BIPOLAR DISORDER – A PORTRAIT OF A COMPLEX MOOD DISORDER

Edited by **Jarrett Barnhill**

Bipolar Disorder – A Portrait of a Complex Mood Disorder
Edited by Jarrett Barnhill

Published by InTech

Janeza Trdine 9, 51000 Rijeka, Croatia

Publishing Process Manager Dragana Manestar

Technical Editor Teodora Smiljanic

Cover Designer InTech Design Team

Additional hard copies can be obtained from orders@intechopen.com

Bipolar Disorder – A Portrait of a Complex Mood Disorder, Edited by Jarrett Barnhill
p. cm.
978-953-51-0002-7

For EU product safety concerns:
IN TECH d.o.o., Prolaz Marije Krucifikse Kozulić 3, 51000 Rijeka, Croatia,
info@intechopen.com or visit our website at intechopen.com.

Meet the editor

Dr L. Jarrett Barnhill, MD is Professor of Psychiatry at the University of North Carolina School of Medicine. He is the Director of the Developmental Neuropharmacology Clinic within the Department of Psychiatry. Dr. Barnhill received his undergraduate degree at the University of North Carolina at Chapel Hill, and his medical degree from the Bowman Gray School of Medicine - Wake Forest University. Dr. Barnhill has completed residencies in General Psychiatry, Child Psychiatry, and Neuropsychiatry. He is a Distinguished Fellow in the American Psychiatric Association and Fellow in the American Academy of Child and Adolescent Psychiatry. Dr. Barnhill has served on the Executive Council of the NCPA for 3 years and is a past president of the NC Council of Child and Adolescent Psychiatry. He is a past President of the North Carolina Psychiatric Association. Dr. Barnhill currently serves on the Governor's Committee on Children with Special Needs. His clinical interests include autism, mental retardation and other developmental disorders, traumatic brain injury, neuropsychiatry of mood and seizure disorders, Tourette's Disorder and other movement disorders.

Contents

Preface

This volume addresses the many faces of Bipolar Disorder (BD). To complete this portrait, we need to explore not only clinical symptoms and treatment responses, but also gene-environment interactions, epigenetics, stress response systems, and psychosocial forces that impact mood disorders in general and BD in particular. From a clinical perspective, the age of onset, severity, and patterns of cycling and symptoms influence clinical course, venue, and treatment decision making. In this edition we address the complexity of BD by focusing on basic neurosciences research, especially the role of neuro-plasticity; multiple neurotransmitter systems; first and second messenger systems; the capacity of several mood stabilizers to alter stress response systems, regulators of circadian rhythms, and the underlying neurophysiology of treating BD. The fundamental processes are addressed in depth by Drs Mahai (bivalent cations in mania); Seong (lithium effects on neuroplasticity); Gino (a novel treatment approach to treatment based on modulating glutaminergic activity treatment with Memantine) and Margarida (the effects of lithium on intracellular mechanisms related to phosphoinositol, GSK3 and other messenger systems).

Clinical heterogeneity is the most challenging feature of major psychiatric disorders. A significant contributor to this heterogeneity is gene-environmental interactions. In keeping with this challenge, Dr Young-Ki addresses the relationship between stress, BD and the role many mood stabilizers pay in modulating the stress response systems (neuro-immunological and neuro-endocrinological systems). Her insights and observations are germane to our understanding of the lifelong changes in the nature and course of BD. One area of intense interest is the boundary and transformation point between high levels of reactivity early in the course of BD to one that seems to be an expression of primarily endogenous events. This "developmental" characteristic of BD is complicated by frequent relapses over time and points us toward kindling, episode sensitization and alterations in neuroplastic responses by recurring symptoms.

In recent years, there is growing interest in the relationship between "soft" forms of BD, namely sub-clinical or sub-syndromal forms (BD. NOS). These subsyndromal forms are grouped under BD spectrum disorder. If we turn our attention to prepubertal onset BD we also confront boundary problems associated with atypical or temperamental presentations of ADHD and severe affective instability, irritability, and explosive behaviors. In this edition Dr Andreea addresses neurobiological features of affect regulation and expression. This chapter dovetails nicely into Dr Peter's review of the problems child psychiatrists encounter, and in some situations create, when dealing with prepubertal onset BD. The boundary between genetic risk for BD and severe temperamental differences, ADHD, abuse and neglect, and affect dysregulation enter into these considerations.

These variations in the presentation of classic BD segue into the markedly increased risk for both attempted and completed suicide in the disorder. Factors such as recurring depression in BP II patients, presence of mixed mood states and subtype early age of onset, substance use, and psychiatric comorbidity contribute to potential suicidal behaviors. For example, many individuals with BD.NOS, Bipolar II (recurring depression), and mixed mood states experience longstanding social and occupational morbidity, prolonged periods of mood-related symptoms and increased risk of suicide. Dr Dagmar provides a thorough analysis of epidemiological research and clinical data that addresses the risks for suicidal behavior and completed suicide in BD.

Dr Levy addresses the complex psychosocial issues that encompass BD. In our era of psychopharmacological treatments, various psychotherapy formats are often overlooked or underemphasized. The role of psychotherapy in enhancing quality of life, relapse prevention, adjuncts for treatment resistant patients, and a methodology of closely monitoring patients for early relapse at key points in their clinical course, can be critical for successful outcomes. Increased mastery of the psychosocial consequences of BD may also play a key role in suicide prevention by enhancing family functioning as well as cognitive behavioral and interpersonal skill sets.

Dr Wendy culminates this volume on a broader medical sociological note. BD is a costly and life-altering disorder that can adversely affect not only quality of life but also occupation, educational, and family functioning. Dr Wendy explores the impact of disclosure on these parameters. In addition, the author investigates the often ambiguous territory of illness disclosure in terms of stigmatization and social role changes due to occupational and family dysfunction and repeated hospitalizations. These psychosocial issues affect one's willingness to accept the diagnosis, comply with treatment recommendations, and levels of subjective distress and morbidity related to a chronic relapsing disorder. The author then turns to the impact of disclosure by celebrities on stigmatization and social acceptance. Even though Dr Wendy focuses on psychosocial factors, these findings can be extrapolated to chapters on neurophysiological parameters associated with occupational distress, social isolation, and sense of alienation, shame/guilt, and other risk factors for both recurrence and treatment resistance.

Bipolar Disorder: A Portrait of a Complex Mood Disorder captures only a portion of the variability and heterogeneity of what we call BD. This book begs the question of approaching BD in terms of a combination of descriptive and etiopathogenic approaches. By doing so, clinicians can apply new research in the neurosciences to the process of differential diagnosis and treatment planning. This approach requires a more detailed understanding of BD, especially the forces underlying differences in typology of mood, polarity, comorbidity, molecular genetics, gender expression, and age of onset. For example, female patients with a family history of BD who experience a single episode of depression remain at risk to some degree for recurring depression, treatment resistance and/or switch potential switch to hypomania. A similar patient victimized by severe physical or sexual abuse and neglect may present with more substantial neuroendocrine dysregulation who have to cope with neuroendocrine changes that influence course and treatment outcome. In contrast, those patients with a classic pattern of well-defined periods of recurring mania alternating with depression; positive family history of lithium responsive BD, male gender, may constitute a more specific endophenotype. Defining such endophenotypes may decrease the level of clinical heterogeneity and help improve diagnosis rigor and treatment response.

Unfortunately, between these two endpoints remain multiple variations of the BD theme. For example, we also grapple with the boundary between psychotic mood and schizophrenia spectrum disorders. These patients can present with mania or depression with psychotic features (severity dimension); combined symptoms associated with both schizophrenia and mood disorder (schizoaffective disorders), and suggest the boundary between mood and schizophrenic disorders is more permeable than previously acknowledged. In the early 20th century, these disorders were considered distinct clinical entities. "Hybrids" between these dichotomized disorders now include schizoaffective disorders. Recent research suggests a possible genetic link based on the presence of psychotic symptoms, there are mood disorders, thought disorders, and then a different risk factor for floridly psychotic forms of each.

Gender and age of onset have significant influences on the clinical presentation and treatment decisions. Among individuals with depression, the gender ratio is nearly 2:1 for females when compared to males. Among individuals with bipolar illness, the gender ratio is nearly equal. Age of onset provides another source of variability. The peak age of onset for unipolar depression is after 40, while the peak onset for bipolar-related mood disorders is often in the late teens and early twenties. The onset of any subtype of mood disorder during childhood tilts the long-term outcome towards chronicity, bipolarity, high rates of comorbid substance use disorders, and comorbidity (including personality disorders. There is a great deal of controversy in child psychiatry about the nature of pre-pubertal early onset bipolar illness, especially since many children may lack clear-cut cyclical changes or present with a mixture of affective symptoms.

In terms of clinical course or longitudinal outcomes, there is a tendency towards increasing frequency and longer duration of episodes; shift towards rapid cycling (more common in females with thyroid dysfunction), and in some circumstances resistance to previously effective treatments. For example, lithium responsive versus valproic responsive individuals may differ in terms of typology, severity, and duration of illness factors. Individuals with comorbid neurological disorders may show similar difference in treatment responsiveness.

Bipolar Disorder: Portrait of a Complex Mood Disorder is a step towards integrating many diverse perspectives on BD. As we shall see, such diversity makes it difficult to clearly define the boundaries of BD. It is helpful to view BD from this perspective, as a final common pathway arises from multiple frames of reference. The integration of epigenetics, molecular pharmacology, and neurophysiology is essential. One solution involves using this diverse data to search for endophenotypes to aid researchers, even though most clinicians prefer broader groupings of symptoms and clinical variables. Our challenge is to consolidate this new information with existing clinical practice in a usable fashion. This need for convergent thinkers who can integrate the findings in this book remains a critical need. This book is a small step in that direction and hopefully guides researchers and clinicians towards a new synthesis of basic neurosciences and clinical psychiatry

Jarrett Barnhill MD, DFAPA, FAACAP
Department of Psychiatry, University of North Carolina School of Medicine,
Chapel Hill, NC,
USA

References

Goodwin FK, Jamison KR. *Manic-Depressive Illness: Bipolar Disorders and Recurring Depression,* 2007 New York: Oxford University Press. ISBN: 13-978-0-19-513579-4.

Knapp P, Mastergeorge AM (2009). Clinical Implications of Current Finding in Neurodevelopment. *Psychiatric Clinics of North America* 32 (1): 177-198.

Ludd LL, Akiskal HS, Schletter DA et al (2003(. The Comparative Clinical Phenotype and Long Term Longitudinal Course of Bipolar I and II: A Clinical Spectrum. *J Affect Disorders* 73(12):19-32.

Mazzarrini L, Vieta E (2010). Toward a Valid Classification of Psychosis: Overcoming the Schizophrenia-Bipolar Dichotomy. *Psych Annals* 40(3): 143-8.

Myers AJ, Nemeroff CB (2010). New Vistas in the Management of Treatment-Refractory Psychiatric Disorders: Genomics and Personalized Medicine. *Focus: J. of Life Long Learning in Psychiatry* 8(4):525-35.

Osuji IJ, Cullum CM (2005). Cognition in Bipolar Disorder. *Psychiatric Clinics North America* 28(2): 427-442.

Scott J, Colom F (2005) Psychosocial Treatments for Bipolar Disorders. *Psychiatric Clinic of North America* 28(2): 371- 84.

Part 1

Basic Science Issues

1

Bivalent Cations in Bipolar Disorders

Mihai Nechifor[1], Cristina Vaideanu[2] and Florina Crivoi[3]
[1]Department of Pharmacology Gr. T. Popa University of Medicine and Pharmacy
[2]Clinical Psychiatric Hospital" Socola " Iasi,
[3]Department of Biophysics Gr. T. Popa University of Medicine and Pharmacy Iasi
Romania

1. Introduction

Bivalent cations play a lot of roles in human brain. Calcium, magnesium zinc ,copper and other cations are involved in normal functions of the brain .The most important mechanisms of action for cations are : modulation of the presynaptic release of some neuromediators , influence on neuronal ionic channels, induction of changes in some receptors activity, influence on the transporters of neuromediators. The disbalance in intracellular or extra cellular concentrations of brain cations are observed in neurological and psychiatric diseases (major depression, schizophrenia, heroine addiction, neurodegenerative diseases, convulsive disorders). The bipolar disorder(BP) is a major public health problem. The economic cost is high and the patients have a high risk of suicide(Antelman et al. 1998) BD also called manic-depressive psychosis is a relative frecvent psychiatric ill. The pathogenic mechanism of the development of BD is unknown yet. The aetiopathogenesis of BD will be better elucidated in the future by experimental and clinical studies.

2. Bipolar cations in bipolar disorders

There are few data , sometimes contradictory regarding the variation of concentration of bivalent cations in the patients with BD. Herzberg & Herzberg 1977 observed a decreased level for plasma magnesium in BD patients compared to healthy control group. Carman et al. 1979 found an increased plasma ratio calcium/ magnesium in BP patients correlated to the intensity and duration of maniacal agitation. The plasma zinc levels were decreased in acute maniacal agitation. Contrary, Frazer et al. 1983 found a higher erythrocyte magnesium level in BD patients not correlated with the severity of clinical symptoms . George et al.1994 showed an increased level of cerebro-spinal fluid magnesium and lack of correlation with the clinical course of mood modulators therapy. Our data (Nechifor et al 2006,2007) show that adult patients with bipolar disorder type I presenting acute maniacal attacks and no previous treatment exhibit lower intracellular magnesium levels than control group. Plasmatic zinc concentrations was significantly lower . There were no significant differences between patients with bipolar disorder type I and control group regarding total magnesium plasma concentration.. An exaggerated intracellular calcium level and an exaggerated of cytosolic calcium concentration in response to serotonin are showed in depressive phase of BD and in major depression. Brown et al .2007 observed clinical maniacal symptoms in the cases of hyperthyroidism in patients with hypercalcemia. The plasma zinc levels were decreased in acute maniacal agitation The existence of many clinical forms of BD contributes to the heterogeneity of obtained data

3. Our studies on bivalent cations level in patients with bipolar disorders

After DSM IV bipolar disorders are classified in four groups. In the bipolar disorder type I the patient has mixed maniac and depressive symptoms with at least one maniac episode. All the patients from our study have been the type I bipolar disorder patients. The mood modulators are the treatment of choice today for BD. Some anticonvulsivant drugs are the main choice drugs today in BD treatment (Goodnick 2006,Muzina et al. 2005, Bowden et al. 2006). Carbamazepine and sodium valproate are usually used in BD therapy(Weisler et al.2005, Walden et al. 1993)The aim of our research was the determination of the plasma and the intracellular levels of some bivalent cations in BD patients before the treatment and also the study of the influence of mood modulators influence on cations concentration during BD therapy. We determined the levels of plasma calcium , magnesium , copper and zinc and the erythrocyte magnesium concentration in adult patients with BD type I (after DSM IV). We worked on adult patients with bipolar disorders (BD)(diagnosed after DSMIV criteria) ageing 21-58 years Admitted into "Socola Psychiatric Hospital" Iasi Romania. In the study were included only type I BD adult patients hospitalized during the maniacal episode .We worked on three groups of patients : I group received carbamazepine 600mg/day p.o. daily 4 weeks ; IInd group received sodium valproate 900mg/day p. o. daily 4 weeks ; IIIrd group received quetiapine 400-600mg/day p.o.daily 4 weeks. A group of 20 healthy adults was control group.

Including criteria: bipolar disorder (diagnosed after DSM IV), at least 4 weeks treatment, absence of anyone BD treatment before admittance in hospital. All patients did not received any treatment for BD before hospitalization . The following non-including criteria were used: pregnancy, lactation, renal failure , heart failure, hepatic failure malabsorbtion syndrome, treatment with bivalent cations containing drugs or diuretics treatment. The plasma levels of total calcium ,total magnesium, zinc and copper and erythrocyte magnesium level were determined by atomic absorbtion spectrofotometry. The determination was performed before the start of treatment and after 4 weeks. In this study were included only patients with at least 4 weeks hospitalization. The results were statistic interpreted.

3.1 Results

The obtained date showed that zinc plasma level was decreased in BD patients before treatment compared to control group. (0.89±0.12mg/l in control group vs. 0.62±0.05mg/l in BD group $p<0.05$). No changes in total calcium plasma concentration(116. 32±9,1 mg/l in BD patients vs. 109.20±12.3mg/l in control group). The erythrocyte magnesium level was decreased in BD patients before treatment (45, 01±1.67mg/l in PD patients vs 59.15 ±2.01mg/l in control group $p< 0.05$). The copper concentration was increased in BD patients before treatment compared to control group). After 4 weeks of treatment the plasma zinc concentration increased in all group of treated patients (ex. after sodium valproate , Zn concentration was 0.83±0.04mg/l $p< 0.05$)The erythrocyte magnesium level increased in all treated patients (ex. in carbamazepine group after treatment Mg concentration was 53.72 ±2.18mg/l $p < 0.05$),but only in quetiapine treated group , the total plasma magnesium concentration significantly increased. In all treated groups copper plasma concentration decreased after the treatment Did non changes in calcium plasma levels after mood modulators treatment. Our results showed that zinc plasma concentration and erythrocyte magnesium level are decreased during the maniacal acute episode of type I BD patients . The obtained data showed that different mood modulator with different mechanism of action increased zinc plasma level and decreased copper plasma concentration. The erythrocyte magnesium is also increased. There are good positive correlations between the improvement of clinical status of patients and the increase of zinc plasma concentration and the augmentation of intracellular magnesium levels.

4. Intractions between mood modulator and bivalent cations

4.1 Magnesium

The mood modulators used in the treatment of BD induced important changes in some bivalent cations concentrations. The increase of magnesium intracellular concentrations were observed after all three mood modulators used. There are data showed that lithium , the oldest effective drug for the treatment of maniac-depressive illness increases the intracellular magnesium levels .Most studies show that the repeated lithium salts administration increases the magnesium concentration.

Lithium increases the intracellular magnesium concentration by competition between magnesium and lithium for some intracellular binding sites(Leyden et al.2000, Mota de Freitas et al.2006)After experimental loading of neuroblastoma cells with 1-2 mM of extracellular Li^{+}, the intracellular free magnesium concentration was significantly higher . Regarding the Li+-Mg2+ competition at some intracellular sites , the existing data indicate the following targets :molecules:inositol monophosphatase, glycogen syntase kinase(GSK 3), fructose 1,6 biphosphatase, biphosphate nucleotidase , ADP and ATP phosphate bindind sites but it is possible to be also and other intracellular binding sites for this competition.(Gould et al. 2004)

The main ways for magnesium action in BD are :a)decreasing the neuronal response to glutamate overstimulation by blocking the calcium channel coupled with NMDA receptors;b) the decreasing presynaptic release of some excitatory neuroaminoacids; c) modulator action at the level of gabaminergic and serotoninergic systems. Gobbi & Janiri ,2006 showed that magnesium-valproate significantly modulates the response induced by NMDA-receptor stimulation. Chuinard et al. 1990, showed that magnesium aspartate administration was effective in stabilizing the mood of rapid-cycling BD which favor the idea that the increase in magnesium concentration is an important factor of lithium and other mood modulators mechanism of action. Magnesium oxide increases the verapamil maintenance therapy in mania (Giannini et al 2000). This fact favors the idea that an increase in magnesium concentration is an important fact, maybe essential for the therapeutic effect of some drugs used in BD treatment. Magnesium-valproate reduces the hyperactivity in an animal model of mania. This effect of magnesium valproate could be abolished by bicuculine.These findings suggest that the action on the postsynaptic GABA effect may be involved also in magnesium-valproate antimaniacal action (Cao & Peng , 1993). The neuroprotective magnesium effect in CNS is important not only for the recovery process after various injury(Vink & Cernak 2000) but also to reduce the maniacal agitation.

Machado-Viera R. et al. 2009 showed that the neurotrophic effect of lithium is very important for neuro protection (Rowe & Chuang 2004) and for prophylaxis of acute mood effects and one of the main targets of lithium intracellular action is GSK-3 . By action at the level of this enzyme, Li increase the intracellular level of magnesium. It is possibil that ,at least in part ,the reduction of apoptosis by lithium is produced by magnesium ions which inhibit the apoptosis. There are few studies about the magnesium alone effect in the treatment of bipolar disorders. Chouinard et al. 1990 showed a moderate effect of magnesium treatment as mood stabilizer for rapid cycling bipolar affective patients .An other way to explain the the effect of magnesium and zinc in BP disorders is the antioxidant effect. The oxidative stress is increased in animal models of mania produced by amphetamine administration. .Chronic amphetamine treatment is associated with an imbalance in SOD and CAT activity. In experimental studies in rats lithium and valproate prevented the excitotoxicity by reducing the oxidative stress (Shao et al. 2005) The both magnesium and zinc have an antioxidant effect.An other target for magnesium involvement in mood stabilizers therapeutic effect is BDNF (brain derived neurotrofic factor).The mood modulators (Frey et al. 2006) but also magnesium increase the BDNF concentration . The ratio between calcium and magnesium and the antagonist effect of magnesium regarding some calcium action are essential for explanation of importance of magnesium in mechanism of mood modulators effects. The transmembrane calcium influx plays an important role in the development of some psychiatric disorders. In BD has been observed dysfunctions in the intracellular signaling transduction ,altered calcium signaling and a elevated protein kinase A activity (Langan & McDonald 2009).

The calcium channels antagonists (verapamil and others) raise carbamazepine effect. Magnesium, acting like a natural calcium antagonist on some ionic channels is a factor which contributes at the pharmacodinamic effect of some mood modulators. Contrary, the calcium ions have an antagonic effect on carbamazepine action. Carbamazepine reduces the neuronal excitability and glutamate release and we consider that this effect is due at least in part by increasing the magnesium concentration. The alteration of calcium homeostasis is involved in the onset or progression of various neurological and psychiatric diseases.such as Parkinson's disease, Alzheimer's and others degenerative diseases, bipolar disorders, Hutington's disease and others (Salvaraj et al. 2010).The hypercalcemia and the change of the ratio calcium/ magnesium could be involved in the pathophysiology of bipolar disorder. Maniacal clinical symptoms from hyperparathyroidism are mediated by hypercalcemia .The patients with maniacal symptoms had have a high level of calcium in the blood and cerebrospinal fluid. The bipolar disorder is not the single disease in which are involved calcium signaling abnormalities in the central nervous system. Cytosolic $Ca2+$ signals are correlated to extracellular calcium enters through plasma membrane channels and to the calcium release from the intracellular stores .In some diseases as seizures, migraine and autism is possibil to be genetic calcium signaling abnormalities(Gargus .2009).

The main mechanism of action of mood modulators in BD is the reduction of the glutamatergic systems activity via NMDA receptors activation. The key common point of magnesium and zinc action in BD is NMDA glutamate receptors. Magnesium ions and some calcium antagonists act also at the level of NMDA receptors coupled calcium channel

The changes in calcium homeostasis are important in the pathophysiology of the bipolar disorder.(Akimoto et al. 2007) . Calcium enhance NMDA receptor activity and stimulate the presynaptic glutamate release. There are evidences of abnormalities in intracellular calcium distribution, concentration and activity and for the involvement of calcium homeostasis disregulation in the molecular mechanism of this disease. Perova T. et al. 2007 showed a hyperactive intracellular calcium dynamics in the lympoblasts from patients with type I bipolar disorder. Acting as a calcium antagonist, magnesium can reduces in part calcium effects in bipolar disorders. Yasuda et al.2009) showed that for mood modulators as valproic acid and lithium ,the GSK-3 inhibition and the stimulation of brain-derived neurotrophic factor(BDNF)synthesis are the primary targets in the action mechanism involved in the suppression of bipolar disorders symptoms. BDNF is strongly involved in the synaptic plasticity, reduces the apoptosis , increase the neuronal survival and regulates the expression of NMDA receptors in brain (Caldeira et al 2007) Magnesium is also involved in neuronal plasticity, inhibits the apoptosis and reduces the neurotoxicity of high glutamate concentration and of overstimulation of NMDA receptors. The kainat-induced neurotoxicity is produced after stimulation of NMDA receptors and lithium decreases this toxicity.Also the NMDA receptors mediated neuronal vacuolization is attenuated by mood modulators(Bown et al. 2003).Crespo-Biel et al.2010 showed that this neuroprotective effect is produced by modulation of calcium entry in neurons. By reducing the calcium penetration in the cell, magnesium can be involved in the neuroprotective lithium effect. The mood modulators have also an important antiapoptotic action (Chuang et al.2005). Magnesium has a important antiapoptotic action and can be involved in this action of mood modulators.We belive that the increase of intracellular magnesium concentration is also a primary step in mood modulators mechanism of action.Bipolar disorder patients have a high rate of relapse(Newberg et al.2008). The Ca^{2+}- permeable melastin related transient receptor potential 2 (TRPM2) channels are important for the entry of calcium ions into the cell.The genetic variations of TRPM 2 increases the risk of devoloving bipolar disorders (Naziroglu 2011)There are data implicating L-type calcium channels disfunctions in the pathophysiology of neuro psychiatric disorders.(Casamassima et al.2010).In mood disorders, L-type calcium channels blockers reduced in clinical practice the intensity of clinical symptoms. In the animal models of depression the calcium channels blockers had have a similar effect. Glutamate stimulates NMDA receptors and increases the entry of calcium in neurons. Verapamil , a calcium channels antagonist augmented the lithium effects in the treatment of mania. In some studies , verapamil alone has shown antimaniac effect. (Mallinger et al. 2008), but the verapamil monotherapy has a reduced efficacy in BD type I patients. In the experimental model of depression induced by forced swim test in rats and in the tail suspension test in mice, zinc administration exerts an antidepressant effect (Kroczka et al. 2001, Rosa et al. 2003). Magnesium oxide increases the verapamil maintenance therapy in mania (Giannini et al 2000). This fact favors the idea that an increase in magnesium concentration is an important fact, maybe essential for the therapeutic effect of some drugs used in BD treatment. Magnesium-valproate reduces the hyperactivity in an animal model of mania. This effect of magnesium valproate could be abolished by bicuculine.These findings suggest that the action on the postsynaptic GABA effect may be involved also in magnesium-valproate antimaniacal action (Cao & Peng , 1993).

Lithium improves the cognitive deficits in animal models of neurodegenerative diseases. Magnesium also increases the memory in experimental studies. It is possibil that , in part , the cognitive effect of lithium be intermediated by the increase of magnesium intracellular concentration.

Magnesium potentiates the effects of anxiolitics. There are experimental evidences of the potentiation of anxiolytic effect of diazepam by magnesium aspartate (Borzeix et al. 1991).

To the extent that glutamatergic signaling *via* NMDA receptors is pathologically upregulated in bipolar disorder patients, (Toro & Deakin 2005, Hashimoto et al. 2007) The modulation of signal transduction at the level of glutamate, serotonin, dopamine and GABA receptor by the mood modulators has major therapeutic involvements (Manji et al. 1999) Valproic acid interferes with glutamatergic function and NMDA receptor signaling., Valproic acid acts at the level of NMDA receptors by different ways (Gean et al.1994). This drug reduces induction of Fos and of activator protein-1 DNA binding activity. By this mean is modulated the transcription of the NMDA receptor subunit, NR2B. Chronic valproate treatment blocks D(2)- receptor-mediated brain signaling via arachidonic acid in rats.By inhibiting NMDA receptors mediated calcium influx ,lithium and magnesium suppresses the calcium dependent way for activation of apoptotic signaling pathways(Chiu & Chuang2011) Lamotrigine is an other mood modulator used in bipolar disorder which has a key point of mechanism of action the blocking effect of NMDA receptor mediated signaling in the brain (Ramadan et al 2011).

Substances that interfere with dopaminergic (Murphy et al. 1971) or glutamatergic (Anand et al. 2000) signaling ameliorate bipolar disorder symptoms. Bipolar symptoms reflect reduced cholinergic(Bymaster et al. 2002), altered serotonergic, and increased dopaminergic and glutamatergic neurotransmission.The mood-stabilization of bipolar patients appear only after 10 days or more time administration. We thing that the retardation of effect appears because needs time for increasing the intracellular magnesium concentration. Our data showed a good positive correlation between therapeutic effect and the increase the intracellular but not with the extracellular magnesium concentration. The acute administration of a single dose of mood modulators don't change the glutamate level but also don't modify the magnesium concentration. The therapeutic effect of carbamazepine in the treatment of epilepsies and in affective disorders was decreased by a low magnesium level in patients effect (Walden et al. 1993).An other possibility for magnesium beneficial effect in the treatment of BD by various mood modulators is action at the leval of GABA-ergic systems from the CNS. Magnesium valproate decreased the hipermotility in experimental rodent model of mania but this effect is diminished by bicuculline , substance which blocks GABA receptors.(Cao & Peng 1993) This fact involves the post synaptic GABA receptors in magnesium effect in BD.Magnesium can modulate the GABA receptors activity and there are evidence for a potentiating effect of magnesium at the level of GABA A receptors (Moykkynen et al .2001).The interactions between magnesium and lithium are very complex because there are possibilities that magnesium increase the lithium entry into the cells.(Rybakowski & Szajnerman 1976).

4.2 Copper

The increase of copper level could be also involved in the pathogenesis of bipolar disorders. Mood disorders are frecvent reported in Wilson disease. Maniac symptoms and depressive behevior are present in patients which high level of copper. (Keller et al. 1999, Akil & Brewer 1995).In 20% of patients, the psychiatric manifestations preceded all other symptoms. (Dening & Barrios 1989 , Machado et al..2008). The mechanism of action of increased copper concentration in mood disorders is not clear but we beleve that the influence on dopamine neuromediation and the possible increase of glutamatergic system activity is important. Studies performed in rats showed that copper has an opposite effect on the compared to magnesium and zinc. Copper-dopamine complex induced mitochondrial autophagy and the neuronal death(Paris et al. 2009). In the same time, the stimulation of synaptic NMDA receptors by glutamate in hippocampus neurons is associated to the release of copper from intracellular stocks .A copper efflux after NMDA receptors activation was observed.. Unlike magnesium and zinc, copper could function as a metal involved in nitrosylation of NMDA receptors. Surely, copper is a neuromodulator factor of some brain area as hippocampus.

GABA is one of the factors involved in the reduction in intensity of opiate dependence, as well as in the reduction in intensity of opiate-withdrawal syndrome signs. At least in some brain areas, copper blocks GABA receptors . Copper and zinc interact al the level of GABA receptors (Sharonova et al. 2000) The copper-GABA complex antagonized diazepam anticonvulsivant effect (Kardas et al 1984). This shows that by forming a complex with GABA, copper ions reduce the efficacy of this amino-acid in antagonizing glutamate effects during the development of morphine dependence.

4.3 Zinc

Zinc allosteric modulates the 5-HT (1A) serotonin receptors(Barrondo & Salles 2009)By this way ,zinc can be an importantfactor influencing the antidepressant and mood modulators therapy.This cation modulates also the neurotransmitters activity including dopamine and serotonin transporters. (Norgaard-Nilsen & Gether 2006) Regarding main ways by which the increase of zinc concentration can be involved in the mechanism of mood modulators action , very important is zinc effect on the NMDA receptors . By stimulation of NMDA receptors glutamate induces agitation and anxiety . Zn^{2+}decreases the NMDA receptor stimulation. Increased zinc concentration in amigdala decreases the fear and the anxiety (Takeda et al 2010). The antidepressant action involved also the serotoninergic system (Szewczyk et al. 2009, Garcia-Colunga et al. 2005).. Serotonin is a big target for antidepressant drugs(Harvey 1997). An important number of very used antidepressant drugs are substances which block serotonin re-uptake. The chronic lithium treatment influences the cortical serotonin uptake and 5-HT1A receptors activity(Carli et al.1997).Zinc modulates the lithium induced biochemical and behavioral changes in rats(Bhalla et al. 2007)

Zinc modulates the serotonin uptake in some areas from the brain . The effects of fluoxetine, imipramine and 6-nitroquipazine on serotonin uptake in rat brain are modulated by zinc (Garcia-Colunga et al. 2005).Serotonin induced platelet calcium mobilization is enhanced in BD patients (Okamoto et al. 1995, Akimoto et al .2007). In the experimental model of depression induced by forced swim test in rats and in the tail suspension test in mice, zinc administration exerts an antidepressant effect (Kroczka et al. 2001, Rosa et al. 2003, Opoka et al. 2009) . There are experimental data about a complex zinc-induced adaptative and modulatory changes in glutamateric and serotoninergic brain systems(Cichy et al. 2009).In our clinical studies (Nechifor et al. 2004, Nechifor 2008) the plasma zinc concentration was decreased and the antidepressant treatment augmented the level of this cation. Chronic lithium administration reduced the NMDA receptors signaling also via arachidonic acid and eicisanoids synthesis(Basselin et al. 2006)but is not clear if zinc or magnesium influence this way.

4.4 Calcium

Calcium ions play a very important role in biological signal transduction, in synthesis and release of neuromediators and neuromodulators and in the several enzymes activity.The calcium neuronal activity in bipolar disorders is increased.There are studies which showed that a elevation of basal calcium intracellular concentration in B D patients (Emamghoreishi et al .1997, Du et al. 2004) .

A way for the involvement of calcium in bipolar disorders pathogeny is PKC activity. This enzyme potentiates the the response after NMDA receptors stimulation by increasing the calcium entry by the channel coupled with NMDA receptors and by reducingthe voltage – dependent magnesium block of this channel (Chen & Huang 1992)PKC also up-regulates the function of L-type calcium channels and the entry of calcium into the cell(McCarty 2006). Magnesium can decrease this mechanism involved in BD. Valproate, lithium and other mood modulators decrease also the PKC activity(Mallinger et al.2008) .The increase of intracellular magnesium concentration can be a common point for the antimaniacal and antidepressive action.

5. Conclusions

The increase of intracellular magnesium concentration by different mood modulators which various chemical structures and different mechanism of action shows that this is important for the therapeutic effect in BD treatment. On the other hand, this magnesium intracellular concentration change is associated with an significant augmentation of plasma zinc level and decrease of plasma copper concentration We consider that there changes is in important part of the mechanism of pharmacotherapeutic action of mood modulators. We believe that the bivalent cations disbalances are involved in the B D relapses and also in the reduction of efficacy of the mood modulators therapy.

6. References

Akil M. & Brewer G. J.(1995) Psychiatric and behavioral abnormalities in Wilson's disease. *Advances in Neurology* 65, pp.171-178 ISSN 0091-3952

Akimoto T.,Kusumi I.,Suzuki K . &Koyama T. Effects of calmodulin and proteine kinase C modulators on transient Ca2+ increase and capacitative Ca2+ entry in human platelets: relevant to pathophysiology of bipolar disordes. Progres in Neuropsychopharmacology and Biological Psychiatry 31(1),pp. 136-141 ISSN 0278-5846

Anand A, Charney DS & Oren DA. (2000) Attenuation of the neuropsychiatric effects of ketamine with lamotrigine: support for hyperglutamatergic effects of N-methyl-D-aspartate receptor antagonists. *Archives of General Psychiatry.* 57(2), pp.270–276. ISSN 0003-990X

Antelman S. M., Caggiula A. R., Kucinski B. J., Fowler H., Gershon S., Edwards D. J., Austin M. C., Stiller R., Kiss S. & Kocan D.(1998).The effects of lithium on a potential cycling model of bipolar disorder. *Progress in Neuro-Psychopharmacology and Biological Psychiatry,* 22(3), pp.495-510, ISSN 0228-5846

Barrondo S & Sallés J Allosteric modulation of 5-HT(1A) receptors by zinc: Binding studies. *Neuropharmacology.;* 56(2), pp.455-462. ISSN0028-3908

Basselin M, Chang L & Bell JM,(2006) Chronic lithium chloride administration attenuates brain NMDA receptor-initiated signaling via arachidonic acid in unanesthetized rats. *Neuropsychopharmacology* 31pp.,1659–1674 ISSN0893- 133X

Bhalla P, Chadha VD & Dhawan DK (2007) Effectiveness of zinc in modulating lithium induced biochemical and behavioral changes in rat brain. *Cellular and Molecular Neurobiology* 27(5) , pp.595-607 ISSN 0272-4340

Borzeix MG, Akimjak JP, Dupont JM, Cahn R & Cahn J (1991). Experimental evidence of a potentiation by alpha,beta magnesium L-aspartate of the anxiolytic effect of diazepam. Four-plate test in mice and EEG study in primates. *Magnesium Research* 4(3-4),pp. 197-200 ISSN 0953-1424

Bowden CL & Karren N.U. (2006)Anticonvulsants in bipolar disorder. *Australian and New Zealand Journal of Psychiatry* 40,pp.386–393. ISSN 1440-1614

Bown CD, Wang JF & Young LT. (2003) Attenuation of N-methyl-D-aspartate-mediated cytoplasmic vacuolization in primary rat hippocampal neurons by mood stabilizers. *Neuroscience.* 117,pp. 949–955 ISSN 0306-4522

Brown S. W., Vyas B. V. & Spiegel D. R.(2007), Mania in a case of hyperparathyroidism. *Psychosomatics* 48, 265-268 ISSN 0033-3182

Bymaster FP & Felder CC.(2002) Role of the cholinergic muscarinic system in bipolar disorder and related mechanism of action of antipsychotic agents. *Molecular Psychiatry.* 7(suppl 1), pp. S57–S63. ISSN 1359-4184

Caldeira MV, Melo CV & Pereira D.B., (2007) BDNF regulates the expression and traffic of NMDA receptors in cultured hippocampal neurons. *Molecular and Cellular Neuroscience* 35,208–219 ISSN 1044-7431

Cao BJ & Peng NA.(1993)Magnesium valproate attenuates hyperactivity induced by dexamphetamine-chlordiazepoxide mixture in rodents.*European Journa l of Pharmaco*logy 237(2-3)pp. 177-181. ISSN 0014-2999

Caraman J. S. & Wyatt R. J.(1979) Calcium: bivalent cations in bivalent psychoses. *Biological Psychiatry* 14(2) ,pp. 295-336 ISSN

Carli M, Afkhami-Dastjerdian S & Reader TA.(1997) Effects of a chronic lithium treatment on cortical serotonin uptake sites and 5-HT1A receptors. *Neurochemical Research* 22,pp. 427–435 ISSN o364-3190

Casamassima F., Hay A. C., Benedetti A., Lattanzi L., Cassano G.B. & PerlisR. H. (2010)L-type calcium channels and psychiatric disorders: A brief review.*American Journal Medical GeneticsB Neuropsychiatriic Genetics* 153B(8)pp. ,1373-1390 ISSN 1552-485X

Chen G, Zeng WZ & Yuan PX, (1999) The mood-stabilizing agents lithium and valproate robustly increase the levels of the neuroprotective protein bcl-2 in the CNS. *Journal of Neurochemistry.* 72, pp. 879–882. ISSN 1471-4159

Chen L. & Huang L., Y.(1992) Protein kinase C reduces Mg 2+ block of NMDA –receptor channels as a mechanism of modulation. *Nature* 356 (6369),521-523 ISSN 0028-0836

Chiu C. T .& Chuang D. M., (2011) Neuroprotective action of lithium in disorders of central nervous system. Zong Nan Da Xue Bao Yi Xue Ban 36(6),pp. 461-476 ISSN 1672-7347

Chouinard G., Beauclair L., Geiser R. & Etienne P. (1990). A pilot study of magnesium aspartate hydrochloride (Magnesiocard) as a mood stabilizer for rapid cycling bipolar affective disorder patients. *Progress in Neuro Psychopharmacology and Biological Psychiatry* 14,pp. 171-180 ISSN 0278-5846

Chuang D.M.(2005) The antiapoptotic actions of mood stabilizers: molecular mechanisms and therapeutic potentials. *Annals of N ew York Acaemie of Sciences* 1053(1),pp. 195–204. ISSN 0077-8923

Cichy A., Sowa-Kuæma M., Legutko B., Pomierny-Chamio L., Siwek A., Piotrowska A.., Poleszak E., Pilc A. & Nowak G. (2009)Zinc-induced adaptive changes in NMDA/glutamatergic and serotonergic receptors. *Pharmacological Reports* 61,1184-1191 ISSN 1734-11140

Crespo-Biel N., Camins A., Canudas A. M. & Pallas M. (2010) Kainat-induced toxicity in the hippocampus : potential role of lithium. (2010).*Bipolar Disorders* 12(4), 425-436 ISSN13995618

DeningT. R. & Barrios G. E. Wilson 's disease :psychiatric symptoms in 195 cases.(1989). *Archive of Generale Psychiatry* 46,pp. 126-134 ISSN 0039-900X

Du J., Quiroz J., Yuan P., Zarate C. & Manji H. K.(2004) Bipolar disorders : involvementof signalingcascades and AMPA receptor trafficking at synapses.*Neuron and Glia Biology* 1(3),231-243 ISSN 0740-925X

Emamghoreishi M., Schlichter L., Li P. P., Parikh S., Sen J. & Karamble A.(1997)High intracellular calcium concentrations in transformed lymphoblasts from patients with bipolar disordes . *American Journal of Psychiatry* 154 (3), 976-982 ISSN 0002-953X

Frazer A, Ramsey TA, Swann A, Bowden C, Brunswick D, Garver D & Secunda S (1983) Plasma and erythrocyte electrolytes in affective disorders. *Journal of Affective Disorders.* 5(2),pp. 103-113 ISSN 0165-0327

Frey BN, Andreazza AC & Cereser KM, (2006) Effects of mood stabilizers on hippocampus BDNF levels in an animal model of mania. *Life Sciences* 79(2)pp. ,281–286 ISSN 0024-3205

García-Colunga J, Reyes-Haro D, Godoy-García IU & Miledi R.(2005)Zinc modulation of serotonin uptake in the adult rat corpus callosum. *Journal of Neuroscience Research.* 80(1) , pp. 145-149

Gargus J. J. (2009) Genetic calcium signaling abnormalities in the central nervous system: seizures , migraine and autism. *Annals of New York Academy of Sciences* 1151, pp. 133-156 ISSN 0077-8923

Gean PW, Huang CC &Hung CR,(1994) Valproic acid suppresses the synaptic response mediated by the NMDA receptors in rat amygdalar slices. *Brain Reseach Bulletin* 33,pp. 333–336 ISSN 0361-9230

George M. S., Rosenstein D.. Rubinow D. E., Kling M. A. & Post R. M. (1994) CFS magnesium in affective disorders -lack of correlationwith clinical course of treatment. *Psychiatry Research* 51(1), pp. 139-146 ISSN 0165-1781

Giannini A. J., Nakoneczie A.M., Melemis S. M., Vetresco J.& Condon M. (2000). Magnesium oxide augmentation of verapamil maintenance therapy in mania.*Psychiatry Research* 93,pp. 83-87 ISSN o165-1781

Gobbi G & Janiri L.(2006) Sodium- and magnesium-valproate in vivo modulate glutamatergic and GABAergic synapses in the medial prefrontal cortex. *Psychopharmacology (Berl)* 185,255–262 ISSN 0033-3158

Goodnick P.J. (2006) Anticonvulsants in the treatment of bipolar mania. *Expert Opinion in Pharmacotherapy.* 7,pp. 401–410 ISSN 1465-6566

Gould T. D., QuirozJ. A., Singh J., Zarate C. A. & Manji H. K. (2004) Emerging experimental therapeutics for bipolar disorder: insights from the molecular and cellular actions of current mood stabilizers.Molecular Psychiatry 9(8),pp. 734-755 ISSN 1359-4184

Harvey B., (1997) The neurobiology and pharmacology of depression. A comparative overview of serotonin selective antidepressants. *South Africa Medical Journal* 87 (suppl. 4),pp. 540-550 ISSN 0256-9574

Hashimoto K, Sawa A & Iyo M.(2007) Increased levels of glutamate in brains from patients with mood disorders. *Biological Psychiatry.* 62. pp. .1310–1316. ISSN 0006-3223

Heiden A., Frey R., Presslich O., Blasbicher T., Smetana R. & Kasper S.(1999).

Herzberg L. & Herzberg B. (1977). Mood changes and magnesium. A possible intewraction between magnesium and lithium? *Journal of Nervous and Mental Diseases* 165 (6),pp. 423-426 ISSN 0022-3018

Kardos J, Samu J, Ujszászi K, Nagy J, Kovács I, Visy J, Maksay G & Simonyi M(1984) Cu^{2+} is the active principle of an endogenous substance from porcine cerebral cortex which antagonizes the anticonvulsant effect of diazepam. *Neurosciences Letters* 52(1),pp. 67-72 ISSN 0304-3940

Keller R., Torta R., Lagget M., Crasto S. & Bergamasco B. (1999). Psychiatric symptoms as late onset of Wilson's disease: neuroradiological findings, clinical features and treatment.*Italian Journal of Neurological Sciences* 20 (1),pp.49-54 ISSN 0022-510X`

Kroczka B., Branski P., Palucha A., Pilc A. & Nowak G., (2001) Antidepressant -like properties of zinc in rodent forced swim test.*Brain Researh Bulletin* 55(3) ,pp. 297-300 ISSN 0361-9230

Langan C. & McDonald C.(2009).Neurobiological trait abnormalities in bipolar disorder.*Moleculat Psychiatry* 14(9), pp.833-846 ISSN 1359-4184

LeydenB., Diven C., Minadeo N., Bryant F. B. & Mota de Freitas D.(2000). Li+/Mg2+competition at therapeutic intracellular Li+ binding sites in human neuroblastoma SH-SYSY cells.*Bipolar Disorders* 2,pp. 200-204 ISSN 1399-5618

Machado A. C., Deguti M. M., Caixeta L., Spitz M., Lucato L. T. & Barbosa E. R.(2008)Mania as the first manifestation of Wilson's disease. *Bipolar Disorders* 10(3),pp. 447-450ISSN 1399-5618

Machado-Vieira R., Manji H. K. & Zarate A. C.(2009). The role of lithium in the treatment of bipolar disorders: convergent evidence for neurotrophic effects as a unifying hypothesis. *Bipolar Disorders* 11(suppl. s2),pp. 92-109 ISSN 1399-5618

Mallinger A. G., Thase M. E., Haskett R., Luckenbaugh D. A., Frank E., Kupfer D. J.& Manji H. K(2008).Verapamil augmentation of lithium treatment improves outcome in mania unresponsive to lithium alone: preliminary findings and discussion of therapeutic mechanism. *Bipolar Disorders* 10(8), pp.856-866 ISSN 1399-5618

Manji H.K., Bebchuk J. M., Moore g. J., Glitz D., Hasanat K. A. & Chen G.(1999).Modulation of CNS signal transduction pathways and gene expression by mood-stabilizing agents:therapeutic implications. *Journal of Clinical Psychiatry* 60(suppl. 2) ,pp.113-116 ISSN 1555-2101

McCarty M., F.(2006). PKC-mediated modulationof L-type calcium channels may contribute to fat-induced insulin resistance. *Medical Hypotheses* 66(4) , pp.824-831 ISSN 0306-9877

Mota de Freitas D, Castro MM & Geraldes CF.(2006)Is competition between Li+ and Mg2+ the underlying theme in the proposed mechanisms for the pharmacological action of lithium salts in bipolar disorder? *Accounts of Chemical Research.* 39(4),pp.283-291 ISSN 0001-4842

Möykkynen T, Uusi-Oukari M, Heikkilä J, Lovinger DM, Lüddens H &Korpi ER. (2001)Magnesium potentiation of the function of native and recombinant GABA(A) receptors.*Neuroreport.* 12(10),2175-2179. ISSN 0959-4965

Murphy DL, Brodie HK & Goodwin FK(1971) Regular induction of hypomania by L-dopa in "bipolar" manic-depressive patients. *Nature.* 229(1) ,pp.135–136 ISSN 0028-0836

Muzina DJ, Elhaj O & Gajwani P. (2005) Lamotrigine and antiepileptic drugs as mood stabilizers in bipolar disorder. *Acta Psychiatrica Neurologica Scandinavica.* Suppl,pp.21–28.ISSN 0365-5598

Nasrallah HA, Ketter TA & Kalali AH. (2006)Carbamazepine and valproate for the treatment of bipolar disorder: a review of the literature. *Journal of Affective Disorders* 95(1), pp.69–78 ISSN 0165-0327

Naziroglu M. (2011)TRPM2 cation channels, oxidative stress and neurological diseases: where are we now? *Neurochemical Research 36*(3), pp. 355-366 ISSN 1573-6903

Nechifor M .(2004)Involvement of some cations in major depression In : Cser M. A. ,Sziklai Laszlo I. , Etienne J. C. , Maymard Y. , Centeno I. , Khassanova I. ,Collery Ph.(Eds.) *Metal Ions in Bioly and Medicine vol.* 8. John Libbey Eurotext Paris pp. 518-521 ISBN 2-7420-522-6

Nechifor M, Vaideanu C, Mândreci I, Palamaru I & Boişteanu P. (2007) .(The influence of bipolar disorders treatment on plasmatic and erythrocyte levels of some catios, Maria Carmen Alpoim, Paula Vasconcellos, Maria Amelia Santos, Armando J Cristovao, Jose A Centero, Philippe Collery (Eds.), *Metal ions in biology and Medicine,* Ed. John Libbey, Eurotext Paris, Vol. 9, pp. 556-560 ISBN 2-7420-0629-X

Nechifor M.(2006) Changement of cations concentration in unipolar and bipolar disorder In ; M. Szilagy, K. Szentmihaly (Eds) *Trace elements in food chain* , Hungarian Central European University Press Budapest pp. 362-366 ISBN 978-963-7067-19-8

Nechifor M.(2008)Interactions between magnesium and psychotropic drugs. *Magnesium Research.* 21(2), pp.97-100.

Nechifor M.,(2007). *Magnesium in psychosis,* Zoshiki Nishizawa, Hirotoshii Morii, Jean Durlach (Eds). in: *New perspectives in magnesium research – Nutrition and Health,* Springer –Verlag London, pp. 369-377 ISBN 978-1-84628-388-8 175, 1-22.

Newberg A. R., Catapano L. A., Zarate C. A. & Manji H. K. (2008) Neurobiology of bipolar disorder. *Expert Review in Neurotherapeutics* 8(1),pp.93-110 ISSN 1473-7175

Nørgaard-Nielsen K & Gether U.(2006)Zn2+ modulation of neurotransmitter transporters. *Handbook of Experimental Pharmacology* 175(1), pp.1-22. ISSN 0171-2004

Okamoto Y. Kagaya A.& Shinno H.(1995) Serotonin -induced platelet intracellular mobilization in mania. *Life Sciences* 56,pp.327-332 ISSN 0024-3205

Perova T., Wasserman M. J., Li P. P. & Warsh J. J. (2008).Hyperactive intracellular calcium dynamics in B lymphoblasts from patients with bipolar I disordes. *International Journal of Neuropsychopharmacology* 11(2),pp.185-196 ISSN 1461-1457

Post R. M., Weise S.R.B.& Chuang D. M. (1992) Mechanisms of action of anticonvulsivant inaffective disorders. Comparison with lithium. *Journal of Clinical Psychopharmacology* 12,pp. 23S-35S ISSN 0268-1315

Ramadan E., Basselin M., Rao J. S., Chen M., Ma K.& Rappoport S. I. (2011)Lamotrigine block NMDA receptor-initiated arachidonic acid signaling in rat brain : implications for its efficacy in bipolar disorder. *International Journal of Neuropsychopharmacology* 28(1) ,pp. 1-13 ISSN 1461-1457

Rosa A. O., Lin J., Calixto J.B., Santos A. R. & Rodrigues A. L.(2003)Involvement of NMDA receptors and L-arginin-nitric oxide pathway in the antidepressant-like effects of zinc in mice.*Brain Research* 144(1) ,pp.87-93 ISSN 0165-0173

Rowe M. K. & Chuang D. M. (2004).Lithium neuroprotection : molecular mechanisms and clinical implications . *Expert Reviews in Molecular Medicine* 6 (21),pp. 1-18 ISSN 1492-3994

Rybakowski J &Szajnerman Z.(1976)Lithium magnesium relationship in red blood cells during lithium profilaxis *Pharmakopsychiatrie und Neuropsychopharmakologie* 9(5),242-246 ISSN 0031-7098

Selvaraj S ., Sun Y. & Singh B. B. (2011) TRPC Channels and their implications for neurological diseases. *CNS Neurological Disorders Drug Targets* 9(1) ,pp. 94-104 ISSN 1871-5273

Shao L, Young LT & Wang JF.(2005) Chronic treatment with mood stabilizers lithium and valproate prevents excitotoxicity by inhibiting oxidative stress in rat cerebral cortical cells. *Biological Psychiatry.* 58,pp.879–884. ISSN0006-3223

Sharonova IN, Vorobjev VS &Haas HL (2000)Interaction between copper and zinc at GABA(A) receptors in acutely isolated cerebellar Purkinje cells of the rat. *British J ournal of Pharmacology* 130(2),pp. 851-856.ISSN 0007-1188

Srinivasan C., ToonJ., Amari L., Abukhdeir A. M., Hamm H., Geraldes C. F., Ho Y. K. & Mota de Freitas D. (2004)Competition between lithium and magnesium ions for the G-protein transducin in the guanosine 5-diphosphate bound conformation. *Journal of inorganic Biochemistry* 98(5),pp. 691-701 ISSN 0162-0134

Szewczyk B., Poleszak E., Sowa- Kucma M., Siwek M., Dudek D., Ryszewska-Pokrasniewicz B., Radziwon-Zaleska M., Opoka W., Czekaj J., Pilc A.B & Nowak G. (2009). Antidepressant activity of zinc and magnesium in view of the current hypotheses of antidepressant action.*Pharmacological Reports* 60(2),pp. 588-599.ISSN 1734-1140

Takeda A.(2000). Movement of zinc and its functional significance in the brain. *Brain Research* 34(3).pp.137-148 ISSN 0165-0173

Thurston JH & Hauhart RE.(1989) Valproate doubles the anoxic survival time of normal developing mice: possible relevance to valproate-induced decreases in cerebral levels of glutamate and aspartate, and increases in taurine. *Life Sciences* 45,(1),pp.59–62 ISSN 0024-3205

Toro C &Deakin JF. (2005)NMDA receptor subunit NRI and postsynaptic protein PSD-95 in hippocampus and orbitofrontal cortex in schizophrenia and mood disorder. *Schizophenia Research* 80,323–330 ISSN 0014-4819

Treatment of severe mania with intravenous magnesium sulphate as a supplementary therapy. *Psychiatry Research* 89(3),.pp. 239-246 ISSN 0165-1781

Ueda Y & Willmore LJ. (2000) Molecular regulation of glutamate and GABA transporter proteins by valproic acid in rat hippocampus during epileptogenesis. *Experimental Brain Research.* 133,334–339. ISSN 0014-4819

Vacheron-Trystram M. N., Braitman A., Cheref S.& Auffray L. (2004) Antipsychotics in bipolar disorders. *Encephale* 30(5),pp. 417-424 ISSN 0013-7006

Vink R.& Cernak I.(2000) .Regulation of brain intracellular free magnesium following traumatic injury to the central nervous system. *Frontiers in Bioscience* 5,pp.656-665 ISSN 1093-9946

Walden J., Grunze H., Mayer A., Dusing R., Schirrmacher K., Liu Z.& Bingmann D.(1993) Calcium antagonistic effect of carbamazepine in epilepsies and affective psychoses. *Neuropsychobiology* 27(3), pp.171-173 ISSN 0960-2011

Weisler R. H., Keck P. E., Swann A. C., Cuttler A. J., Keller T. A.& Kalali A. H.,.(2005).Extended-release carbamazepine capsules as monotherapy for acute mania in bipolar disorder: a multicenter randomized, double-blind, placebo-controlled trial. *Journal of Clinical Psychiatry* 66(3),pp. 323-330. ISSN 0160-689

Yasuda S., Liang M.H., Marinova Z., Yahyavi A.& Chuang D. M.(2009).The mood stabilizers lithium and valproate selectivity activate the promoter IV of brain-derived neurotrophic factor in neurons. *Molecular Psychiatry* 14(1),pp. 51-59 ISSN 1359-4184

2

Anti-Stress Effects of Mood Stabilizers and Relevance to Their Therapeutic Actions

Young-Ki Chung and Seungmin Yoo
Department of Psychiatry, Ajou University School of Medicine
South Korea

1. Introduction

Bipolar disorder proves to be a far more serious mental illness than previously thought due to high suicide rate, great functional impairments, and chronicity (Belmaker, 2004; Post et al., 1996). In patients with bipolar disorder, reduction of volume in specific brain areas has consistently been found (Drevets et al., 1997). Although pathophysiology of bipolar disorder is far from being clear, alterations in neuroprotective and neurotrophic signaling cascades are implicated in pathophysiology of bipolar disorder (Schloesser et al., 2008).

The original concept of the mood stabilizer was the agent that was effective in treating acute manic episodes and preventing their relapses, thus "stabilizing" the manic pole of bipolar disorder (Stahl, 2008). More recently, the concept of mood stabilizer has been broadened to include any drug to treat bipolar disorder (Stahl, 2008). Lithium and valproate are two of the most prominent drugs approved by the United States Federal Food and Drug Administration (FDA) used for the treatment of bipolar disorder. Since John Cade discovered the efficacy of lithium in the control of acute manic episode in 1948 (Cade, 1949), lithium has been a prototypic mood stabilizer. Lithium has also been found to be effective for the prevention of recurrent manic and depressive episodes and for augmenting the activity of classical antidepressants in some depressed patients (Baldessarini et al., 1999). Anticonvulsant mood stabilizers, of which structures are totally unrelated to lithium, are also efficacious for the treatment of bipolar disorder. However, despite the long history of use for the treatment of bipolar disorder, the mechanism of actions of mood stabilizers still remains obscure.

Mood stabilizers are known to possess neuroprotective and neurotrophic properties. It is well known that representative mood stabilizers, lithium and valproate regulate multiple target molecules in neuroprotective and neurotrophic signaling cascades. Major signaling cascades regulated by mood stabilizers are brain derived neurotrophic factor (BDNF)and extracellular signal-regulated kinase (ERK) pathways, glycogen synthase kinase-3 (GSK-3) mediated pathway, and regulating pathways for bcl-2 expression (Manji et al., 2000; Shaltiel et al., 2007). In addition, regulation of arachidonic acid cascade by mood stabilizers is considered to be another common mechanism of action of mood stabilizers (Rao et al., 2008).

Psychosocial stress is a predisposing and precipitating factor to mood disorders like depression and bipolar disorder. Psychosocial stress affects brain functions profoundly and stress response is mediated through interactions between brain areas implicated in pathophysiology of mood disorders (McEwen, 2004). Psychosocial stress, for example, contributes to depressive and anxious symptoms in patients with affective illness (Caspi et al., 2003; Hammen, 2005; Hammen et al., 2004; Melchior et al., 2007). Environmental and psychological stressors can recapitulate biochemical, structural, and behavioral aspects of depressive illness in laboratory animals (Pittenger & Duman, 2008). Substantial evidence indicates that neural plasticity induced by depression and chronic stress share common features. However, antidepressant treatment has shown to produce opposing effects to those induced by stress (Pittenger & Duman, 2008). Mood stabilizers also prevent stress-induced neural plasticity (Bachmann et al., 2005; Bourin et al., 2009). Moreover, external stress enhances apoptosis by decreasing antiapoptotic and increasing proapoptotic molecules (Kubera et al., 2011). These molecules are also the targets of mood stabilizers.

In the context of these overlapping signaling cascades involved in bipolar disorder, actions of mood stabilizers, and stress effect it is conceivable to infer that protection of stress-induced neural plasticity contributes to the mood stabilizing effect. In this chapter, protective effects of mood stabilizers against stress-induced neural plasticity (anti-stress effects) and their relevance to therapeutic effects will be discussed. For neuroprotective and neurotrophic properties of mood stabilizers, readers can refer to numerous excellent papers elsewhere.

2. Preventive effect of mood stabilizers on stress induced structural plasticity

Preventive effects on stress-induced structural plasticity seems to be centered on lithium, so in this section, stress-prevention effect will be focused on lithium.

2.1 The effect of lithium on structural plasticity in the hippocampus and prefrontal cortex

Stress effects on the brain have been most extensively studied in the hippocampus. The hippocampus is one of the key structures of the brain mediating the stress response. It has the highest density of stress hormone receptors, glucocorticoids and mineralocorticoids receptors in the brain with potent inhibitory influence on the activity of the hypothalamic-pituitary-adrenal axis. Morphology and functions of the hippocampus are altered by acute or chronic stress (Brunson et al., 2003; Buwalda et al., 2005; Mirescu & Gould, 2006). The morphological alterations and functional impairments of the hippocampus are observed in a number of mental illnesses such as mood disorders, posttraumatic stress disorder, schizophrenia, and Alzheimer's disease (DeCarolis & Eisch, 2010). Sustained exposure to stress or glucocorticoids (GCs) leads to structural remodeling of the hippocampus with impairments in the performance of hippocampal-dependent cognitive functions (de Quervain et al., 1998; Magarinos & McEwen, 1995).

This structural remodeling effect is best studied in the CA3 region of the hippocampus. Sousa et al. (2000) reported that chronic stress or sustained GCs treatment induces structural remodeling in the CA1 as well as CA3 region with impairments in hippocampal-dependent learning and memory functions. Reduction in the length and branch point of apicall dendrites in the CA3 was observed following various types of stress. Daily restraint stress for 21 days or GCs exposure led to atrophy and the retraction of apical dendrites of CA3 pyramidal neurons in the rodent. Exposure to chronic psychosocial stress also resulted in structural remodeling of pyramidal neurons in the CA3 (Magarinos et al., 1996). This type of stress-induced structural plasticity is reversible because disrupted cellular morphology is slowly restored after cessation of stress without actual cell loss. Stress-induced dendritic atrophy in CA3 is mediated through decreased glucose uptake, increased glutamate and Ca^{2+}excitotoxicity, and decreased BDNF expression (Duman et al., 1999).

Chronic restraint stress results in the similar structural remodeling in the medial prefrontal cortex (mPFC) to that observed in the CA3 region of the hippocampus (Cook & Wellman, 2004; Radley et al., 2004). However, the apical dendritic atrophy following chronic stress is observed in the mPFC, but not in the orbital cortex (Liston et al., 2006), which suggests regional specificity of stress effect. Stress induces dendritic spine loss and altered patterns of spine morphology in the PFC in young rats, but not in aged rats (Bloss et al., 2011).

Lithium was found to prevent stress-induced dendritic atrophy of CA3 pyramidal neurons. Wood et al. (2004) reported that concomitant lithium treatment to rats prevented apical dendritic atrophy of CA3 pyramidal neurons caused by the exposure to restraint stress for 21 days with prevention of stress induced glutamatergic activation such as increase in glial glutamate transporter-1 mRNA expression and phosphorylation of cAMP response element binding protein (CREB) in the hippocampus. Alterations in cAMP signaling are implicated in mood disorders including bipolar disorder (Blendy, 2006; Karege et al., 2004). However, lithium treatment had no effect on dendritic morphology in non-stressed rats. Taken together, it is conceivable that lithium's ability in preventing stress-induced dendritic remodelling and increased phosphorylation of CREB may contribute to its mood stabilizing efficacy.

2.2 The effect of lithium on structural plasticity in amygdala

Prolonged stress exposure induced increase in dendritic length and branching of principal cells in the amygdala, which is in contrast to dendritic remodeling found in the hippocampus and medial prefrontal cortex (Vyas et al., 2002). This stress-induced dendritic hypertrophy in the amygdala is associated with increased anxiety-like behavior (Vyas et al., 2006). Increase in the size and function of the amygdala was demonstrated in depression (Drevets, 2003). Functional and structural abnormalities in the amygdala have been implicated in bipolar disorder (Garrett & Chang, 2008). Increased amygdala activity was observed in manic and euthymic bipolar patients during emotional discrimination tasks (Yurgelun-Todd et al., 2000). The preventive effect of lithium on stress-induced dendritic remodeling was demonstrated in the amygdala as well. Johnson et al. (2009) investigated the stress effect and lithium's effect on the dendritic morphology in the amygdala using the same study paradigm of Wood et al. (2004). They found that chronic lithium treatment prevented the stress-induced (restraint stress for 21 days) increase in dendritic length and branching of principal pyramidal neurons in the basolateral amygdala. However, like pyramidal neurons in CA3, lithium had no effect on dendritic morphology of the principal cells in the amygdala in non-stressed animals. This finding also suggests a specificity of neuroprotective action of lithium against stress.

2.3 The protective effect of lithium on stress-induced structural remodeling involves stabilization of glutamatergic activation

Lithium's preventive effect on dendritic morphology is thought to be mediated at least by the decrease and stabilization of the stress induced glutamatergic activation. Several lines of evidence support this notion. First, the dendritic remodeling by chronic stress in the CA3 area was prevented with N-methyl-D-aspartate (NMDA) antagonist (Magarinos & McEwen, 1995) or dilantin which reduces glutamate release (Watanabe et al., 1992). In addition, dendritic atrophy of pyramidal neurons in the mPFC which occurs during restraint stress was prevented by the administration of a competitive NMDA antagonist, ±3-(2-carboxypiperazin-4yl) propyl-1-phosphonic acid (CPP) (Martin & Wellman, 2011). Second, stress elevates extracellular glutamate level, thus facilitating glutamate signaling in the limbic brain. Chronic restraint stress, not acute stress, increases glutamate release and uptake in the hippocampus (Fontella et al., 2004; Lowy et al., 1993). Acute stress elevates extracellular glutamate level in the amygdala (Reznikov et al., 2007)and in the medial PFC (Moghaddam et al., 1994). Third, lithium is known to suppress glutamatergic signaling. Lithium treatment increases glutamate uptake into synaptosomes (Dixon & Hokin, 1998) and stimulates glutamate metabolism (Marcus et al., 1986). Fourth, lithium decreases the level of phosphorylation of NR2B subunit of NMDA receptor, thus suppressing the activity of NMDA receptor (Hashimoto et al., 2002). Lamotrigine which is an effective anticonvulsant mood stabilizer reduces glutamate release (E.S. Brown et al., 2010). These findings suggest that one of the key mechanisms of the therapeutic effect of lithium is the prevention of stress-induced dendritic remodelling by the stabilization of glutamate levels (Johnson et al., 2009).

3. The preventive effects of mood stabilizers on stress induced impairment in neurogenesis

3.1 Neurogenesis in the hippocampus

Neurogenesis in the adult brain has been found in most vertebrates including humans (Eriksson et al., 1998; Gould et al., 1999). Adult neurogenesis mainly occurs in the subgranular zone of the dentate gyrus in the hippocampus and subventricular zone of the ventricle wall (DeCarolis & Eisch, 2010). However, adult neurogenesis in the hippocampus is restricted to the subgranular zone of the dentate gyrus. Some researchers (Henn & Vollmayr, 2004) argue that impaired adult neurogenesis in the dentate gyrus may contribute to the reduced hippocampal volume in depressed patients although the contribution of neurogenesis in the dentate gyrus to the total volume of the hippocampus is considered very low in rodents and primates (Cameron & McKay, 2001; Kornack & Rakic, 1999). Adult neurogenesis is a multistep process. A brief description of adult neurogenesis is presented in this section (see Boku et al., 2010; Lucassen et al., 2010).

In the precursor cell phase, radial glia-like stem cells in the subgranular zone are, through transit amplifying cells, transformed to neural progenitor cells. Cell proliferation increases the number of neural progenitor cells. Cell fate determination is also believed to occur in transit amplifying cells. In the postmitotic cell phase progenitor cells are transformed to immature neurons in which axonal and dendritic extensions are initiated. These immature neurons are to be eliminated dramatically. This elimination process is apoptotic, NMDA receptor-mediated and input dependent (Biebl et al., 2000; Kuhn et al., 2005; Tashiro et al., 2006). Surviving immature neurons migrate to the granular cell layer. Immature neurons mature with the maturation of dendritic spines. Newly formed mature neurons are finally incorporated to the hippocampal circuitry with projections to the CA3 neurons through mossy fiber pathway. These newborn granule cells have lower threshold for the induction of long term potentiation (LTP) and their survival depends on input dependent activity (Schmidt-Hieber et al., 2004). If integration to the pre-existing network does not occur, newborn cells are rapidly eliminated by apoptosis (Dayer et al., 2005).

3.2 The effect of mood stabilizers on stress-induced inhibition of neurogenesis in the hippocampus

Adult neurogenesis in the hippocampus is regulated by various environmental factors and age (J. Brown et al., 2003; Kempermann et al., 1997). With increasing age, adult neurogenesis rapidly declines (Manganas et al., 2007). Factors regulating adult neurogenesis include stress, exercise, dietary restriction, and an enriched environment. Stress is a potent negative regulator of adult neurogenesis (Boku et al., 2010). Exposure to both psychosocial and physical stressors inhibits one or more subphases of adult neurogenesis (Czeh et al., 2002; Gould et al., 1997; Malberg & Duman, 2003; Pham et al., 2003). Various types of stress induce impairments of neurogenesis. Results obtained from several studies on the stress effects on adult neurogenesis are to be presented. Thomas et al. (2007) demonstrated that acute social dominance stress inhibited cell survival but not cell proliferation. Chronic intermittent restraint stress, but not acute stress, reduced progenitor cell proliferation without affecting levels of expression in BDNF, growth associated protein-43 (GAP-43), and synaptophysin (Rosenbrock et al., 2005). In learned helplessness, an animal model of depression induced by stress, stress acutely inhibits cell proliferation regardless of the induction of learned helplessness (Heine et al., 2004). Suppression of cell proliferation and enhanced apoptosis were noticed in the dentate gyrus of pups with maternal separation, which was reversed by fluoxetine treatment (Lee et al., 2001).

Mood stabilizers are known to prevent stress-induced reduction in neurogenesis, which requires multiple target molecules involving neuroprotective and neurotrophic signaling pathways. Chronic psychosocial stress in tree shrews reduced cell proliferation in the dentate gyrus and reduced hippocampal volume that were prevented by chronic treatment with an antidepressant, tianeptine (Czeh et al., 2001). Chronic mild stress (chronic mild stress) resulted in decrease in cell proliferation and differentiation, which was paralleled by depression-like behavior in forced swim test in rats. This chronic mild stress-induced decrease in neurogenesis was prevented by chronic lithium treatment. In addition, this effect was mediated at least through inhibition of glycogen synthase kinase-3β by lithium (Silva et al., 2008).

Chronic restraint stress induced suppression of cell proliferation which was accompanied by decreased expression of BDNF in the hippocampus (H. Xu et al., 2006). Boku et al. (2011) demonstrated that lithium and valproate, but not carbamazepine and lamotrigine, prevented the decrease in dentate gyrus-derived neural precursor cell proliferation induced by dexamethasone. However, all four mood stabilizers blocked apoptosis of dentate gyrus-derived neural precursor cells. This suggests that effects of mood stabilizers on adult neurogenesis in the dentate gyrus contribute to their therapeutic actions. This group also showed that lithium reversed glucocorticoids-induced decrease in cell proliferation using dentate gyrus-derived neural precursor cells (Boku et al., 2010).

Interestingly, a study showed that treatment with valproate reduced cell proliferation in the subgranular zone of the dentate gyrus within the hippocampus, which was linked to significant impairment in their ability to perform a hippocampal-dependent spatial memory test. Contrary to expectation, valproate treatment caused a significant reduction in BDNF (Umka et al., 2010). But this study did not examine effects of valproate on neurogenesis under the stressed condition.

Stress-induced reduction in neurogenesis is also associated with reduction in vascular endothelial growth factor. Both angiogenesis and neurogenesis can be modulated by similar stimuli (Fabel et al., 2003). Vascular endothelial growth factor (VEGF) also has neurogenic effect (During & Cao, 2006; Silva et al., 2007). Chronic stress induces reduction in VEGF expression in the hippocampus (Heine et al., 2005; Silva et al., 2007) with concomitant reduction in proliferating cells (Heine et al., 2005). On the other hand, lithium treatment attenuated stress-induced reduction in VEGF expression and prevented stress-induced upregulation of GSK-3β and stress-induced β-catenin. These results suggest that protection by lithium against stress-induced impairment of neurogenesis can be mediated through the GSK-3β/β-catenin/VEGF signaling pathway (Silva et al., 2007).

Early life stress causes decrease and increase in cell proliferation and apoptosis respectively. Rat pups with maternal separation stress showed decreased cell proliferation and increased apoptosis, which was attenuated by concomitant fluoxetine treatment (Lee et al., 2001). In line with this, a study showed that early life stress effects could be counteracted by lithium treatment. Maternal deprivation induced stress decreased neuropeptide Y-like immunoreactivity in the hippocampus and striatum and increased neuropeptide Y-like immunoreactivity and corticototrophin releasing hormone-like immunoreactivity in the hypothalamus. Lithium treatment counteracted maternal deprivation effects by increasing neuropeptide Y-like immunoreactivity in the hippocampus and striatum and decreasing corticototrophin releasing hormone-like immunoreactivity in the hypothalamus (Husum & Mathe, 2002).

In addition to their preventive effect of lithium against stress-induced impairment of neurogenesis, lithium and valproate by themselves increase neurogenesis in the dentate gyrus of adult animals (G. Chen et al., 2000; Hao et al., 2004; Son et al., 2003). Neurogenesis promoting effect of lithium and valproate is ascribed to their capability to activate extracellular signal regulated kinase (ERK) signaling cascade (G. Chen et al., 2000; Hao et al., 2004).

4. The effect of mood stabilizers on stress-induced impairments in synaptic plasticity

Induction of hippocampal long term synaptic plasticity is profoundly affected by stress (Yang et al., 2004). It is well known that mild and transient stress enhances hippocampal-dependent learning and memory (Luine et al., 1996). However, exposure to more prolonged stress or severe stress definitely impairs hippocampal-dependent cognition (Sapolsky, 2003). Hippocampal-dependent learning and memory is also disrupted following chronic exposure to GCs or exposure to high dose of GCs (Joels, 2001). Stress-induced impairment of LTP is considered to contribute to impaired hippocampal-dependent memory. Rats exposed to inescapable stress shows impairment in hippocampal-dependent memory and LTP and this impairment is mimicked by NMDA antagonist, (2R)-amino-5-phosphonovalenic acid (APV), injection into the dorsal hippocampus (Baker & Kim, 2002). Thus, the relationship between hippocampal-dependent cognitive performance and level of stress or GCs takes the inverted U pattern. Stress or GCs disrupts LTP in the various hippocampal subfields (Diamond & Park, 2000; Foy et al., 1987; Shors & Dryver, 1994; Shors et al., 1989). But similar stress paradigms resulting in impairment of LTP enhance long term depression (LTD) (L. Xu et al., 1997).

Inverted U pattern can also be applied to LTP, i.e. milder level of exposure to stress or GCs enhances LTP while more severe level of exposure to stress or GCs impairs LTP (Diamond et al., 1992; Pavlides et al., 1994). The mechanism underlying this inverted U pattern is ascribed to differential occupancy of mineralocorticoid receptors (MR) and glucocorticoid receptors (GR). Under mild stress or low GCs condition, heavy occupancy or saturation of MR by GCs is believed to be responsible to enhancement of cognition and LTP (Sapolsky, 2003). Previous studies showed that MR occupancy enhances LTP (Pavlides et al., 1995; Vaher et al., 1994) and hippocampal-dependent spatial memory tasks (Oitzl et al., 1998).

The mechanism of LTP will be briefly described according the review by Pittenger and Duman (2008). Local elevation of cAMP and Ca^{2+} induces short term synaptic plasticity, i.e. early LTP. Ca^{2+} influx through NMDA receptors or voltage gated calcium channel activates Ca^{2+}/calmodulin dependent kinases (CaMK). Among them, Ca^{2+}/calmodulin dependent kinase II (CaMK II) is critical for early LTP. CaMK II and other CaMK phosphorylate the GluR 1 subunit of AMPA receptors, which renders AMPA receptors on the postsynaptic membrane potentiated in function. In addition, phosphorylation of AMPA receptors promotes insertion of AMPA receptors on the postsynaptic membrane. Once activated by local increase in Ca^{2+}, CaMK II phosphorylates itself persistently even after fall in local Ca^{2+} level. Thus, this property of CaMK II makes CaMK II a suitable molecule for short term synaptic change. However, long term synaptic change requires other signaling pathways: cAMP-dependent protein kinase (PKA) and mitogen-activated protein kinase (MAPK) pathways. Stimulation of PKA and MAPK pathways results in activation of regulated transcription factors which induce new genes for late LTP (L-LTP). CaMK IV is also an important molecule for L-LTP which activates regulated transcription factors like CREB.

Stress-induced synaptic plasticity seems to result at least from activation of ERK1/2 pathway by stress (Yang et al., 2004). Yang et al. (2004) showed that stress-induced suppression of LTP and enhancement of LTD in the CA1 was blocked by pretreatment with GR antagonist, 11β, 17β-11[4-(dimethylamino)phenyl]-17-hydroxy-17-(1-propynyl)-estra-4-9-dien-3-one. Further, this stress effect in LTP and LTD was blocked by specific inhibitors of MEK1/2, protein kinase C, tyrosine kinase, and BDNF antisense oligonucleotides, suggesting the involvement of corticosterone-induced sustained activation of ERK1/2-coupled signaling cascades in the stress effect on LTP and LTD.

CREB is also involved in the regulation of numerous types of synaptic changes in the hippocampus, amygdala, and cortex. Since stress and mood stabilizers are known to regulate these signaling pathways involved in the mediation of LTP, it is conceivable that stress-induced synaptic alterations can be remedied by the treatment with mood stabilizers. There is substantial evidence supporting this notion.

Lithium by itself facilitates LTP in hippocampal subregions. Chronic treatment with lithium (28 days) (Nocjar et al., 2007) and subchronic treatment with lithium (14 days) enhanced LTP in the CA1 and in the dentate gyrus (Shim et al., 2007), respectively. Lithium is believed to regulate signaling cascades mediating the induction of LTP (Shim et al., 2007). Son et al. (2003) showed that acute and chronic lithium treatment enhanced LTP in the dentate gyrus of the rat. They argued that LTP enhancing effect of lithium was mediated through upregulation of signaling molecules like BDNF, phosphorylated MAPK, phosphorylated CREB, CaMK II, phosphorylated Elk, and TrkB, but not through increased neurogenesis. Silva et al. (2008) showed that lithium prevented chronic mild stress-induced upregulation GSK-3β and downregulation of its downstream molecules bcl-2 associated athanogenes-1 and synapsin-I. Lithium also blocked depressive-like behavior resulting from chronic mild stress via inhibition of GSK-3β. A study showed that acute immobilization stress impaired LTP induction in the CA1 region, which was restored by addition of lithium of therapeutic concentration to artificial CSF in brain slices. However, the addition of lithium to slices from non-stressed animals had no effect (Lim et al., 2005).

5. The protective effect of mood stabilizers against stress-induced apoptosis and molecular changes

As mentioned above, exposure to stress results in increase in proapoptotic molecules and decrease in antiapoptotic molecules in laboratory animals (for review, see Kubera et al., 2011). On the other hand chronic stress reduces the expression of cell surviving molecules such as bcl-2 family antiapoptotic proteins, bcl-2 and BAG-1, brain derived neurotrophic factor (BDNF), and vascular growth factor (VGF) (Nair et al., 2007; Thomas et al., 2007). It is well known that primary mood stabilizers, lithium and valproate, increase antiapoptotic proteins of bcl-2 family such as bcl-2 and bcl-xl. In addition, chronic stress activates proinflammatory cytokines and induces neuroinflammatory changes (Kubera et al., 2011). Increased apoptosis following chronic stress was observed in the hippocampal subregions and entorhinal cortex of the tree shrew (Lucassen et al., 2001).

Several previous studies suggest that mood stabilizers protect or counteract these stress-induced apoptosis or molecular changes. Bachis et al. (2008) demonstrated that chronic unpredictable mild stress for five weeks promoted neuronal apoptosis by demonstrating increased caspase-3 positive neurons in the rat cortex. This effect was reversed by treatment with desipramine. Given the fact that the most prominent mood stabilizers, lithium and valproate upregulate anti-apoptotic proteins like bcl-2 (G. Chen et al., 1999; R.W. Chen & Chuang, 1999) and suppress proapoptotic proteins, p53 and Bax (R.W. Chen & Chuang, 1999) and other anticonvulsant mood stabilizers share this neuroprotective property (X. Li et al., 2002), it is conceivable that mood stabilizers protect against stress-induced apoptosis or molecular changes.

Actually, several lines of evidence support this notion. Miller and Mathé (1997) reported that chronic lithium treatment attenuated c-fos mRNA induction by acute injection stress in the rat frontal cortex and hippocampus. The same group reported that chronic lithium treatment attenuated AP-1 DNA binding induction by acute restraint stress in the frontal cortex and hippocampus and CREB binding in the frontal cortex (Miller et al., 2007). In line with these findings, chronic stress-induced increase in CREB phosphorylation and CREB transcriptional activity was reversed with concomitant chronic lithium treatment (Boer et al., 2008) and imipramine treatment (Boer et al., 2007).

Kosten et al. (2008) demonstrated that chronic stress reduced the expression of antiapoptotic protein genes (bcl-2 and bcl-xl) in the limbic brain, which was prevented by the treatment with different classes of antidepressant. They also showed that a major proaptotic protein, Bax gene expression was repressed by fluoxetine. Lithium is also known to block chronic mild stress-induced depressive-like behavior. This blockade of depressive-like behavior is accompanied by prevention of mild stress-induced increase of GSK-3β and decrease in its downstream targets, BAG-1, and synapsin- I (Silva et al., 2008). This finding suggests that neuroprotective action of lithium against chronic stress effect involves the suppression of GSK-3β.

6. Stress-induced neuroinflammatory reactions are likely to be prevented by mood stabilizers

Exposure to psychosocial stressors leads to increased proinflammatory molecules as well as behavioral or mood disturbances in humans and animals. Stress-induced depressive-like behavior is associated with increase in interleukin 1-β (IL-1β), tumor necrosis factor-α (TNF-α), IL-6, nuclear factor κB (NFκB), cyclooxygenase-2(COX-2), expression of Toll-like receptors and lipid peroxidation in animals (Kubera et al., 2011). In humans, various chronic psychosocial stressors like burnout, low socioeconomic status, childhood adversity and maltreatment, and loneliness also affect adversely stress response and induce increase in proinflammatory molecules (Hansel et al., 2010). Increased level of proinflammatory cytokines was observed in patients with stress-related disorders like depression (Maes, 1995) and posttraumatic stress disorder (Hoge et al., 2009). Depressed and manic states in bipolar disorder are considered to be a proinflammatory state. A study about cytokine levels in euthymic bipolar patients suggested that proinflammatory state resolved in euthymic state (Guloksuz et al., 2010). On the other hand, antidepressant therapy decreases proinflammatory cytokines, IL-β, IL-6, and TNF-α (Kubera et al., 2011). This line of studies suggests that pathophysiology of bipolar disorder and stress effects share the mechanism leading to proinflammatory state.

Chronic stress-induced increase in proinflammatory cytokines is associated with anhedonia (decreased sucrose intake) in laboratory animals (Grippo et al., 2005). Decrease in hippocampal IL-1 was observed following chronic stress and only chronic IL-1 injection caused the similar behaviors which were observed in animals exposed to chronic stress (Goshen & Yirmiya, 2009). There is evidence to support that IL-1β is a mediator of stress-induced behaviors (Badowska-Szalewska et al., 2009). A study suggested IL-1β activity is coupled to phospholipase A_2 (PLA_2) by showing inhibition of IL-1β activity with nonspecific PLA_2 antagonist, quinacrine (Song et al., 2007). Besides, IL-β, TNF-α and NFκB increase in the brain after chronic stress (Grippo et al., 2005; Gu et al., 2009; Madrigal et al., 2002; Madrigal et al., 2001). NFκB recognizes specific DNA sequences in the promoters of genes encoding proinflammatory factors including cyclooxygenase-2 (COX-2) and inducible nitric oxide synthase(iNOS) (Kubera et al., 2011). Stress also increases reactive oxygen species and iNOS (Madrigal et al., 2001; Olivenza et al., 2000).

On the other hand, it is known that chronic treatment with lithium, carbamazepine, and valproate at therapeutically relevant doses decrease arachidonic acid turnover and the level of COX-2 and prostaglandin E_2 (PGE_2) in the rat brain (Chang et al., 1996; Rao et al., 2008). Lithium and carbamazepine reduce the AP-2 binding activity, which leads to decreased transcription, translation and activity of arachidonic acid-specific and calcium dependent phospholipase A_2 ($cPLA_2$) (Bosetti et al., 2003; Rao et al., 2008), a PLA_2 isoform specific for release of arachidonic acid from membrane phospholipids of the brain, whereas valproate inhibits long chain acyl-CoA synthetase and thus decreases the arachidonic acid turnover (Rao et al., 2008). However, chronic treatment with lithium does not affect COX-1 or Ca^{2+}-independent PLA_2 (Bosetti et al., 2002). Chronic treatment with valproate also decreases DNA biding activity of NFκB in the rat frontal cortex (Rao et al., 2007a). Chronic administration of lamotrigine, an effective mood stabilizer blocking the relapse of depressive episode, decreases the level of COX-2 protein and mRNA expression in the rat frontal cortex without affecting the protein level of PLA_2 (Lee et al., 2008).

In addition, decreased turnover by chronic treatment with mood stabilizers is associated with increased expression of bcl-2 and BDNF in the brain. Chronic deprivation of dietary essential n-3 polyunsaturated fatty acids results in bipolar disorder-like symptoms in rats (DeMar et al., 2006) and increased expression of $cPLA_2$ and COX-2 (DeMar et al., 2006; Rao et al., 2007b). Deprivation of essential n-3 polyunsaturated fatty acids also leads to decreased expression of phosphorylated CREB and BDNF mRNA and proteins (Rao et al., 2007c).

7. Mood stabilizers attenuate stress-induced oxidative stress

Chronic stress increases the level of oxidative stress. Lucca et al. (2009) reported that chronic stress increased thiobarbituric acid reactive substances (TBARS) in the brain, a parameter of increased oxidative stress. In line with this study, lithium and valproate reversed amphetamine treatment-induced elevation of TBARS and prevented hyperactivity in an animal model of mania (Frey et al., 2006). These studies suggest that mood stabilizers, lithium and valproate, share the capability to prevent stress-induced increase in oxidative stress and that this anti-oxidant effect may contribute to their therapeutic actions. Lithium and valproate at therapeutically relevant concentrations in humans inhibited glutamate-induced increase in intracellular free Ca2+ concentration, lipid peroxidation, and protein oxidation in cultured rat cerebral cortical cells, which suggests that lithium and valproate inhibit glutamate-induced excitotoxicity by inhibiting oxidative stress (Shao et al., 2005).

8. Are mood stabilizers efficacious for treating traumatic stress-related disorders?

Experiences can modify gene functions without DNA sequence changes and these mechanisms are called epigenetic modification. Epigenetic modification encompasses DNA methylation and histone acetylation and methylation (Krishnan & Nestler, 2008). Deprivation of maternal care in rat pups resulted in increased methylation of promoter region of GR gene in the hippocampus (Tsankova et al., 2007), and thus repressed gene expression. Early maltreatment also led to lasting increased methylation and decreased expression of BDNF gene in the prefrontal cortex of adult rats (Roth et al., 2009). A human study reported the association of higher level of methylation in serotonin transporter promoter with increased risk to unresolved loss or other trauma among adoptees (van et al., 2010). In an animal model of posttraumatic stress disorder (PTSD), traumatized rats showed hypermethylation of BDNF gene in the dorsal CA1 with reduced expression of BDNF mRNA. However, decreased methylation of BDNF gene was observed in the CA3, which suggests region-specificity (Roth et al., 2011). Epigenetic modification is involved in stress-related disorders like depression, bipolar disorder, and PTSD (Mill & Petronis, 2007; Petronis, 2003). On the other hand, among mood stabilizers, valproate is a histone deacetylase (HDAC) inhibitor and HDAC is a direct target of valproate (Phiel & Klein, 2001). Neuroprotective action of valproate is associated with hyperacetylation of histone (Chuang, 2005). Valproate was found to produce DNA demethylation and hyperacetylation in glutamate transporter-1 gene promoter in cultured primary astrocytes (Perisic et al., 2010).

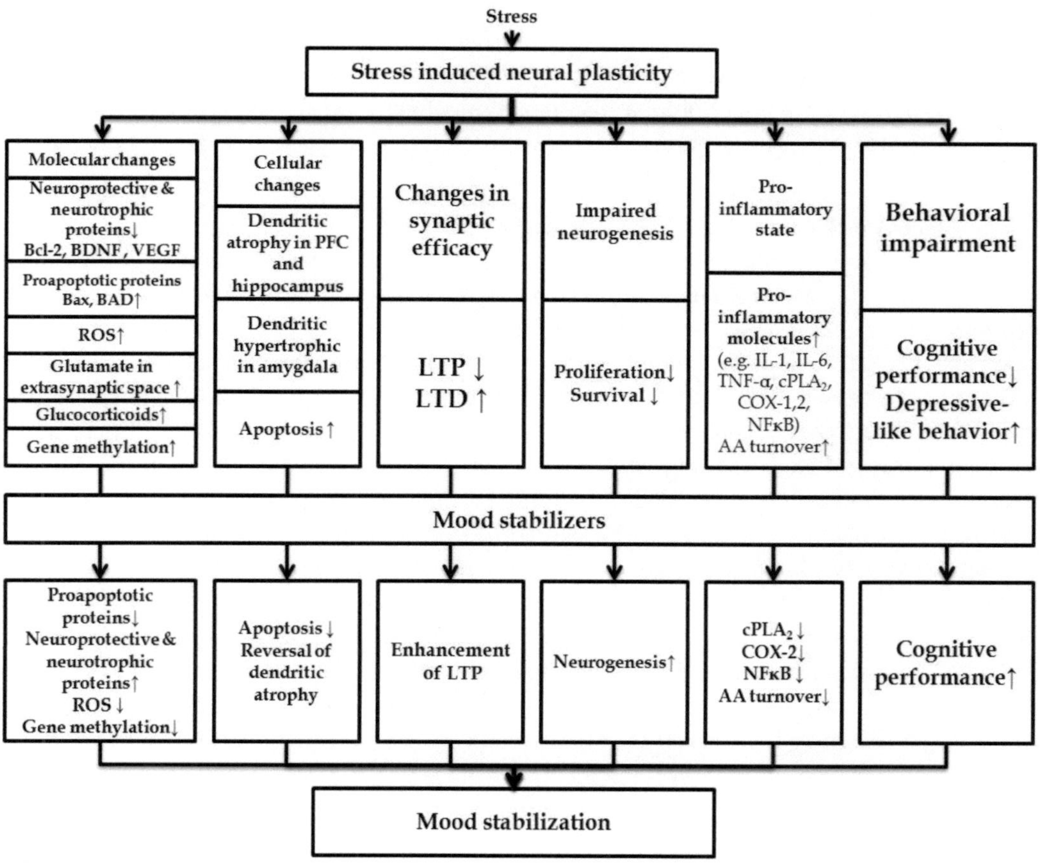

Fig. 1. Schematic representation of stress induced neural changes antistress effects of mood stabilizers. Abbreviations used: Bcl-2: B-cell leukemia/lymphoma-2; BDNF: brain-derived neurotrophic factor; VEGF: vascular endothelial growth factor; Bax: bcl-2-associated X protein; BAD: bcl-associated death protein; ROS: reactive oxygen species; PFC: prefrontal cortex; LTP: long-term potentiation; LTD: long-term depression; IL-1,6: interleukin-1,6; TNF-α: tumor necrosis factor-α; $cPLA_2$: cytosolic phospholipase A_2; COX-1,2: cyclooxgenase-1,2; NFκB: nuclear transcription factor kappa B; AA: arachidonic acid

Taken together, it is suggested that mood stabilizers, especially valproate might be efficacious for treating traumatic stress-related disorders like PTSD. A systemic review suggest that valproate can be effective for the treatment of PTSD by reducing hyperarousal, improving irritability, anger outburst, and mood although it remains that more double blind studies are needed (Adamou et al., 2007). Lamotrigine significantly suppressed stress responses to public speech including suppression of stress hormones such as growth hormones and cortisol (Makatsori et al., 2004).

9. Conclusion

We live in a world full of stress. Inescapable changes in our life styles are imposing tremendous stress on us. Stress profoundly affects morphology and functions of the brain which is the central organ mediating stress response (McEwen, 2004). In the mental health area, we are increasingly faced with heavy burden of stress-related mental disorders like depression and bipolar disorder. Even though bipolar disorder is a serious mental disorder, pharmacological treatment of bipolar disorder is not always satisfactory. Pathophysiology of bipolar disorder and the mechanism of therapeutic actions of mood stabilizers remain to be elucidated.

Stress effects on the brain i.e. altered structural plasticity, impaired neurogenesis, altered synaptic efficacy, accelerated apoptosis, cognitive impairment, and proclivity to neuroinflammation are mediated through overlapping cascades and molecules involved in cell survival and death. For example, exposure to chronic mild stress in mice produces cognitive impairment accompanied by increased proinflammatory cytokines and stress hormones, decreased BDNF, and enhanced cell damage (S. Li et al., 2008). Furthermore, mood stabilizers have capabilities to reverse stress-induced behavioral impairments. Chronic stress-induced spatial memory impairment is attenuated by chronic lithium treatment (Vasconcellos et al., 2003). Chronic or subchronic treatment with lamotrigine ameliorates behavioral parameters in stress-induced animal models of depression with concomitant restoration of stress-induced downregulation of neuroprotective proteins (N. Li et al., 2010; N. Li et al., 2011). Interestingly, protective effects against stress-induced deleterious effects on the brain (anti-stress effects) seem to be shared by most mood stabilizers. As described in the preceding sections, effects of stress are mediated through signaling cascades that are also the targets of mood stabilizers. In this context, mood stabilizing effects may be ascribed to anti-stress effects of mood stabilizers.

10. Acknowledgment

The authors would like to thank Dr. Jeewon Lee and Dr. Hye-won Baek for their contribution to the revision process of this article.

11. References

Adamou, M., Puchalska, S., Plummer, W.&Hale, A. S. (2007). Valproate in the Treatment of Ptsd: Systematic Review and Meta Analysis. *Curr Med Res Opin,* Vol.23, No.6, (Jun, 2007), pp. (1285-1291), ISSN. 1473-4877 (Electronic)

Bachis, A., Cruz, M. I., Nosheny, R. L.&Mocchetti, I. (2008). Chronic Unpredictable Stress Promotes Neuronal Apoptosis in the Cerebral Cortex. *Neurosci Lett,* Vol.442, No.2, (Sep 12, 2008), pp. (104-108), ISSN. 0304-3940 (Print)

Bachmann, R. F., Schloesser, R. J., Gould, T. D.&Manji, H. K. (2005). Mood Stabilizers Target Cellular Plasticity and Resilience Cascades: Implications for the Development of Novel Therapeutics. *Mol Neurobiol,* Vol.32, No.2, (Oct, 2005), pp. (173-202), ISSN. 0893-7648 (Print)

Badowska-Szalewska, E., Klejbor, I., Sidor-Kaczmarek, J., Cecot, T., Lietzau, G., Spodnik, J. H.&Morys, J. (2009). Stress-Induced Changes of Interleukin-1beta within the Limbic System in the Rat. *Folia Morphol (Warsz),* Vol.68, No.3, (Aug, 2009), pp. (119-128), ISSN. 0015-5659 (Print)

Baker, K. B.&Kim, J. J. (2002). Effects of Stress and Hippocampal Nmda Receptor Antagonism on Recognition Memory in Rats. *Learn Mem,* Vol.9, No.2, (Mar-Apr, 2002), pp. (58-65), ISSN. 1072-0502 (Print)

Baldessarini, R. J., Tondo, L.&Hennen, J. (1999). Effects of Lithium Treatment and Its Discontinuation on Suicidal Behavior in Bipolar Manic-Depressive Disorders. *J Clin Psychiatry,* Vol.60 Suppl 2, 1999), pp. (77-84; discussion 111-116), ISSN. 0160-6689 (Print)

Belmaker, R. H. (2004). Bipolar Disorder. *N Engl J Med,* Vol.351, No.5, (Jul 29, 2004), pp. (476-486), ISSN. 1533-4406 (Electronic)

Biebl, M., Cooper, C. M., Winkler, J.&Kuhn, H. G. (2000). Analysis of Neurogenesis and Programmed Cell Death Reveals a Self-Renewing Capacity in the Adult Rat Brain. *Neurosci Lett,* Vol.291, No.1, (Sep 8, 2000), pp. (17-20), ISSN. 0304-3940 (Print)

Blendy, J. A. (2006). The Role of Creb in Depression and Antidepressant Treatment. *Biol Psychiatry,* Vol.59, No.12, (Jun 15, 2006), pp. (1144-1150), ISSN. 0006-3223 (Print)

Bloss, E. B., Janssen, W. G., Ohm, D. T., Yuk, F. J., Wadsworth, S., Saardi, K. M., McEwen, B. S.&Morrison, J. H. (2011). Evidence for Reduced Experience-Dependent Dendritic Spine Plasticity in the Aging Prefrontal Cortex. *J Neurosci,* Vol.31, No.21, (May 25, 2011), pp. (7831-7839), ISSN. 1529-2401 (Electronic)

Boer, U., Alejel, T., Beimesche, S., Cierny, I., Krause, D., Knepel, W.&Flugge, G. (2007). Cre/Creb-Driven up-Regulation of Gene Expression by Chronic Social Stress in Cre-Luciferase Transgenic Mice: Reversal by Antidepressant Treatment. *PLoS One,* Vol.2, No.5, 2007), pp. (e431), ISSN. 1932-6203 (Electronic)

Boer, U., Cierny, I., Krause, D., Heinrich, A., Lin, H., Mayr, G., Hiemke, C.&Knepel, W. (2008). Chronic Lithium Salt Treatment Reduces Cre/Creb-Directed Gene Transcription and Reverses Its Upregulation by Chronic Psychosocial Stress in Transgenic Reporter Gene Mice. *Neuropsychopharmacology,* Vol.33, No.10, (Sep, 2008), pp. (2407-2415), ISSN. 1740-634X (Electronic)

Boku, S., Nakagawa, S.&Koyama, T. (2010). Glucocorticoids and Lithium in Adult Hippocampal Neurogenesis. *Vitam Horm,* Vol.82, 2010), pp. (421-431), ISSN. 0083-6729 (Print)

Boku, S., Nakagawa, S., Masuda, T., Nishikawa, H., Kato, A., Toda, H., Song, N., Kitaichi, Y., Inoue, T.&Koyama, T. (2011). Effects of Mood Stabilizers on Adult Dentate Gyrus-Derived Neural Precursor Cells. *Prog Neuropsychopharmacol Biol Psychiatry,* Vol.35, No.1, (Jan 15, 2011), pp. (111-117), ISSN. 1878-4216 (Electronic)

Bosetti, F., Rintala, J., Seemann, R., Rosenberger, T. A., Contreras, M. A., Rapoport, S. I.&Chang, M. C. (2002). Chronic Lithium Downregulates Cyclooxygenase-2 Activity and Prostaglandin E(2) Concentration in Rat Brain. *Mol Psychiatry,* Vol.7, No.8, 2002), pp. (845-850), ISSN. 1359-4184 (Print)

Bosetti, F., Weerasinghe, G. R., Rosenberger, T. A.&Rapoport, S. I. (2003). Valproic Acid Down-Regulates the Conversion of Arachidonic Acid to Eicosanoids Via Cyclooxygenase-1 and -2 in Rat Brain. *J Neurochem,* Vol.85, No.3, (May, 2003), pp. (690-696), ISSN. 0022-3042 (Print)

Bourin, M., Chenu, F.&Hascoet, M. (2009). The Role of Sodium Channels in the Mechanism of Action of Antidepressants and Mood Stabilizers. *Curr Drug Targets,* Vol.10, No.11, (Nov, 2009), pp. (1052-1060), ISSN. 1873-5592 (Electronic)

Brown, E. S., Zaidel, L., Allen, G., McColl, R., Vazquez, M.&Ringe, W. K. (2010). Effects of Lamotrigine on Hippocampal Activation in Corticosteroid-Treated Patients. *J Affect Disord,* Vol.126, No.3, (Nov, 2010), pp. (415-419), ISSN. 1573-2517 (Electronic)

Brown, J., Cooper-Kuhn, C. M., Kempermann, G., Van Praag, H., Winkler, J., Gage, F. H.&Kuhn, H. G. (2003). Enriched Environment and Physical Activity Stimulate Hippocampal but Not Olfactory Bulb Neurogenesis. *Eur J Neurosci,* Vol.17, No.10, (May, 2003), pp. (2042-2046), ISSN. 0953-816X (Print)

Brunson, K. L., Chen, Y., Avishai-Eliner, S.&Baram, T. Z. (2003). Stress and the Developing Hippocampus: A Double-Edged Sword? *Mol Neurobiol,* Vol.27, No.2, (Apr, 2003), pp. (121-136), ISSN. 0893-7648 (Print)

Buwalda, B., Kole, M. H., Veenema, A. H., Huininga, M., de Boer, S. F., Korte, S. M.&Koolhaas, J. M. (2005). Long-Term Effects of Social Stress on Brain and Behavior: A Focus on Hippocampal Functioning. *Neurosci Biobehav Rev,* Vol.29, No.1, (Feb, 2005), pp. (83-97), ISSN. 0149-7634 (Print)

Cade, J. F. (1949). Lithium Salts in the Treatment of Psychotic Excitement. *Med J Aust,* Vol.2, No.10, (Sep 3, 1949), pp. (349-352), ISSN. 0025-729X (Print)

Cameron, H. A.&McKay, R. D. (2001). Adult Neurogenesis Produces a Large Pool of New Granule Cells in the Dentate Gyrus. *J Comp Neurol,* Vol.435, No.4, (Jul 9, 2001), pp. (406-417), ISSN. 0021-9967 (Print)

Caspi, A., Sugden, K., Moffitt, T. E., Taylor, A., Craig, I. W., Harrington, H., McClay, J., Mill, J., Martin, J., Braithwaite, A.&Poulton, R. (2003). Influence of Life Stress on Depression: Moderation by a Polymorphism in the 5-Htt Gene. *Science,* Vol.301, No.5631, (Jul 18, 2003), pp. (386-389), ISSN. 1095-9203 (Electronic)

Chang, M. C., Grange, E., Rabin, O., Bell, J. M., Allen, D. D.&Rapoport, S. I. (1996). Lithium Decreases Turnover of Arachidonate in Several Brain Phospholipids. *Neurosci Lett,* Vol.220, No.3, (Dec 20, 1996), pp. (171-174), ISSN. 0304-3940 (Print)

Chen, G., Rajkowska, G., Du, F., Seraji-Bozorgzad, N.&Manji, H. K. (2000). Enhancement of Hippocampal Neurogenesis by Lithium. *J Neurochem,* Vol.75, No.4, (Oct, 2000), pp. (1729-1734), ISSN. 0022-3042 (Print)

Chen, G., Zeng, W. Z., Yuan, P. X., Huang, L. D., Jiang, Y. M., Zhao, Z. H.&Manji, H. K. (1999). The Mood-Stabilizing Agents Lithium and Valproate Robustly Increase the Levels of the Neuroprotective Protein Bcl-2 in the Cns. *J Neurochem,* Vol.72, No.2, (Feb, 1999), pp. (879-882), ISSN. 0022-3042 (Print)

Chen, R. W.&Chuang, D. M. (1999). Long Term Lithium Treatment Suppresses P53 and Bax Expression but Increases Bcl-2 Expression. A Prominent Role in Neuroprotection against Excitotoxicity. *J Biol Chem,* Vol.274, No.10, (Mar 5, 1999), pp. (6039-6042), ISSN. 0021-9258 (Print)

Chuang, D. M. (2005). The Antiapoptotic Actions of Mood Stabilizers: Molecular Mechanisms and Therapeutic Potentials. *Ann N Y Acad Sci,* Vol.1053, (Aug, 2005), pp. (195-204), ISSN. 0077-8923 (Print)

Cook, S. C.&Wellman, C. L. (2004). Chronic Stress Alters Dendritic Morphology in Rat Medial Prefrontal Cortex. *J Neurobiol,* Vol.60, No.2, (Aug, 2004), pp. (236-248), ISSN. 0022-3034 (Print)

Czeh, B., Michaelis, T., Watanabe, T., Frahm, J., de Biurrun, G., van Kampen, M., Bartolomucci, A.&Fuchs, E. (2001). Stress-Induced Changes in Cerebral Metabolites, Hippocampal Volume, and Cell Proliferation Are Prevented by Antidepressant Treatment with Tianeptine. *Proc Natl Acad Sci U S A,* Vol.98, No.22, (Oct 23, 2001), pp. (12796-12801), ISSN. 0027-8424 (Print)

Czeh, B., Welt, T., Fischer, A. K., Erhardt, A., Schmitt, W., Muller, M. B., Toschi, N., Fuchs, E.&Keck, M. E. (2002). Chronic Psychosocial Stress and Concomitant Repetitive Transcranial Magnetic Stimulation: Effects on Stress Hormone Levels and Adult Hippocampal Neurogenesis. *Biol Psychiatry,* Vol.52, No.11, (Dec 1, 2002), pp. (1057-1065), ISSN. 0006-3223 (Print)

Dayer, A. G., Cleaver, K. M., Abouantoun, T.&Cameron, H. A. (2005). New Gabaergic Interneurons in the Adult Neocortex and Striatum Are Generated from Different Precursors. *J Cell Biol,* Vol.168, No.3, (Jan 31, 2005), pp. (415-427), ISSN. 0021-9525 (Print)

de Quervain, D. J., Roozendaal, B.&McGaugh, J. L. (1998). Stress and Glucocorticoids Impair Retrieval of Long-Term Spatial Memory. *Nature,* Vol.394, No.6695, (Aug 20, 1998), pp. (787-790), ISSN. 0028-0836 (Print)

DeCarolis, N. A.&Eisch, A. J. (2010). Hippocampal Neurogenesis as a Target for the Treatment of Mental Illness: A Critical Evaluation. *Neuropharmacology,* Vol.58, No.6, (May, 2010), pp. (884-893), ISSN. 1873-7064 (Electronic)

DeMar, J. C., Jr., Ma, K., Bell, J. M., Igarashi, M., Greenstein, D.&Rapoport, S. I. (2006). One Generation of N-3 Polyunsaturated Fatty Acid Deprivation Increases Depression and Aggression Test Scores in Rats. *J Lipid Res,* Vol.47, No.1, (Jan, 2006), pp. (172-180), ISSN. 0022-2275 (Print)

Diamond, D. M., Bennett, M. C., Fleshner, M.&Rose, G. M. (1992). Inverted-U Relationship between the Level of Peripheral Corticosterone and the Magnitude of Hippocampal Primed Burst Potentiation. *Hippocampus,* Vol.2, No.4, (Oct, 1992), pp. (421-430), ISSN. 1050-9631 (Print)

Diamond, D. M.&Park, C. R. (2000). Predator Exposure Produces Retrograde Amnesia and Blocks Synaptic Plasticity. Progress toward Understanding How the Hippocampus Is Affected by Stress. *Ann N Y Acad Sci,* Vol.911, (Jun, 2000), pp. (453-455), ISSN. 0077-8923 (Print)

Dixon, J. F.&Hokin, L. E. (1998). Lithium Acutely Inhibits and Chronically up-Regulates and Stabilizes Glutamate Uptake by Presynaptic Nerve Endings in Mouse Cerebral Cortex. *Proc Natl Acad Sci U S A,* Vol.95, No.14, (Jul 7, 1998), pp. (8363-8368), ISSN. 0027-8424 (Print)

Drevets, W. C. (2003). Neuroimaging Abnormalities in the Amygdala in Mood Disorders. *Ann N Y Acad Sci,* Vol.985, (Apr, 2003), pp. (420-444), ISSN. 0077-8923 (Print)

Drevets, W. C., Price, J. L., Simpson, J. R., Jr., Todd, R. D., Reich, T., Vannier, M.&Raichle, M. E. (1997). Subgenual Prefrontal Cortex Abnormalities in Mood Disorders. *Nature,* Vol.386, No.6627, (Apr 24, 1997), pp. (824-827), ISSN. 0028-0836 (Print)

Duman, R. S., Malberg, J.&Thome, J. (1999). Neural Plasticity to Stress and Antidepressant Treatment. *Biol Psychiatry,* Vol.46, No.9, (Nov 1, 1999), pp. (1181-1191), ISSN. 0006-3223 (Print)

During, M. J.&Cao, L. (2006). Vegf, a Mediator of the Effect of Experience on Hippocampal Neurogenesis. *Curr Alzheimer Res,* Vol.3, No.1, (Feb, 2006), pp. (29-33), ISSN. 1567-2050 (Print)

Eriksson, P. S., Perfilieva, E., Bjork-Eriksson, T., Alborn, A. M., Nordborg, C., Peterson, D. A.&Gage, F. H. (1998). Neurogenesis in the Adult Human Hippocampus. *Nat Med,* Vol.4, No.11, (Nov, 1998), pp. (1313-1317), ISSN. 1078-8956 (Print)

Fabel, K., Tam, B., Kaufer, D., Baiker, A., Simmons, N., Kuo, C. J.&Palmer, T. D. (2003). Vegf Is Necessary for Exercise-Induced Adult Hippocampal Neurogenesis. *Eur J Neurosci,* Vol.18, No.10, (Nov, 2003), pp. (2803-2812), ISSN. 0953-816X (Print)

Fontella, F. U., Vendite, D. A., Tabajara, A. S., Porciuncula, L. O., da Silva Torres, I. L., Jardim, F. M., Martini, L., Souza, D. O., Netto, C. A.&Dalmaz, C. (2004). Repeated Restraint Stress Alters Hippocampal Glutamate Uptake and Release in the Rat. *Neurochem Res,* Vol.29, No.9, (Sep, 2004), pp. (1703-1709), ISSN. 0364-3190 (Print)

Foy, M. R., Stanton, M. E., Levine, S.&Thompson, R. F. (1987). Behavioral Stress Impairs Long-Term Potentiation in Rodent Hippocampus. *Behav Neural Biol,* Vol.48, No.1, (Jul, 1987), pp. (138-149), ISSN. 0163-1047 (Print)

Frey, B. N., Valvassori, S. S., Reus, G. Z., Martins, M. R., Petronilho, F. C., Bardini, K., Dal-Pizzol, F., Kapczinski, F.&Quevedo, J. (2006). Changes in Antioxidant Defense Enzymes after D-Amphetamine Exposure: Implications as an Animal Model of Mania. *Neurochem Res,* Vol.31, No.5, (May, 2006), pp. (699-703), ISSN. 0364-3190 (Print)

Garrett, A.&Chang, K. (2008). The Role of the Amygdala in Bipolar Disorder Development. *Dev Psychopathol,* Vol.20, No.4, (Fall, 2008), pp. (1285-1296), ISSN. 1469-2198 (Electronic)

Goshen, I.&Yirmiya, R. (2009). Interleukin-1 (Il-1): A Central Regulator of Stress Responses. *Front Neuroendocrinol,* Vol.30, No.1, (Jan, 2009), pp. (30-45), ISSN. 1095-6808 (Electronic)

Gould, E., McEwen, B. S., Tanapat, P., Galea, L. A.&Fuchs, E. (1997). Neurogenesis in the Dentate Gyrus of the Adult Tree Shrew Is Regulated by Psychosocial Stress and Nmda Receptor Activation. *J Neurosci,* Vol.17, No.7, (Apr 1, 1997), pp. (2492-2498), ISSN. 0270-6474 (Print)

Gould, E., Reeves, A. J., Graziano, M. S.&Gross, C. G. (1999). Neurogenesis in the Neocortex of Adult Primates. *Science,* Vol.286, No.5439, (Oct 15, 1999), pp. (548-552), ISSN. 0036-8075 (Print)

Grippo, A. J., Francis, J., Beltz, T. G., Felder, R. B.&Johnson, A. K. (2005). Neuroendocrine and Cytokine Profile of Chronic Mild Stress-Induced Anhedonia. *Physiol Behav,* Vol.84, No.5, (Apr 13, 2005), pp. (697-706), ISSN. 0031-9384 (Print)

Gu, H., Tang, C., Peng, K., Sun, H.&Yang, Y. (2009). Effects of Chronic Mild Stress on the Development of Atherosclerosis and Expression of Toll-Like Receptor 4 Signaling

Pathway in Adolescent Apolipoprotein E Knockout Mice. *J Biomed Biotechnol,* Vol.2009, 2009), pp. (613879), ISSN. 1110-7251 (Electronic)

Guloksuz, S., Cetin, E. A., Cetin, T., Deniz, G., Oral, E. T.&Nutt, D. J. (2010). Cytokine Levels in Euthymic Bipolar Patients. *J Affect Disord,* Vol.126, No.3, (Nov, 2010), pp. (458-462), ISSN. 1573-2517 (Electronic)

Hammen, C. (2005). Stress and Depression. *Annu Rev Clin Psychol,* Vol.1, 2005), pp. (293-319), ISSN. 1548-5943 (Print)

Hammen, C., Brennan, P. A.&Shih, J. H. (2004). Family Discord and Stress Predictors of Depression and Other Disorders in Adolescent Children of Depressed and Nondepressed Women. *J Am Acad Child Adolesc Psychiatry,* Vol.43, No.8, (Aug, 2004), pp. (994-1002), ISSN. 0890-8567 (Print)

Hansel, A., Hong, S., Camara, R. J.&von Kanel, R. (2010). Inflammation as a Psychophysiological Biomarker in Chronic Psychosocial Stress. *Neurosci Biobehav Rev,* Vol.35, No.1, (Sep, 2010), pp. (115-121), ISSN. 1873-7528 (Electronic)

Hao, Y., Creson, T., Zhang, L., Li, P., Du, F., Yuan, P., Gould, T. D., Manji, H. K.&Chen, G. (2004). Mood Stabilizer Valproate Promotes Erk Pathway-Dependent Cortical Neuronal Growth and Neurogenesis. *J Neurosci,* Vol.24, No.29, (Jul 21, 2004), pp. (6590-6599), ISSN. 1529-2401 (Electronic)

Hashimoto, R., Hough, C., Nakazawa, T., Yamamoto, T.&Chuang, D. M. (2002). Lithium Protection against Glutamate Excitotoxicity in Rat Cerebral Cortical Neurons: Involvement of Nmda Receptor Inhibition Possibly by Decreasing Nr2b Tyrosine Phosphorylation. *J Neurochem,* Vol.80, No.4, (Feb, 2002), pp. (589-597), ISSN. 0022-3042 (Print)

Heine, V. M., Maslam, S., Zareno, J., Joels, M.&Lucassen, P. J. (2004). Suppressed Proliferation and Apoptotic Changes in the Rat Dentate Gyrus after Acute and Chronic Stress Are Reversible. *Eur J Neurosci,* Vol.19, No.1, (Jan, 2004), pp. (131-144), ISSN. 0953-816X (Print)

Heine, V. M., Zareno, J., Maslam, S., Joels, M.&Lucassen, P. J. (2005). Chronic Stress in the Adult Dentate Gyrus Reduces Cell Proliferation near the Vasculature and Vegf and Flk-1 Protein Expression. *Eur J Neurosci,* Vol.21, No.5, (Mar, 2005), pp. (1304-1314), ISSN. 0953-816X (Print)

Henn, F. A.&Vollmayr, B. (2004). Neurogenesis and Depression: Etiology or Epiphenomenon? *Biol Psychiatry,* Vol.56, No.3, (Aug 1, 2004), pp. (146-150), ISSN. 0006-3223 (Print)

Hoge, E. A., Brandstetter, K., Moshier, S., Pollack, M. H., Wong, K. K.&Simon, N. M. (2009). Broad Spectrum of Cytokine Abnormalities in Panic Disorder and Posttraumatic Stress Disorder. *Depress Anxiety,* Vol.26, No.5, 2009), pp. (447-455), ISSN. 1520-6394 (Electronic)

Husum, H.&Mathe, A. A. (2002). Early Life Stress Changes Concentrations of Neuropeptide Y and Corticotropin-Releasing Hormone in Adult Rat Brain. Lithium Treatment Modifies These Changes. *Neuropsychopharmacology,* Vol.27, No.5, (Nov, 2002), pp. (756-764), ISSN. 0893-133X (Print)

Joels, M. (2001). Corticosteroid Actions in the Hippocampus. *J Neuroendocrinol,* Vol.13, No.8, (Aug, 2001), pp. (657-669), ISSN. 0953-8194 (Print)

Johnson, S. A., Wang, J. F., Sun, X., McEwen, B. S., Chattarji, S.&Young, L. T. (2009). Lithium Treatment Prevents Stress-Induced Dendritic Remodeling in the Rodent Amygdala. *Neuroscience,* Vol.163, No.1, (Sep 29, 2009), pp. (34-39), ISSN. 1873-7544 (Electronic)

Karege, F., Schwald, M., Papadimitriou, P., Lachausse, C.&Cisse, M. (2004). The Camp-Dependent Protein Kinase a and Brain-Derived Neurotrophic Factor Expression in Lymphoblast Cells of Bipolar Affective Disorder. *J Affect Disord,* Vol.79, No.1-3, (Apr, 2004), pp. (187-192), ISSN. 0165-0327 (Print)

Kempermann, G., Kuhn, H. G.&Gage, F. H. (1997). More Hippocampal Neurons in Adult Mice Living in an Enriched Environment. *Nature,* Vol.386, No.6624, (Apr 3, 1997), pp. (493-495), ISSN. 0028-0836 (Print)

Kornack, D. R.&Rakic, P. (1999). Continuation of Neurogenesis in the Hippocampus of the Adult Macaque Monkey. *Proc Natl Acad Sci U S A,* Vol.96, No.10, (May 11, 1999), pp. (5768-5773), ISSN. 0027-8424 (Print)

Kosten, T. A., Galloway, M. P., Duman, R. S., Russell, D. S.&D'Sa, C. (2008). Repeated Unpredictable Stress and Antidepressants Differentially Regulate Expression of the Bcl-2 Family of Apoptotic Genes in Rat Cortical, Hippocampal, and Limbic Brain Structures. *Neuropsychopharmacology,* Vol.33, No.7, (Jun, 2008), pp. (1545-1558), ISSN. 0893-133X (Print)

Krishnan, V.&Nestler, E. J. (2008). The Molecular Neurobiology of Depression. *Nature,* Vol.455, No.7215, (Oct 16, 2008), pp. (894-902), ISSN. 1476-4687 (Electronic)

Kubera, M., Obuchowicz, E., Goehler, L., Brzeszcz, J.&Maes, M. (2011). In Animal Models, Psychosocial Stress-Induced (Neuro)Inflammation, Apoptosis and Reduced Neurogenesis Are Associated to the Onset of Depression. *Prog Neuropsychopharmacol Biol Psychiatry,* Vol.35, No.3, (Apr 29, 2011), pp. (744-759), ISSN. 1878-4216 (Electronic)

Kuhn, H. G., Biebl, M., Wilhelm, D., Li, M., Friedlander, R. M.&Winkler, J. (2005). Increased Generation of Granule Cells in Adult Bcl-2-Overexpressing Mice: A Role for Cell Death During Continued Hippocampal Neurogenesis. *Eur J Neurosci,* Vol.22, No.8, (Oct, 2005), pp. (1907-1915), ISSN. 0953-816X (Print)

Lee, H. J., Ertley, R. N., Rapoport, S. I., Bazinet, R. P.&Rao, J. S. (2008). Chronic Administration of Lamotrigine Downregulates Cox-2 Mrna and Protein in Rat Frontal Cortex. *Neurochem Res,* Vol.33, No.5, (May, 2008), pp. (861-866), ISSN. 0364-3190 (Print)

Lee, H. J., Kim, J. W., Yim, S. V., Kim, M. J., Kim, S. A., Kim, Y. J., Kim, C. J.&Chung, J. H. (2001). Fluoxetine Enhances Cell Proliferation and Prevents Apoptosis in Dentate Gyrus of Maternally Separated Rats. *Mol Psychiatry,* Vol.6, No.6, (Nov, 2001), pp. (610, 725-618), ISSN. 1359-4184 (Print)

Li, N., He, X., Qi, X., Zhang, Y.&He, S. (2010). The Mood Stabilizer Lamotrigine Produces Antidepressant Behavioral Effects in Rats: Role of Brain-Derived Neurotrophic Factor. *J Psychopharmacol,* Vol.24, No.12, (Dec, 2010), pp. (1772-1778), ISSN. 1461-7285 (Electronic)

Li, N., He, X., Zhang, Y., Qi, X., Li, H., Zhu, X.&He, S. (2011). Brain-Derived Neurotrophic Factor Signalling Mediates Antidepressant Effects of Lamotrigine. *Int J Neuropsychopharmacol,* Vol.14, No.8, (Sep, 2011), pp. (1091-1098), ISSN. 1469-5111 (Electronic)

Li, S., Wang, C., Wang, W., Dong, H., Hou, P.&Tang, Y. (2008). Chronic Mild Stress Impairs Cognition in Mice: From Brain Homeostasis to Behavior. *Life Sci,* Vol.82, No.17-18, (Apr 23, 2008), pp. (934-942), ISSN. 0024-3205 (Print)

Li, X., Ketter, T. A.&Frye, M. A. (2002). Synaptic, Intracellular, and Neuroprotective Mechanisms of Anticonvulsants: Are They Relevant for the Treatment and Course of Bipolar Disorders? *J Affect Disord,* Vol.69, No.1-3, (May, 2002), pp. (1-14), ISSN. 0165-0327 (Print)

Lim, K. Y., Yang, J. J., Lee, D. S., Noh, J. S., Jung, M. W.&Chung, Y. K. (2005). Lithium Attenuates Stress-Induced Impairment of Long-Term Potentiation Induction. *Neuroreport,* Vol.16, No.14, (Sep 28, 2005), pp. (1605-1608), ISSN. 0959-4965 (Print)

Liston, C., Miller, M. M., Goldwater, D. S., Radley, J. J., Rocher, A. B., Hof, P. R., Morrison, J. H.&McEwen, B. S. (2006). Stress-Induced Alterations in Prefrontal Cortical Dendritic Morphology Predict Selective Impairments in Perceptual Attentional Set-Shifting. *J Neurosci,* Vol.26, No.30, (Jul 26, 2006), pp. (7870-7874), ISSN. 1529-2401 (Electronic)

Lowy, M. T., Gault, L.&Yamamoto, B. K. (1993). Adrenalectomy Attenuates Stress-Induced Elevations in Extracellular Glutamate Concentrations in the Hippocampus. *J Neurochem,* Vol.61, No.5, (Nov, 1993), pp. (1957-1960), ISSN. 0022-3042 (Print)

Lucassen, P. J., Vollmann-Honsdorf, G. K., Gleisberg, M., Czeh, B., De Kloet, E. R.&Fuchs, E. (2001). Chronic Psychosocial Stress Differentially Affects Apoptosis in Hippocampal Subregions and Cortex of the Adult Tree Shrew. *Eur J Neurosci,* Vol.14, No.1, (Jul, 2001), pp. (161-166), ISSN. 0953-816X (Print)

Lucca, G., Comim, C. M., Valvassori, S. S., Reus, G. Z., Vuolo, F., Petronilho, F., Dal-Pizzol, F., Gavioli, E. C.&Quevedo, J. (2009). Effects of Chronic Mild Stress on the Oxidative Parameters in the Rat Brain. *Neurochem Int,* Vol.54, No.5-6, (May-Jun, 2009), pp. (358-362), ISSN. 1872-9754 (Electronic)

Luine, V., Martinez, C., Villegas, M., Magarinos, A. M.&McEwen, B. S. (1996). Restraint Stress Reversibly Enhances Spatial Memory Performance. *Physiol Behav,* Vol.59, No.1, (Jan, 1996), pp. (27-32), ISSN. 0031-9384 (Print)

Madrigal, J. L., Hurtado, O., Moro, M. A., Lizasoain, I., Lorenzo, P., Castrillo, A., Bosca, L.&Leza, J. C. (2002). The Increase in Tnf-Alpha Levels Is Implicated in Nf-Kappab Activation and Inducible Nitric Oxide Synthase Expression in Brain Cortex after Immobilization Stress. *Neuropsychopharmacology,* Vol.26, No.2, (Feb, 2002), pp. (155-163), ISSN. 0893-133X (Print)

Madrigal, J. L., Moro, M. A., Lizasoain, I., Lorenzo, P., Castrillo, A., Bosca, L.&Leza, J. C. (2001). Inducible Nitric Oxide Synthase Expression in Brain Cortex after Acute Restraint Stress Is Regulated by Nuclear Factor Kappab-Mediated Mechanisms. *J Neurochem,* Vol.76, No.2, (Jan, 2001), pp. (532-538), ISSN. 0022-3042 (Print)

Maes, M. (1995). Evidence for an Immune Response in Major Depression: A Review and Hypothesis. *Prog Neuropsychopharmacol Biol Psychiatry,* Vol.19, No.1, (Jan, 1995), pp. (11-38), ISSN. 0278-5846 (Print)

Magarinos, A. M.&McEwen, B. S. (1995). Stress-Induced Atrophy of Apical Dendrites of Hippocampal Ca3c Neurons: Involvement of Glucocorticoid Secretion and Excitatory Amino Acid Receptors. *Neuroscience,* Vol.69, No.1, (Nov, 1995), pp. (89-98), ISSN. 0306-4522 (Print)

Magarinos, A. M., McEwen, B. S., Flugge, G.&Fuchs, E. (1996). Chronic Psychosocial Stress Causes Apical Dendritic Atrophy of Hippocampal Ca3 Pyramidal Neurons in Subordinate Tree Shrews. *J Neurosci,* Vol.16, No.10, (May 15, 1996), pp. (3534-3540), ISSN. 0270-6474 (Print)

Makatsori, A., Duncko, R., Moncek, F., Loder, I., Katina, S.&Jezova, D. (2004). Modulation of Neuroendocrine Response and Non-Verbal Behavior During Psychosocial Stress in Healthy Volunteers by the Glutamate Release-Inhibiting Drug Lamotrigine. *Neuroendocrinology,* Vol.79, No.1, (Jan, 2004), pp. (34-42), ISSN. 0028-3835 (Print)

Malberg, J. E.&Duman, R. S. (2003). Cell Proliferation in Adult Hippocampus Is Decreased by Inescapable Stress: Reversal by Fluoxetine Treatment. *Neuropsychopharmacology,* Vol.28, No.9, (Sep, 2003), pp. (1562-1571), ISSN. 0893-133X (Print)

Manganas, L. N., Zhang, X., Li, Y., Hazel, R. D., Smith, S. D., Wagshul, M. E., Henn, F., Benveniste, H., Djuric, P. M., Enikolopov, G.&Maletic-Savatic, M. (2007). Magnetic Resonance Spectroscopy Identifies Neural Progenitor Cells in the Live Human Brain. *Science,* Vol.318, No.5852, (Nov 9, 2007), pp. (980-985), ISSN. 1095-9203 (Electronic)

Manji, H. K., Moore, G. J.&Chen, G. (2000). Clinical and Preclinical Evidence for the Neurotrophic Effects of Mood Stabilizers: Implications for the Pathophysiology and Treatment of Manic-Depressive Illness. *Biol Psychiatry,* Vol.48, No.8, (Oct 15, 2000), pp. (740-754), ISSN. 0006-3223 (Print)

Marcus, S. R., Nadiger, H. A., Chandrakala, M. V., Rao, T. I.&Sadasivudu, B. (1986). Acute and Short-Term Effects of Lithium on Glutamate Metabolism in Rat Brain. *Biochem Pharmacol,* Vol.35, No.3, (Feb 1, 1986), pp. (365-369), ISSN. 0006-2952 (Print)

Martin, K. P.&Wellman, C. L. (2011). Nmda Receptor Blockade Alters Stress-Induced Dendritic Remodeling in Medial Prefrontal Cortex. *Cereb Cortex,* (Mar 7, 2011), pp. ISSN. 1460-2199 (Electronic)

McEwen, B. S. (2004). Protection and Damage from Acute and Chronic Stress: Allostasis and Allostatic Overload and Relevance to the Pathophysiology of Psychiatric Disorders. *Ann N Y Acad Sci,* Vol.1032, (Dec, 2004), pp. (1-7), ISSN. 0077-8923 (Print)

Melchior, M., Caspi, A., Milne, B. J., Danese, A., Poulton, R.&Moffitt, T. E. (2007). Work Stress Precipitates Depression and Anxiety in Young, Working Women and Men. *Psychol Med,* Vol.37, No.8, (Aug, 2007), pp. (1119-1129), ISSN. 0033-2917 (Print)

Mill, J.&Petronis, A. (2007). Molecular Studies of Major Depressive Disorder: The Epigenetic Perspective. *Mol Psychiatry,* Vol.12, No.9, (Sep, 2007), pp. (799-814), ISSN. 1359-4184 (Print)

Miller, J. C., Jimenez, P.&Mathe, A. A. (2007). Restraint Stress Influences Ap-1 and Creb DNA-Binding Activity Induced by Chronic Lithium Treatment in the Rat Frontal Cortex and Hippocampus. *Int J Neuropsychopharmacol,* Vol.10, No.5, (Oct, 2007), pp. (609-619), ISSN. 1461-1457 (Print)

Miller, J. C.&Mathe, A. A. (1997). Basal and Stimulated C-Fos Mrna Expression in the Rat Brain: Effect of Chronic Dietary Lithium. *Neuropsychopharmacology,* Vol.16, No.6, (Jun, 1997), pp. (408-418), ISSN. 0893-133X (Print)

Mirescu, C.&Gould, E. (2006). Stress and Adult Neurogenesis. *Hippocampus,* Vol.16, No.3, 2006), pp. (233-238), ISSN. 1050-9631 (Print)

Moghaddam, B., Bolinao, M. L., Stein-Behrens, B.&Sapolsky, R. (1994). Glucocorticoids Mediate the Stress-Induced Extracellular Accumulation of Glutamate. *Brain Res,* Vol.655, No.1-2, (Aug 29, 1994), pp. (251-254), ISSN. 0006-8993 (Print)

Nair, A., Vadodaria, K. C., Banerjee, S. B., Benekareddy, M., Dias, B. G., Duman, R. S.&Vaidya, V. A. (2007). Stressor-Specific Regulation of Distinct Brain-Derived Neurotrophic Factor Transcripts and Cyclic Amp Response Element-Binding Protein Expression in the Postnatal and Adult Rat Hippocampus. *Neuropsychopharmacology,* Vol.32, No.7, (Jul, 2007), pp. (1504-1519), ISSN. 0893-133X (Print)

Nocjar, C., Hammonds, M. D.&Shim, S. S. (2007). Chronic Lithium Treatment Magnifies Learning in Rats. *Neuroscience,* Vol.150, No.4, (Dec 19, 2007), pp. (774-788), ISSN. 0306-4522 (Print)

Oitzl, M. S., Fluttert, M., Sutanto, W.&de Kloet, E. R. (1998). Continuous Blockade of Brain Glucocorticoid Receptors Facilitates Spatial Learning and Memory in Rats. *Eur J Neurosci,* Vol.10, No.12, (Dec, 1998), pp. (3759-3766), ISSN. 0953-816X (Print)

Olivenza, R., Moro, M. A., Lizasoain, I., Lorenzo, P., Fernandez, A. P., Rodrigo, J., Bosca, L.&Leza, J. C. (2000). Chronic Stress Induces the Expression of Inducible Nitric Oxide Synthase in Rat Brain Cortex. *J Neurochem,* Vol.74, No.2, (Feb, 2000), pp. (785-791), ISSN. 0022-3042 (Print)

Pavlides, C., Kimura, A., Magarinos, A. M.&McEwen, B. S. (1994). Type I Adrenal Steroid Receptors Prolong Hippocampal Long-Term Potentiation. *Neuroreport,* Vol.5, No.18, (Dec 20, 1994), pp. (2673-2677), ISSN. 0959-4965 (Print)

Pavlides, C., Kimura, A., Magarinos, A. M.&McEwen, B. S. (1995). Hippocampal Homosynaptic Long-Term Depression/Depotentiation Induced by Adrenal Steroids. *Neuroscience,* Vol.68, No.2, (Sep, 1995), pp. (379-385), ISSN. 0306-4522 (Print)

Perisic, T., Zimmermann, N., Kirmeier, T., Asmus, M., Tuorto, F., Uhr, M., Holsboer, F., Rein, T.&Zschocke, J. (2010). Valproate and Amitriptyline Exert Common and Divergent Influences on Global and Gene Promoter-Specific Chromatin Modifications in Rat Primary Astrocytes. *Neuropsychopharmacology,* Vol.35, No.3, (Feb, 2010), pp. (792-805), ISSN. 1740-634X (Electronic)

Petronis, A. (2003). Epigenetics and Bipolar Disorder: New Opportunities and Challenges. *Am J Med Genet C Semin Med Genet,* Vol.123C, No.1, (Nov 15, 2003), pp. (65-75), ISSN. 1552-4868 (Print)

Pham, K., Nacher, J., Hof, P. R.&McEwen, B. S. (2003). Repeated Restraint Stress Suppresses Neurogenesis and Induces Biphasic Psa-Ncam Expression in the Adult Rat Dentate Gyrus. *Eur J Neurosci,* Vol.17, No.4, (Feb, 2003), pp. (879-886), ISSN. 0953-816X (Print)

Phiel, C. J.&Klein, P. S. (2001). Molecular Targets of Lithium Action. *Annu Rev Pharmacol Toxicol,* Vol.41, 2001), pp. (789-813), ISSN. 0362-1642 (Print)

Pittenger, C.&Duman, R. S. (2008). Stress, Depression, and Neuroplasticity: A Convergence of Mechanisms. *Neuropsychopharmacology,* Vol.33, No.1, (Jan, 2008), pp. (88-109), ISSN. 0893-133X (Print)

Post, R. M., Ketter, T. A., Pazzaglia, P. J., Denicoff, K., George, M. S., Callahan, A., Leverich, G.&Frye, M. (1996). Rational Polypharmacy in the Bipolar Affective Disorders. *Epilepsy Res Suppl,* Vol.11, 1996), pp. (153-180), ISSN. 0922-9833 (Print)

Radley, J. J., Sisti, H. M., Hao, J., Rocher, A. B., McCall, T., Hof, P. R., McEwen, B. S.&Morrison, J. H. (2004). Chronic Behavioral Stress Induces Apical Dendritic Reorganization in Pyramidal Neurons of the Medial Prefrontal Cortex. *Neuroscience,* Vol.125, No.1, 2004), pp. (1-6), ISSN. 0306-4522 (Print)

Rao, J. S., Bazinet, R. P., Rapoport, S. I.&Lee, H. J. (2007a). Chronic Treatment of Rats with Sodium Valproate Downregulates Frontal Cortex Nf-Kappab DNA Binding Activity and Cox-2 Mrna. *Bipolar Disord,* Vol.9, No.5, (Aug, 2007a), pp. (513-520), ISSN. 1398-5647 (Print)

Rao, J. S., Ertley, R. N., DeMar, J. C., Jr., Rapoport, S. I., Bazinet, R. P.&Lee, H. J. (2007b). Dietary N-3 Pufa Deprivation Alters Expression of Enzymes of the Arachidonic and Docosahexaenoic Acid Cascades in Rat Frontal Cortex. *Mol Psychiatry,* Vol.12, No.2, (Feb, 2007b), pp. (151-157), ISSN. 1359-4184 (Print)

Rao, J. S., Ertley, R. N., Lee, H. J., DeMar, J. C., Jr., Arnold, J. T., Rapoport, S. I.&Bazinet, R. P. (2007c). N-3 Polyunsaturated Fatty Acid Deprivation in Rats Decreases Frontal Cortex Bdnf Via a P38 Mapk-Dependent Mechanism. *Mol Psychiatry,* Vol.12, No.1, (Jan, 2007c), pp. (36-46), ISSN. 1359-4184 (Print)

Rao, J. S., Lee, H. J., Rapoport, S. I.&Bazinet, R. P. (2008). Mode of Action of Mood Stabilizers: Is the Arachidonic Acid Cascade a Common Target? *Mol Psychiatry,* Vol.13, No.6, (Jun, 2008), pp. (585-596), ISSN. 1476-5578 (Electronic)

Reznikov, L. R., Grillo, C. A., Piroli, G. G., Pasumarthi, R. K., Reagan, L. P.&Fadel, J. (2007). Acute Stress-Mediated Increases in Extracellular Glutamate Levels in the Rat Amygdala: Differential Effects of Antidepressant Treatment. *Eur J Neurosci,* Vol.25, No.10, (May, 2007), pp. (3109-3114), ISSN. 0953-816X (Print)

Rosenbrock, H., Koros, E., Bloching, A., Podhorna, J.&Borsini, F. (2005). Effect of Chronic Intermittent Restraint Stress on Hippocampal Expression of Marker Proteins for Synaptic Plasticity and Progenitor Cell Proliferation in Rats. *Brain Res,* Vol.1040, No.1-2, (Apr 8, 2005), pp. (55-63), ISSN. 0006-8993 (Print)

Roth, T. L., Lubin, F. D., Funk, A. J.&Sweatt, J. D. (2009). Lasting Epigenetic Influence of Early-Life Adversity on the Bdnf Gene. *Biol Psychiatry,* Vol.65, No.9, (May 1, 2009), pp. (760-769), ISSN. 1873-2402 (Electronic)

Roth, T. L., Zoladz, P. R., Sweatt, J. D.&Diamond, D. M. (2011). Epigenetic Modification of Hippocampal Bdnf DNA in Adult Rats in an Animal Model of Post-Traumatic Stress Disorder. *J Psychiatr Res,* Vol.45, No.7, (Jul, 2011), pp. (919-926), ISSN. 1879-1379 (Electronic)

Sapolsky, R. M. (2003). Stress and Plasticity in the Limbic System. *Neurochem Res,* Vol.28, No.11, (Nov, 2003), pp. (1735-1742), ISSN. 0364-3190 (Print)

Schloesser, R. J., Huang, J., Klein, P. S.&Manji, H. K. (2008). Cellular Plasticity Cascades in the Pathophysiology and Treatment of Bipolar Disorder. *Neuropsychopharmacology,* Vol.33, No.1, (Jan, 2008), pp. (110-133), ISSN. 0893-133X (Print)

Schmidt-Hieber, C., Jonas, P.&Bischofberger, J. (2004). Enhanced Synaptic Plasticity in Newly Generated Granule Cells of the Adult Hippocampus. *Nature,* Vol.429, No.6988, (May 13, 2004), pp. (184-187), ISSN. 1476-4687 (Electronic)

Shaltiel, G., Chen, G.&Manji, H. K. (2007). Neurotrophic Signaling Cascades in the Pathophysiology and Treatment of Bipolar Disorder. *Curr Opin Pharmacol,* Vol.7, No.1, (Feb, 2007), pp. (22-26), ISSN. 1471-4892 (Print)

Shao, L., Young, L. T.&Wang, J. F. (2005). Chronic Treatment with Mood Stabilizers Lithium and Valproate Prevents Excitotoxicity by Inhibiting Oxidative Stress in Rat Cerebral Cortical Cells. *Biol Psychiatry,* Vol.58, No.11, (Dec 1, 2005), pp. (879-884), ISSN. 0006-3223 (Print)

Shim, S. S., Hammonds, M. D., Ganocy, S. J.&Calabrese, J. R. (2007). Effects of Sub-Chronic Lithium Treatment on Synaptic Plasticity in the Dentate Gyrus of Rat Hippocampal Slices. *Prog Neuropsychopharmacol Biol Psychiatry,* Vol.31, No.2, (Mar 30, 2007), pp. (343-347), ISSN. 0278-5846 (Print)

Shors, T. J.&Dryver, E. (1994). Effect of Stress and Long-Term Potentiation (Ltp) on Subsequent Ltp and the Theta Burst Response in the Dentate Gyrus. *Brain Res,* Vol.666, No.2, (Dec 15, 1994), pp. (232-238), ISSN. 0006-8993 (Print)

Shors, T. J., Seib, T. B., Levine, S.&Thompson, R. F. (1989). Inescapable Versus Escapable Shock Modulates Long-Term Potentiation in the Rat Hippocampus. *Science,* Vol.244, No.4901, (Apr 14, 1989), pp. (224-226), ISSN. 0036-8075 (Print)

Silva, R., Martins, L., Longatto-Filho, A., Almeida, O. F.&Sousa, N. (2007). Lithium Prevents Stress-Induced Reduction of Vascular Endothelium Growth Factor Levels. *Neurosci Lett,* Vol.429, No.1, (Dec 11, 2007), pp. (33-38), ISSN. 0304-3940 (Print)

Silva, R., Mesquita, A. R., Bessa, J., Sousa, J. C., Sotiropoulos, I., Leao, P., Almeida, O. F.&Sousa, N. (2008). Lithium Blocks Stress-Induced Changes in Depressive-Like Behavior and Hippocampal Cell Fate: The Role of Glycogen-Synthase-Kinase-3beta. *Neuroscience,* Vol.152, No.3, (Mar 27, 2008), pp. (656-669), ISSN. 0306-4522 (Print)

Son, H., Yu, I. T., Hwang, S. J., Kim, J. S., Lee, S. H., Lee, Y. S.&Kaang, B. K. (2003). Lithium Enhances Long-Term Potentiation Independently of Hippocampal Neurogenesis in the Rat Dentate Gyrus. *J Neurochem,* Vol.85, No.4, (May, 2003), pp. (872-881), ISSN. 0022-3042 (Print)

Song, C., Li, X., Kang, Z.&Kadotomi, Y. (2007). Omega-3 Fatty Acid Ethyl-Eicosapentaenoate Attenuates Il-1beta-Induced Changes in Dopamine and Metabolites in the Shell of the Nucleus Accumbens: Involved with Pla2 Activity and Corticosterone Secretion. *Neuropsychopharmacology,* Vol.32, No.3, (Mar, 2007), pp. (736-744), ISSN. 0893-133X (Print)

Sousa, N., Lukoyanov, N. V., Madeira, M. D., Almeida, O. F.&Paula-Barbosa, M. M. (2000). Reorganization of the Morphology of Hippocampal Neurites and Synapses after Stress-Induced Damage Correlates with Behavioral Improvement. *Neuroscience,* Vol.97, No.2, 2000), pp. (253-266), ISSN. 0306-4522 (Print)

Stahl, S. M. *Stahl's Essential Psychopharmacology : Neuroscientific Basis and Practical Applications* (3rd ed, Fully rev. and expanded.), Cambridge University Press, ISBN. 9780521857024 (hardcover), Cambridge ; New York

Tashiro, A., Sandler, V. M., Toni, N., Zhao, C.&Gage, F. H. (2006). Nmda-Receptor-Mediated, Cell-Specific Integration of New Neurons in Adult Dentate Gyrus. *Nature,* Vol.442, No.7105, (Aug 24, 2006), pp. (929-933), ISSN. 1476-4687 (Electronic)

Thomas, R. M., Hotsenpiller, G.&Peterson, D. A. (2007). Acute Psychosocial Stress Reduces Cell Survival in Adult Hippocampal Neurogenesis without Altering Proliferation. *J Neurosci,* Vol.27, No.11, (Mar 14, 2007), pp. (2734-2743), ISSN. 1529-2401 (Electronic)

Tsankova, N., Renthal, W., Kumar, A.&Nestler, E. J. (2007). Epigenetic Regulation in Psychiatric Disorders. *Nat Rev Neurosci,* Vol.8, No.5, (May, 2007), pp. (355-367), ISSN. 1471-003X (Print)

Umka, J., Mustafa, S., ElBeltagy, M., Thorpe, A., Latif, L., Bennett, G.&Wigmore, P. M. (2010). Valproic Acid Reduces Spatial Working Memory and Cell Proliferation in the Hippocampus. *Neuroscience,* Vol.166, No.1, (Mar 10, 2010), pp. (15-22), ISSN. 1873-7544 (Electronic)

Vaher, P. R., Luine, V. N., Gould, E.&McEwen, B. S. (1994). Effects of Adrenalectomy on Spatial Memory Performance and Dentate Gyrus Morphology. *Brain Res,* Vol.656, No.1, (Sep 5, 1994), pp. (71-78), ISSN. 0006-8993 (Print)

van, I. M. H., Caspers, K., Bakermans-Kranenburg, M. J., Beach, S. R.&Philibert, R. (2010). Methylation Matters: Interaction between Methylation Density and Serotonin Transporter Genotype Predicts Unresolved Loss or Trauma. *Biol Psychiatry,* Vol.68, No.5, (Sep 1, 2010), pp. (405-407), ISSN. 1873-2402 (Electronic)

Vasconcellos, A. P., Tabajara, A. S., Ferrari, C., Rocha, E.&Dalmaz, C. (2003). Effect of Chronic Stress on Spatial Memory in Rats Is Attenuated by Lithium Treatment. *Physiol Behav,* Vol.79, No.2, (Jul, 2003), pp. (143-149), ISSN. 0031-9384 (Print)

Vyas, A., Jadhav, S.&Chattarji, S. (2006). Prolonged Behavioral Stress Enhances Synaptic Connectivity in the Basolateral Amygdala. *Neuroscience,* Vol.143, No.2, (Dec 1, 2006), pp. (387-393), ISSN. 0306-4522 (Print)

Vyas, A., Mitra, R., Shankaranarayana Rao, B. S.&Chattarji, S. (2002). Chronic Stress Induces Contrasting Patterns of Dendritic Remodeling in Hippocampal and Amygdaloid Neurons. *J Neurosci,* Vol.22, No.15, (Aug 1, 2002), pp. (6810-6818), ISSN. 1529-2401 (Electronic)

Watanabe, Y., Gould, E.&McEwen, B. S. (1992). Stress Induces Atrophy of Apical Dendrites of Hippocampal Ca3 Pyramidal Neurons. *Brain Res,* Vol.588, No.2, (Aug 21, 1992), pp. (341-345), ISSN. 0006-8993 (Print)

Wood, G. E., Young, L. T., Reagan, L. P., Chen, B.&McEwen, B. S. (2004). Stress-Induced Structural Remodeling in Hippocampus: Prevention by Lithium Treatment. *Proc Natl Acad Sci U S A,* Vol.101, No.11, (Mar 16, 2004), pp. (3973-3978), ISSN. 0027-8424 (Print)

Xu, H., Chen, Z., He, J., Haimanot, S., Li, X., Dyck, L.&Li, X. M. (2006). Synergetic Effects of Quetiapine and Venlafaxine in Preventing the Chronic Restraint Stress-Induced Decrease in Cell Proliferation and Bdnf Expression in Rat Hippocampus. *Hippocampus,* Vol.16, No.6, 2006), pp. (551-559), ISSN. 1050-9631 (Print)

Xu, L., Anwyl, R.&Rowan, M. J. (1997). Behavioural Stress Facilitates the Induction of Long-Term Depression in the Hippocampus. *Nature,* Vol.387, No.6632, (May 29, 1997), pp. (497-500), ISSN. 0028-0836 (Print)

Yang, C. H., Huang, C. C.&Hsu, K. S. (2004). Behavioral Stress Modifies Hippocampal Synaptic Plasticity through Corticosterone-Induced Sustained Extracellular Signal-Regulated Kinase/Mitogen-Activated Protein Kinase Activation. *J Neurosci,* Vol.24, No.49, (Dec 8, 2004), pp. (11029-11034), ISSN. 1529-2401 (Electronic)

Yurgelun-Todd, D. A., Gruber, S. A., Kanayama, G., Killgore, W. D., Baird, A. A.&Young, A. D. (2000). Fmri During Affect Discrimination in Bipolar Affective Disorder. *Bipolar Disord,* Vol.2, No.3 Pt 2, (Sep, 2000), pp. (237-248), ISSN. 1398-5647 (Print)

3

Lithium Enhances Synaptic Plasticity: Implication for Treatment of Bipolar Disorder

Seong S. Shim
Department of Psychiatry, Case Western Reserve University School of Medicine, USA

1. Introduction

Although lithium has remained as the drug of choice in the treatment of bipolar disorder over the past 50 years, the mechanism by which it exerts its therapeutic effects is not well understood. A large body of evidence from molecular, cellular and clinical studies proposes that lithium have positive actions in enhancing neuroplasticity and synaptic plasticity, and these actions are associated with its efficacy in the treatment of bipolar disorder and other psychiatric disorders (Manji et al. 1999; Manji and Duman 2001; Rowe and Chuang 2004; Tsaltas et al. 2009). Lithium modulates intracellular signal transduction pathways involved in the activation of transcription factors (PEBP-2β, P53, and CREB) and the gene expression of diverse neurotrophic and neuroprotective factors (Chalecka-Franaszek and Chuang 1999; Grimes and Jope 2001; Jope and Roh 2006; Manji et al. 1999; Manji and Duman 2001). Among these factors, brain derived neurotrophic factor (BDNF), B-cell CLL/lymphoma 2 (Bcl-2) and cyclic adenosine monophosphate response element-binding protein (CREB) have been the most extensively studied (Angelucci et al. 2003; Chuang et al. 2002; Fukumoto et al. 2001; Manji and Chen 2002). It is well known that BDNF, Bcl-2 and CREB play important roles in maintaining normal synaptic plasticity in diverse ways (Adams and Cory 1998; McAllister et al. 1999; ; Shaywitz and Greenberg 1999; Silva et al. 1998). Evidence from brain imaging and postmortem studies and studies with animal models suggests that in patients with bipolar disorder, diverse pathological changes in neuroplastic processes lead to impairment in synaptic communications in neuronal circuits involved in the pathophysiology of bipolar disorder. Lithium may enhance synaptic plasticity and thereby restore normal synaptic communications in the circuits by up-regulating neurotrophic and neuroprotective factor such as BDNF, Bcl-2 and activating CREB, the major transcription factor of gene expression of BDNF and Bcl-2, and this action may be associated with the efficacy of the drug. However, this theory has been developed primarily based on molecular biological and clinical studies. This theory has never been fully tested by directly examining the effects of lithium exposure on synaptic plasticity. Only few studies have specifically focused on the effects of lithium exposure on synaptic plasticity.

This study investigated the effects of subchronic and chronic exposure to lithium on synaptic plasticity in the hippocampus. First, the study examined whether 2 weeks and 4 weeks lithium treatment alters functional synaptic plasticity by examining the effects of lithium treatment on input and output (I/O) responses and long-term potentiation (LTP) of field excitatory postsynaptic potential (fEPSP) of the principal neurons in the dentate gyrus (DG) and area CA1 in hippocampal slices. Second, the study examined the effects of 2 and 4 weeks lithium treatment on structural synaptic plasticity by examining effects of lithium treatment on the density of dendrites of the principal neurons in DG and area CA1, using the Golgi staining and Sholl analysis. The study also examined the effects of the same lithium treatment on the levels of phosphorylated CREB (pCREB), Bcl-2 and BDNF, which play critical roles in developing and maintaining normal synaptic plasticity, in the DG and area CA1.

2. Methods

2.1 Animals and lithium treatment

Adult male Sprague-Dawley rats weighing 200g to 250g at procurement were housed three per cage with food and water available ad libitum and housed in a temperature-controlled room with a light/dark cycle of 12/12 hours. For the 4 weeks lithium study, animals were divided as lithium treated group and control groups. Lithium chows (0.24% Li_2CO_3, Harlan Teklad®, Madison, WI) or control chow was fed for 4 weeks. For the 2 weeks lithium study, animals were treated and fed the same way as the 4 weeks lithium study. At the end of lithium treatment, the blood levels of lithium were confirmed to be within its therapeutic range in human. Animal treatment and experiments were conducted in accordance with the principles and procedures of the National Institutes of Health Guide for the Care and Use of Laboratory Animals.

2.2 Effects of lithium treatment on LTP of the principal neurons in the hippocampus

Animals were deeply anesthetized with isofluorane inhalation anesthesia. The brains were removed then hemi-sected, the hippocampus was separated from the rest of the hemisphere and placed on the stage of a tissue chopper, and 400 micrometer-thick slices were harvested. Hippocampal slices for electrophysiological recording were incubated in artificial cerebrospinal fluid (aCSF) bubbled with $95\%O_2/5\%CO_2$ at room temperature until used. Hippocampal slices selected for use were placed in an interface-type chamber and perfused with oxygenated ($95\%O_2/5\%CO_2$) aCSF and allowed to equilibrate for 30 min. The aCSF was exchanged at a flow rate of 1-1.5ml/min and was composed of (in mM) 125 NaCl, 3.35 KCl, 1.25 NaH_2PO_4, 2.0 $CaCl_2$, 2.0 $MgSO_4$, 25 $NaHCO_3$, and 10 glucose.

For fEPSP recordings in the DG, a bipolar stimulating electrode was placed in the medial perforant pathway fibers, and a glass micropipette recording electrode (2-4 MΩ) filled with aCSF placed in the granule cell dendritic layer using microelectrode manipulators under an illuminated upright microscope. The same protocol was followed for fEPSP recording in hippocampal area CA1. For LTP recording in hippocampal area CA1, the bipolar stimulating electrode was placed in the Schafer collaterals and the recording electrode in the apical CA1 dendritic layer for fEPSP. To determine the fEPSP I/O function, stimulations (1/20 sec.) at an intensity ranging from just below threshold (~10 uA) until response has reached asymptote (no greater than ~900uA) were applied, and responses recorded onto a PC utilizing Clampex software. Five stimulations at each intensity level were given during ascending intensities at 10, 30, 50 and 100uA, then increased by 100□uA until the maximum (900uA) was reached.

Prior to induction of LTP of the fEPSP, a 15 minute baseline was recorded with stimulations at 1/20 Hz at a stimulation intensity of 500 uA. LTP was elicited with tetanus of 2 trains of 100Hz stimulations for 0.5sec with a 5 sec interval between trains at the same intensity. LTP was recorded at least for 60 minutes following tetanus with 1/20 Hz stimulations, and LTP responses only for the last 10 min. (50-60 min) are used for analyses to avoid the contamination of noises and other neuronal activities. All measures were made utilizing the same Clampex software used in data collection. The magnitude of all the fEPSPs recorded was expressed by measuring the initial slope of the postsynaptic potential waveforms. For each value, mean ± SEM was determined, and the resulting values between the lithium treated and control groups were compared statistically.

2.3 Effects of lithium treatment on the density of dendrites in the hippocampus

2.3.1 Golgi staining

The Rapid Golgi method was used for this study (Gabbot and Somogyi, 1984). Animals were administered with lithium as described above. At the end of lithium treatment, the chest of animals was opened under deep anesthesia with Nembutal® (75 mg/kg, IP), the left ventricle was punctured, and the right atrium incised to allow the blood exit during saline injection. 0.9% saline was perfused into the vascular system to flush out the blood from the brain. Then, fixative solution (phosphate buffered 10% formalin) was infused. Animals were then decapitated, and the brains were removed. The blocks of hippocampal tissue were separated and fixed in 10% paraformaldeyde and 2.5% glutaraldehyde dissolved in 0.1 M phosphate buffer overnight. The blocks of fixed tissue were kept in staining solutions of potassium dichromate and osmium tetraoxide followed by immersion in a solution of silver nitrate. Hippocampal neurons were stained en bloc and the stained blocks were then dehydrated in an graded series of increasing ethanol concentrations with 1% uranyl acetate and embedded in an ascending series of low viscosity nitrocellulose solutions. The nitrocellulose was hardened by exposure to chloroform. This hardened nitrocellulose was affixed to a sectioning block, and sections are cut (at 120 μM) on an AO sliding microtome. The sections were cleared in alpha-terpineol and mounted on slides under Permount.

2.3.2 Analysis of dendritic materials (Sholl analysis)

The slides of Golgi-impregnated neurons were mounted on a Zeiss microscope, which was equipped with drawing tubes for preparing camera lucida drawings and digital photomicroscopy capabilities. The analysis defined the amount of dendritic material and its spatial distribution. A template of a series of enlarging concentric circles was centered on the soma of each camera lucida drawing of the dendritic trees of neurons. The number of intersections of the dendritic trees with each successive shell was quantified. This generated a profile defining the distribution of the dendritic material at equidistant intervals from the soma (every 10 micron increased from the soma). For each value, mean ± SEM was determined, and the resulting values between the lithium treated and control groups were analyzed statistically.

2.4 Effects of lithium treatment on levels of BDNF, Bcl-2 and pCREB in the hippocampus

At the end of the lithium treatment, animals were sacrificed, the brains were hemisected, and the hippocampi were separated. Tissues of DA and the areas of CA1 were collected and weight. A volume of protein/peptide extraction buffer equal to 4-9 times the tissue weight was added. Then, tissue blocks were homogenized with an ultrasonic cell dismemberator (Fisher Scientific), boiled for 5 minutes and centrifuged at 15000 for 20 minutes. Thereafter, supernatant was transferred to another set of microtubes for further processing immediately or stored in -80 °C. BDNF, pCREB and Bcl-2 levels were determined by the commercially available ELISA kits with BDNF from Promega ((Madison, WI) and the kits for p-CREB and Bcl-2 from Aldrich-Sigma (St. Louis, MO). Their protocols of ELISA were followed exactly. After completing the entire steps of processing, optical densities of wells microplate were read by an assay reader, FLUOstar (BMG LabTech, Germany). Peptide value of each were converted to and presented as the value of pg/mg wet tissue. The peptide/protein value of each was as the value of pg/mg wet tissue or units of activated enzyme/ml. For each value, mean ± SEM was determined, and the resulting values between the lithium treated and control groups were statistically analyzed.

3. Results

This study has obtained results for the effect of lithium treatment for two and four weeks lithium on functional (LTP) and structural synaptic plasticity (density of dendrites) and on the levels of BDNF, Bcl-2 and pCREB in the DG and hippocampal area CA1 (Table 1).

	2 Week Lithium Treatment		**4 Week Lithium Treatment**	
	DG	Area CA1	DG	Area CA1
Lithium Effects on Synaptic Plasticity				
LTP	↑	-	↑*	↑
I/O	↑	-	-	→
Lithium Effects on Density of Dendrites and Dendritic Spines				
Dendrites	↓	↓	↑	↑
Lithium Effects on Levels of BDNF, Bcl-2 and P-CREB				
BDNF	→	→	↑	→
Bcl-2	→	→	↑	↑
p-CREB	→	→	↑	↑

Table 1. Effects of 2 weeks and 4 weeks lithium treatments on functional and structural synaptic plasticity and the levels of BDNF, Bcl-2 and pCREB in the hippocampus. Shim and Russell 2004; Hammonds *et al.* 2007; Hammonds and Shim 2009; Shim *et al.* 2007; Shim and Hammonds 2009; Shim and Hammonds 2010; *Son *et al.* 2003

3.1 Effects of lithium treatment on LTP of the principal neurons in the hippocampus

Four weeks lithium treatment did not enhance the magnitude of I/O responses (Fig. 2), but significantly enhanced the magnitude of LTP of fEPSP of CA1 pyramidal cells (Figs. 1, 3). Two weeks lithium treatment significantly enhanced the magnitude of I/O responses and LTP of fEPSP of the DG granule cells (Figs. 4, 5). (Shim and Hammonds 2009; Shim et al. 2007).

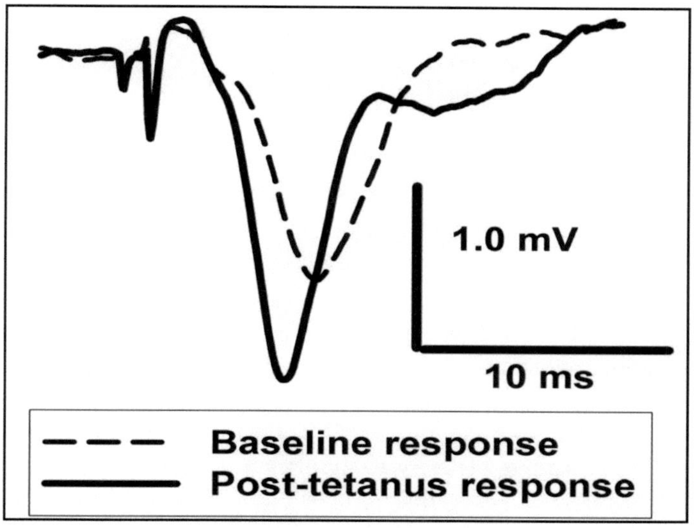

Fig. 1. Typical wave forms of LTP of *f*EPSP of the granule cells in DG.

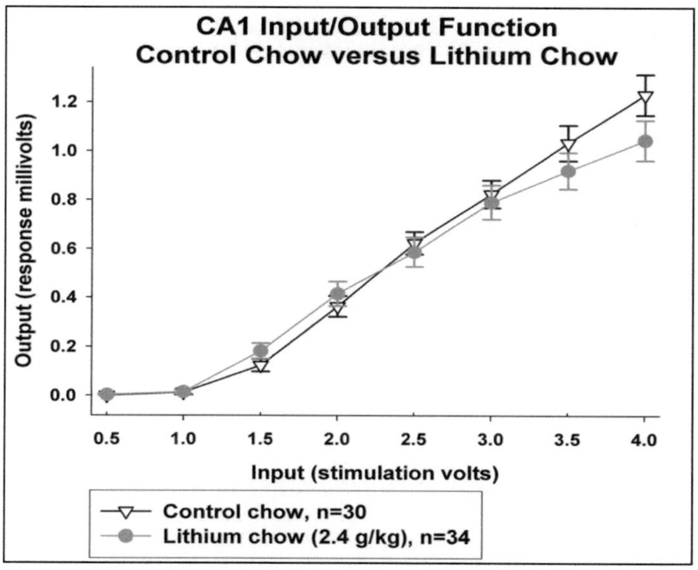

Fig. 2. Effects of 4 weeks lithium treatment on I/O responses of *f*EPSP in area CA1. There was no significant differences in I/O response between lithium (number of animals (N) =19; number of slices (n) = 34) and control (N =16, n = 30) chow treatment [$p > 0.05$, independent Student's *t*-tests

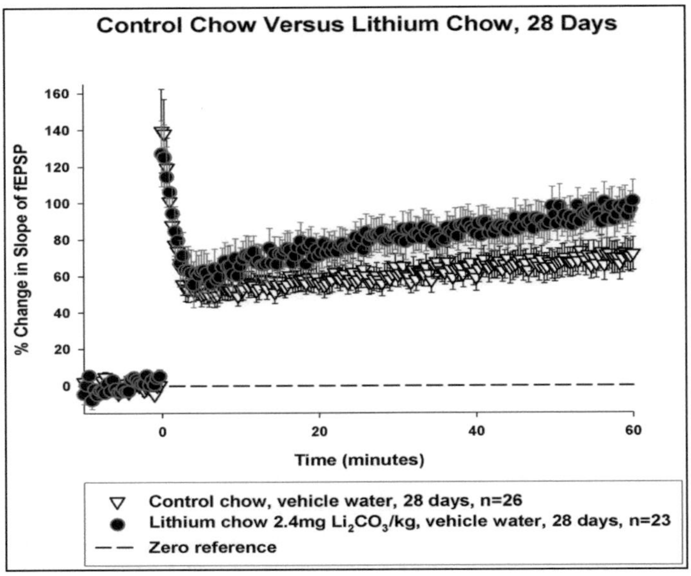

Fig. 3. Effects of 4 weeks lithium treatment on LTP of *f*EPSP in area CA1. Two trains of 100 Hz tetanus stimuli significantly magnified *f*EPSP of CA1 pyramidal cells as determined for the last 10 min. of the recordings. There was significant differences in LTP responses between lithium (N = 14, n = 26) and control (N =11, n = 23) treated animals in area CA1 ($p <$ 0.01].

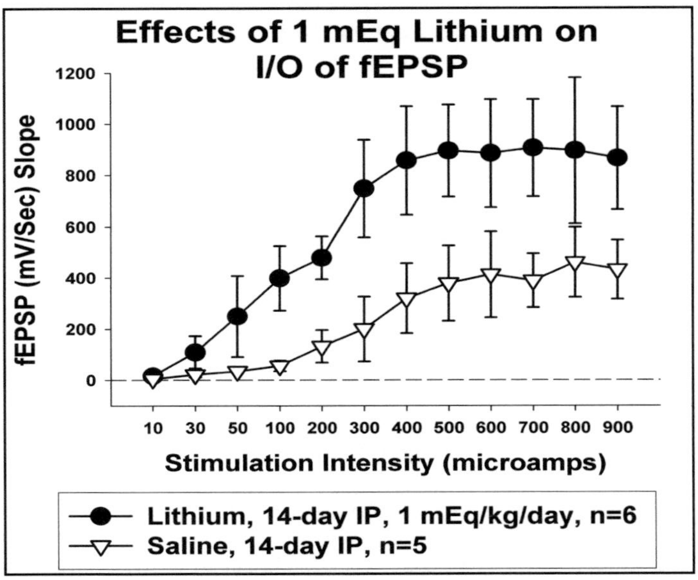

Fig. 4. Effects of 2 weeks lithium treatment on I/O responses of *f*EPSP in the DG. There was significant differences in I/O response between lithium (N = 5; n = 6) and control (N = 4, n = 5) chow treatment [($p < 0.01$, independent Student's *t*-test).

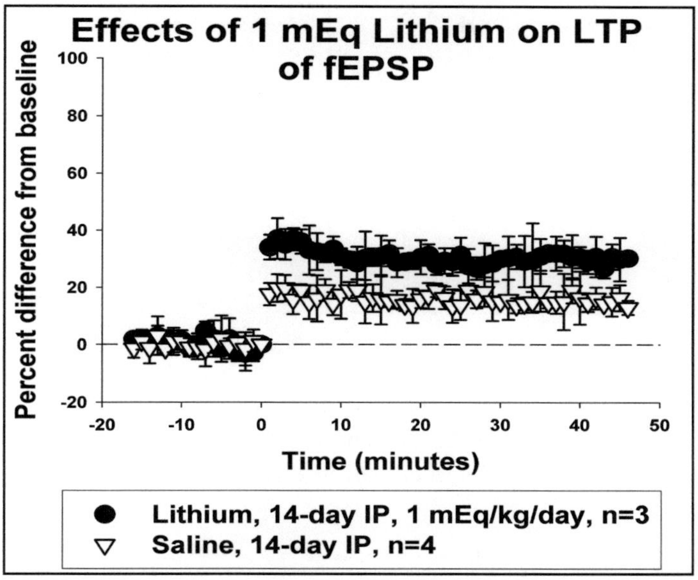

Fig. 5. Effects of 2 weeks lithium treatment on LTP of *f*EPSP in the DG. Two trains of 100 Hz tetanus stimuli significantly magnified fEPSP of the DG granule cells as determined for the last 10 min. of the recordings. There was significant differences in LTP responses between lithium treated (N = 4, n = 5) and control (N = 5, n = 6) animals in the DG [$p < 0.01$, independent Student's *t*-test].

3.2 Effects of lithium treatment on density of dendrites in the hippocampus

Sholl analysis shows that 4 weeks and 2 weeks lithium treatments redistributed dendritic branches in different manners. In the DG granule cells, 4 weeks lithium treatment increased the number of dendritic branches in nearly the inner 2/3 of dendritic trees, but reduced the number of the branches in outer 1/3 of dendritic trees (Figs. 6, 7). In hippocampal area CA1, the same lithium treatment increased the number of apical dendritic branches in the inner 2/3 of the dendritic trees, whereas, in the distal 1/3 of the apical tree, the treatment reduced dendritic branching (Fig. 8). In contrast, 2 weeks lithium treatment reduced the number of dendritic branches approximately in the outer 2/3 of the dendritic trees of the DG granule cells (Fig. 9). The same treatment reduced the apical dendritic branches of CA1 pyramidal cells throughout dendritic trees except a small proximal region of the dendrites (Fig. 10). Both 4 and 2 weeks lithium treatments redistributed dendritic branches without significant changes in the total number of them in the DG and area CA1 (Shim and Hammonds 2010; Shim and Russell 2004).

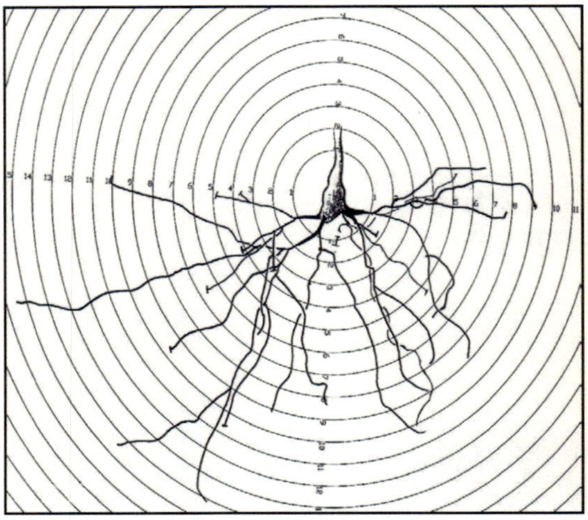

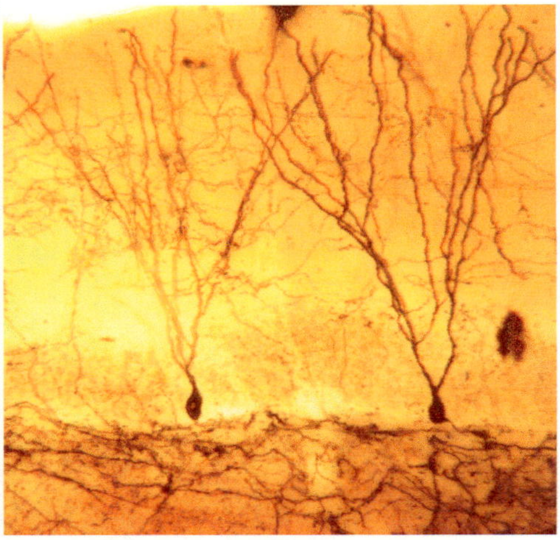

A B

Fig. 6. (A) Golgi impregnated granule cells with well branched and bifurcated dendrites extending into the molecular layers in the DG. (B) Camera lucida drawing of the dendritic trees of hippocampal CA1 pyramidal cells. The dendritic branches of a CA1 pyramidal cell under a concentric shell template for Sholl dendrite density analysis are shown.

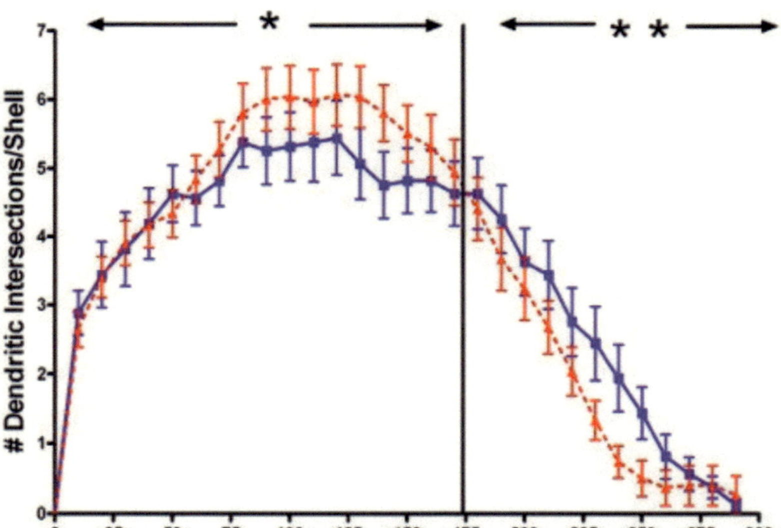

Fig. 7. Effects of 4 weeks lithium treatment on dendrites of the DG granule cells. Lithium-treated animals (lithium-treated animals N = 6, n = 30; controls N = 4, n = 16) have significantly more dendritic branches in the inner 2/3 of dendritic trees (Wilcoxonn test, p = 0.0038). In the outer 1/3 of dendrtic trees, lithium-treated animals have less dendritic branches (Wilsoxon test, p = 0.0024) broken line: lithium treated group; full line: control group. The over-all dendritic material of granule cells is not significantly different between the two groups.

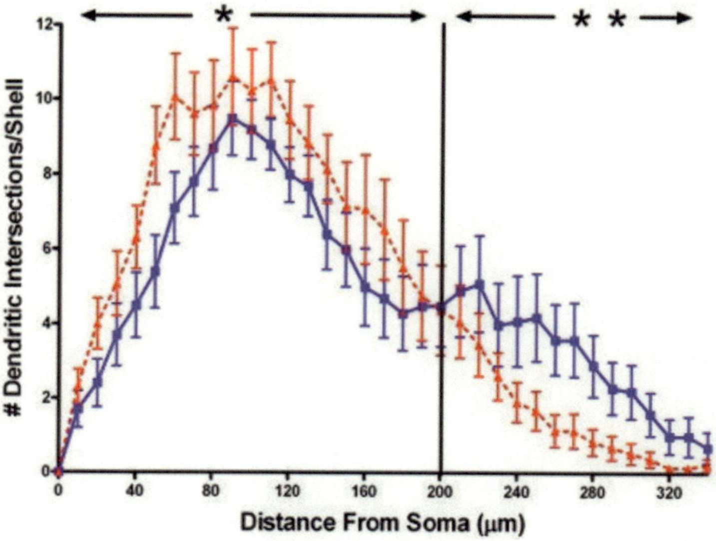

Fig. 8. Effects of 4 weeks lithium treatment on apical dendrites of the CA1 pyramidal cells. Lithium-treated animals (lithium-treated animals N = 3, n = 8; controls N=4, n=10) have significantly more dendritic branches in the inner 2/3 of dendritic branches (Wilcoxonn test, p = 0.0005). In the outer 1/3 of dendrtic trees, lithium-treated animals have less dendritic branches of dendrites (Wilcoxon test, $p < 0.0009$). The total amount of dendritic material of CA1 pyramidal cells is not significantly different between the two groups.

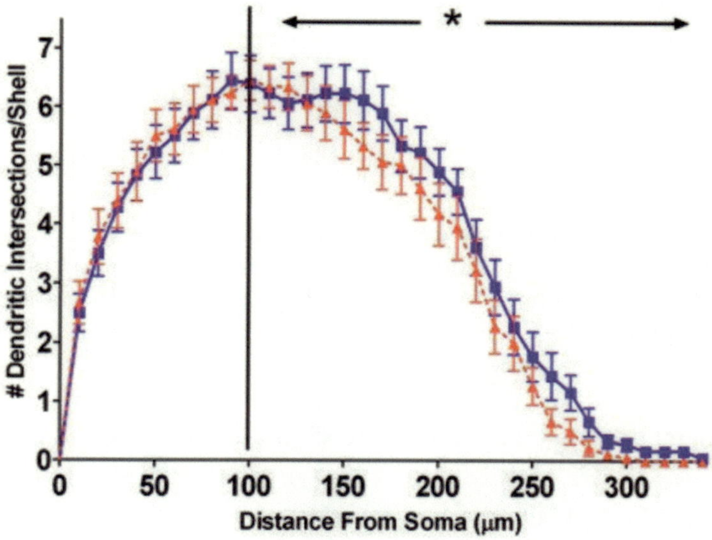

Fig. 9. Effects of 2 weeks lithium treatment on dendritic branches of the DG granule cells. The lithium treated animals (N = 3; n = 18) had significantly less dendritic branches on the outer 2/3 of dendritic tree comparing to those in the control group (N = 3, n = 18) ($p = 0.001$, Wilcoxon test). The over-all dendritic branches of granule cells were not statistically significant between the two groups.

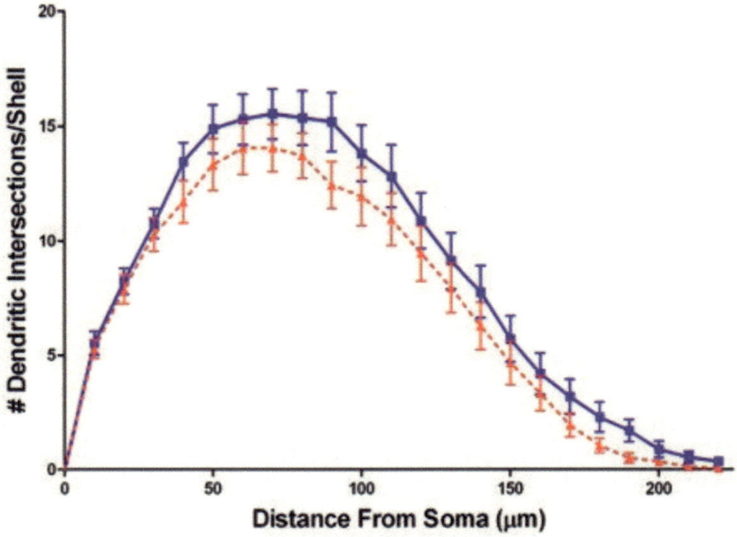

Fig. 10. Effects of 2 weeks lithium treatment on apical dendritic branches of area CA1 pyramidal neurons. The loss of dendritic branches was found in more than the outer 2/3 of the dendritic arbor in the lithium treated group ($p < 0.0001$, Wilcoxon test)(lithium -treated group N =3, n=18; Controls N = 3, n=17). The total amount of dendritic branches was not significantly different between the two groups.

3.3 Effects of lithium treatment on levels of BDNF, Bcl-2 and pCREB in the hippocampus

This study have also examined the effects of 4 weeks and 2 weeks lithium treatments on the levels of BDNF, Bcl-2 and p-CREB in the DG and area CA1 using ELISA. Four weeks lithium treatment increased the levels of Bcl-2 and pCREB in the DG and area CA1 (Fig 11B, 11C) (Hammonds and Shim 2009). This treatment increased the level of BDNF in the DG, but did not increase the level of BDNF, in area CA1 (Fig. 11A). However, 2 weeks lithium treatment did not alter the levels of BDNF, Bcl-2 and pCREB in either the DG or area CA1 (Figures are not shown) (Hammmonds and Shim 2007).

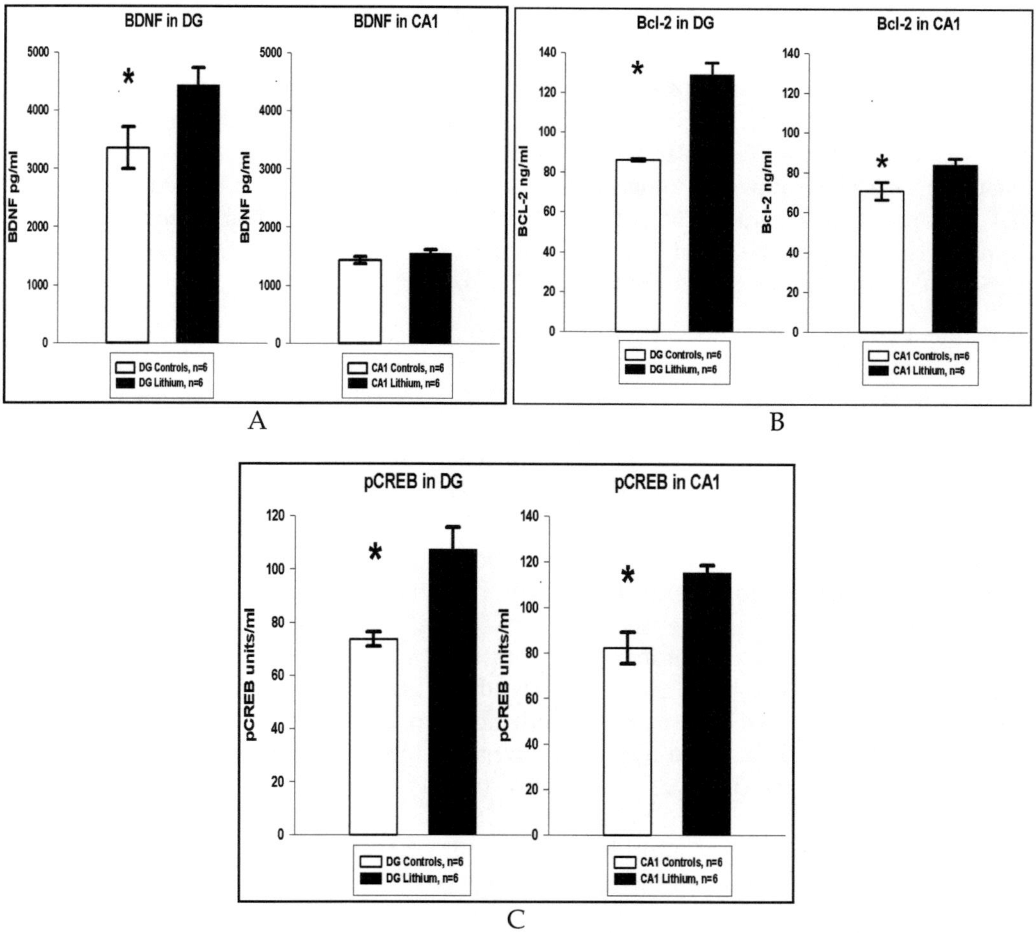

Fig. 11. Effects of 4-weeks lithium treatment on levels of BDNF, Bcl-2 and pCREB in the DG and area CA1. (A) There was significant difference in the levels of BDNF between the two groups (N = 6 of each group) in DG ($p = 0.016$, independent Student t-test), but no significant difference between the two groups in area CA1 ($p = 0.227$). There was significant difference in the level of total free Bcl-2 ($p < 0.001$ in DG; $p < 0.025$ in CA1) (B) and pCREB ($p = 0.001$ in DG; $p < 0.001$ in CA1) (C) between the lithium-treated (N = 6) control groups (N = 6) in both areas Two weeks lithium treatment did not change the levels of BDNF, Bcl-2 and pCREB in the DG and area CA1 (Hammonds et al, 2007..

4. Discussion

Our study found that chronic lithium treatment (CLT) magnified the expression of LTP of hippocampal area CA1 pyramidal cells (Fig. 3) without enhancing baseline synaptic response (Fig. 2), and subchronic lithium treatment (SCLT) magnified LTP and I/O function of the DG granule cells (Figs. 4, 5). Son et al. (2003) also reported that 4 weeks lithium treatment magnified LTP in the DG. Previously, we reported that 2 weeks lithium treatment magnified LTP in the DG (Shim et al. 2007). Our findings combined with others suggest that long-term lithium treatment increases synaptic plasticity in the hippocampus. We recently observed that acute in vitro and in vivo (one day) lithium treatment decreased the magnitude of LTP in area CA1 (a manuscript in preparation). These findings together supports our hypothesis that prolonged lithium treatment upregulates functional synaptic plasticity in the hippocampus, and this effect may not be associated with the direct chemical effect of lithium, but is likely to be associated with the molecular actions of lithium at genetic levels. Since hippocampal synaptic plasticity is known to underlie memory and learning, we investigated the effects of CLT on learning and memory (Nocjar et al 2007), which showed CLT increases learning and memory. This suggests the possibility that lithium can have therapeutic effects in patients with Alzheimer's disease and other dementia. (Engel et al. 2008).

Our sholl analyses of the effects of lithium treatment on dendrites suggest that prolonged lithium treatment produces a remodeling in the distribution of dendrites of the principal neurons in the hippocampus. Interestingly, CLT and SCLT remodels in distribution of dendrites in different manners. CLT produces an increase in the distribution of dendritic branches with more branches in the segment of highly populated branches of dendritic trees and reduces dendritic branches in the segment of less populated branches (Figs. 7, 8). The consequence of this redistribution is unknown. We hypothesize that CLT produces more branches in the segment of denritic trees, where synapses with major excitatory input occurs form, and the most active synaptic transmission occurs, whereas CLT reduces branches in the segment where less synapses form. This structural remodeling may lead to an enhancement in synaptic transmission in a more efficient manner. SCLT remodels in distribution of dendritic branches in a manner opposite to the manner CLT does. SCLT decreases dendritic branches in dendrite segments, where branches are highly distributed and active synaptic interactions occur (Fig. 9, 10). This unexpected finding is not due to lithium toxicity. In our studies, lithium was administered twice daily (1 mEq/kg a day) for the two weeks lithium study, and no animals died or showed signs of lithium toxicity. It was confirmed that the blood levels of lithium were within the human therapeutic range of 0.51 to 0.78 mEq/L (Baldessarini and Tarazi, 2001). Furthermore, In our 4 weeks lithium treatment studies, animals were fed lithium chow for 4 weeks, and their lithium blood levels were again confirmed to be within the therapeutic range (0.68 to 0.89 mEq/L). Thus, lithium toxicity is not supported as a potential cause for the findings in the current studies. The dendritic segments where SCLT reduces dendritic branching are the region where dendrites receive major excitatory input and form main synapses with the input. Our study shows that SCLT increase baseline synaptic transmission and LTP, which was recorded approximately in the same region of the DG as the region where SCLT reduced dendritic branches (Figs. 4, 5). It is possible that the lithium-induced loss of dendritic branches in these regions could dampen excited synaptic activity as a homeostatic adaptive response to maintain physiologically efficacious synaptic transmission.

Our results show that CLT upregulates Bcl-2 and CREB in the DG and hippocampal area CA1 and upregulates BDNF in the DG (Fig. 11). However, SCLT does not upregulate BDNF, Bcl-2 or CREB in these hippocampal subregions (Hammonds et al. 2007). These findings suggest that CLT, not SCLT, upregulates these neuroplatic proteins, which are actively involved in synaptic plasticity.

Our results from electrophysiological, morphological and molecular studies suggest that SCLT may increase functional synaptic plasticity. SCLT, however, increases neither the density of dendrites in the active synaptic regions of dendritic trees of the principal neurons nor the levels of BDNF, Bcl-2 and pCREB in the hippocampus. In contrast, CLT increases functional synaptic strength, which is accompanied with remodels in distribution of dendritic branches in such a way that more dendritic material is available in dendritic regions where the most active synaptic activity occurs. These cellular changes are consistent with the up-regulation of BDNF, Bcl-2 and pCREB in the hippocampus. Thus, CLT may produce more stable and persistent changes in synaptic plasticity in the hippocampus. In contrast, SCLT may increase the functional aspect of synaptic plasticity. These changes may not be accompanied with the increased density of dendrites at active synaptic regions or the up-regulation of neuroplastic proteins. These findings are consistent with the clinical observation that lithium takes 3-4 weeks to show its efficacy in treatment of bipolar disorder and other mental illness (Baldessarinni and Tarazi 2001).

5. References

Adams JM, Cory S: The Bcl-2 protein family: arbiters of cell survival. Science 281;1998:1322-1326

Angelucci F, Aloe L, Jimenez-Vasquez P, Mathe AA: Lithium treatment alters brain concentrations of nerve growth factor, brain-derived neurotrophic factor and glial cell line-derived neurotrophic factor in a rat model of depression. Int J Neuropsychopharmacol 2003; 6:225-231

Baldessarini RJ, Tarazi FI (2001) Drugs and treatment of psychotic disorders: psychosis and mania. In: The pharmacological basis of therapeutics (Hardman JG, Limbird LE, eds), pp 485–520. New York: McGraw-Hill

Chalecka-Franaszek E, Chuang D-M: Lithium activates the serine /threonine kinase Akt-1 and suppresses glutamate-induced inhibition of Akt-1 activity in neurons. Proc Natl Acad Sci U S A 1999; 96:8745–8750

Chuang DM, Chen R, Chalecka-Franaszek E, Ren M, Hashimoto R, Senatorov V et al.: Neuroprotective effects of lithium in cultured cells and animal models of diseases. Bipolar Disord 2002; 4:129-136

Engel T, Goñi-Oliver P, Gómez de Barreda E, Lucas JJ, Hernández F, Avila J.: Lithium, a potential protective drug in Alzheimer's disease. Neurodegener Dis 2008; 5:247-249

Fukumoto T, Morinobu S, Okamoto Y, Kagaya A, Yamawaki S. Chronic lithium treatment increases the expression of brain-derived neurotrophic factor in the rat brain. Psychopharmacol 2001; 158:100-106

Gabbot PL, Somogyi J: The "single" section Golgi impregnation procedure: methodological description. J Neurosci Methods 1984;11:221-230

Gould TD, Manji HK: Signaling networks in the pathophysiology and treatment of mood disorders. J Psychosom Res 2002; 53:687-697

Grimes CA, Jope RS: CREB DNA binding activity is inhibited by glycogen synthase kinase-3β and facilitated by lithium. J Neurochem 2001; 78: 1219–1232

Hammonds M.D., Shim, S.S.: Effects of 4 weeks lithium and olanzapine treatment on levels of BDNF, Bcl-2 and phosphorylated CREB in the hippocampus Basic Clin Pharmacol Toxicol 2009;105:113–119

Hammonds M.D., Shim S.S., Pingfu F., Calabrese J.R.: Effects of subchronic lithium treatment on levels of BDNF, Bcl-2 and phospho-CREB in the rat hippocampus. Basic Clin Pharmacol Toxicol 2007; 100:356-359

Jope RS and Roh M-S: Glycogen Synthase Kinase-3 (GSK3) in Psychiatric Diseases and Therapeutic Interventions. Curr Drug Targets 2006; 7:1421-1434

Manji HK, Chen G: PKC, MAP kinases and the bcl-2 family of proteins as long-term targets for mood stabilizers. Mol Psychiatry 2002; 1:46-56

Manji HK, Duman RS: Impairments of neuroplasticity and cellular resilience in severe mood disorders: implications for the development of novel therapeutics. Psychopharmacol Bull 2001; 35:5-49

Manji HK, Moore GJ, Chen G: Lithium at 50: have the neuroprotective effects of this unique cation been overlooked? Biol Psychiatry 1999; 46:929-940

McAllister AK, Katz LC, Lo DC: Neurotrophins and synaptic plasticity. Annu Rev Neurosci 1999;22:295-318

Nocjar C, Hammonds MD, Shim SS: Chronic lithium treatment magnifies learning in rats Neuroscience 2007; 774:774-788

Rowe MK and Chuang D-M: Lithium neuroprotection: molecular mechanisms and clinical implications. Expert Rev Mol Med 2004; 6:1-18

Shaywitz AJ, Greenberg ME: CREB: A stimulus-induced transcription factor activated by a diverse array of extracellular signals. Annu Rev Biochem 1999;68:821-861

Shim SS, Hammonds MD, Ganocy SJ, Calabrese JR: Effects of lithium exposure on synaptic plasticity in the dentate gyrus of the hippocampus. Prog Neuropsychopharmacol Biol Psychiat 2007;31:343-347

Shim S.S., Hammonds, M.D.: Effects of treatment with olanzapine and lithium on synaptic plasticity and levels of BDNF, Bcl-2 and phosphorylated CREB in the hippocampus. Biol Psychiatry Abstr 2009;64:89S

Shim S.S., Hammonds M.D.: Effect of chronic lithium treatment on morphology of dendrites in hippocampus and prefrontal cortex. Psychiatry Abstr 2010;67:238S

Shim S.S., Russell R.: Effects of exposure to lithium on morphology of dendrites in the hippocampus and the parietal cortex. Neuropsychopharmacol 2004; 29:S208

Silva AJ, Kogon JH, Frankland PW: Creb and memory Annu Rev Neurosci 1998;21:127-148

Son H, Yu IT, Hwang SJ, Kim JS, Lee SH, Lee YS et al: Lithium enhances long-term potentiation independently of hippocampal neurogenesis in the rat dentate gyrus. J Neurochem 2003; 85:872-881

Tsaltas E, Kontis D, Boulougouris V, Papadimitriou GN: Lithium and cognitive enhancement: leave it or take it? Psychopharmacol 2009; 202:457-476

Wood GE, Young LT, Reagan LP, Chen B, McEwen BS: Stress-induced structural remodeling in hippocampus: prevention by lithium treatment. Proc Natl Acad Sci USA 2004;101:3973

Part 2

Neuropharmacological Challenges

4

Li^+ in Bipolar Disorder – Possible Mechanisms of Its Pharmacological Mode of Action

Carla P. Fonseca[1,2], Liliana P. Montezinho[1,3] and
M. Margarida C.A. Castro[1*]
[1]*Dept. of Life Sciences, Faculty of Sciences and Technology, University of Coimbra; Center for Neurosciences and Cell Biology (CNC) of Coimbra, Coimbra,*
[2]*CICS-UBI–Health Sciences Research Centre, University of Beira Interior, Covilhã,*
[3]*Dept. of Neurodegeneration 1, H. Lundbeck A/S, Valby,*
[1,2]*Portugal*
[3]*Denmark*

1. Introduction

Bipolar disorder is a severe psychiatric illness characterized by cyclic episodes of mania and depression that affects approximately 1 % of the world population, and has a great economic and social impact. Lithium (Li^+), in the form of lithium salts, has been used for more than five decades and is still the drug of choice in the treatment of this pathology. The anticonvulsants carbamazepine and valproate, originally used to treat epileptic seizures, are alternative or adjunctive therapies to lithium, representing the first-line therapy for lithium-resistant or lithium intolerant patients. Bipolar disorder is associated with structural, functional and physiological alterations in the brain of bipolar patients, which reflect chemical, neurochemical and metabolic changes, specifically at the levels of brain metabolites and neurotransmitters, as already detected by different techniques (Silverstone et al., 2005). Abnormalities in signal transduction pathways, in particular G proteins, adenylate cyclase (AC) and phosphatidylinositol (PI) signalling cascade, as well as protein kinase C (PKC) were related with the pathophysiology of bipolar disorder (Berns & Nemeroff, 2003, Brunello & Tascedda, 2003; Manji et al., 2001; Manji & Lenox, 2000a, 2000b).

Despite the widespread clinical use of lithium salts, the pharmacological mode of action underlying Li^+ mood stabilizing effects is still unclear and several hypotheses have been advanced. Once inside the cells, Li^+ has been proposed to compete with Na^+ and Mg^{2+} for these ions intracellular binding sites in biomolecules, to affect intracellular Ca^{2+} concentration and to have an important role on the activity of G proteins, AC and inositol monophosphatase (IMPase) thus interfering with the levels of neurotransmitters and other substances in brain. Li^+ has also been proposed to modulate the activity of certain glycolytic and tricarboxylic acid (TCA) cycle enzymes affecting several metabolic pathways and

*Corresponding author

altering the concentrations of intermediary metabolites. Recent research has been focused on how Li^+ changes the activity of cellular signal transduction systems, in particular those involving AC. The present chapter is a review of data published in literature from studies carried out to test some of these hypotheses, which, in most cases, are inter-related. The objective is to give an overview of what is known about possible mechanisms of Li^+ therapeutic action in bipolar disorder, at the molecular and cellular levels, using different approaches and techniques.

Li^+ effects have been shown to be highly cell-type specific and so the study of Li^+ transport processes across cell membranes is pertinent. Studies of the kinetics of Li^+ influx, intracellular immobilisation and Li^+/Mg^{2+} competition in different types of cells are presented (Amari et al., 1999a; Castro et al, 1996; Fonseca et al., 2000, 2004; Layden et al., 2003; Montezinho et al., 2002; Mota de Freitas et al., 2006; Nikolakopoulos et al., 1998). Li^+ influx rate constants were determined by atomic absorption spectrophotometry (experiments performed in cell suspensions) and ^7Li nuclear magnetic resonance (NMR) spectroscopy (cells immobilized in agarose), in the presence and absence of different activators and inhibitors of transport pathways present in these cells membrane. L-type voltage-sensitive Ca^{2+} channels were found to have an important role in Li^+ uptake under depolarising conditions in excitable cells. Once inside the cells, Li^+ was found to be bound to intracellular structures, as demonstrated by the ratio between the intracellular ^7Li NMR longitudinal and transversal relaxation times (T_1/T_2 ratio). The degree of intracellular Li^+ immobilisation is cell type dependent. Intracellular Mg^{2+} was found to be significantly displaced by Li^+ from its binding sites, as demonstrated by ^{31}P NMR and fluorescence spectroscopy (Fonseca et al., 2004)

The study of the effects of Li^+ on metabolic pathways is also referred. Results for Li^+ effects on glucose metabolism and on the competitive metabolism of glucose and lactate in a cell line, the human neuroblastoma SH-SY5Y cells, using ^{13}C NMR spectroscopy are presented (Fonseca et al., 2005). A relatively simple metabolic network was proposed for these cells, based on the computer program tcaCALC best fitting solutions. The results obtained suggested that cell energetic metabolism might be an important target for Li^+ action. ^{13}C NMR spectroscopy was also used to investigate Li^+ effects on glucose and acetate metabolism in primary cell cultures, rat cortical neurons and astrocytes, as well as in rat brain (Fonseca et al., 2009). It was proposed that Li^+ has an important role on the GABAergic neurotransmitter system as detected in cortical neurons when ^{13}C-glucose was used as substrate, as well as in rat brain after infusion with [1-^{13}C]glucose.

Since it has been suggested that cyclic adenosine 3',5'monophosphate (cAMP) levels are abnormal in bipolar patients and are regulated by mood stabilizing agents, it is of utmost importance to know whether this second messenger regulates Li^+ transport into neurons. It is also relevant to determine Li^+ effects on the homeostasis of intracellular cAMP levels. The effect of different intracellular cAMP levels on Li^+ uptake, at therapeutic plasma concentrations, was studied using pertinent cellular models and a radioactive assay (Montezinho et al., 2004). The data obtained demonstrated that intracellular cAMP levels regulate the uptake of Li^+ in a Ca^{2+} dependent manner, and that Li^+ plays an important role in the homeostasis of this second messenger in neuronal cells .

Second messenger-mediated pathways are targets for Li^+ action, thus, it is important to investigate whether other mood stabilizing agents exert similar effects on the same signalling pathways. Bipolar disorder seems to be associated with an enhanced signalling activity of the cAMP cascade and most of its events have been implicated in the action of mood stabilizing agents. Therefore the effects of Li^+, carbamazepine and valproate on basal and forskolin-evoked cAMP accumulation and the capacity of dopamine D_2-like receptors stimulation, with quinpirole, to block the increase of forskolin-stimulated cAMP levels were studied both *in vitro*, in cultured cortical neurons, and *in vivo* in the rat prefrontal cortex using microdialysis in freely moving animals, under control conditions and after treatment with the mood stabilizing drugs (Montezinho et al., 2006).

Several studies have suggested the involvement of biogenic monoaminergic neurotransmission in bipolar disorder and in the therapies used for this disease. The effects of the mood stabilizing drugs Li^+, carbamazepine or valproate on the dopaminergic and adrenergic systems, particularly on dopamine D_2-like and β-adrenergic receptors, were studied both in cultured rat cortical neurons and in the rat prefrontal cortex using microdialysis in freely moving animas (Montezinho et al., 2007). It was observed that these receptors have a regulatory role on AC activity and each drug acts by a unique mechanism. Dopamine D_2 and β_1-adrenergic receptors were found to be co-localized in cells of the prefrontal cortex, as determined by immunohistochemistry and were differentially affected by treatment with the three mood stabilizers, as determined by Western blot experiments. Data showed that the mood stabilizers studied affected dopamine D_2 receptors.

Figure 1 summarizes the possible targets for Li^+ action.

2. Characterisation of Li^+ transport pathways and intracellular binding: Effects on Li^+/Mg^{2+} competition in cellular models

Li^+ transport, intracellular immobilisation and Li^+/Mg^{2+} competition were studied in chromaffin cells, isolated from the bovine adrenal medulla, which are good neuronal models (Trifaró, 1982), and human neuroblastoma SH-SY5Y cells, a clonal derivative of the SK-N-SH cell line that provides a suitable model of neurons due to its exclusive neuroblast phenotype (Biedler et al, 1973). The results obtained and the main conclusions are presented in the following sections.

2.1 Li^+ membrane transport studies in neuronal models

Atomic Absorption (AA) spectrophotometry was used to investigate the membrane transport pathways involved in the uptake of Li^+ by chromaffin cells (Fonseca et al., 2004). Figure 2 shows the kinetics of Li^+ influx in the control situation, in the presence of ouabain (a (Na^+, K^+)-ATPase inhibitor), and under continuous depolarising conditions in the absence and presence of nitrendipine (a specific blocker of the L-type voltage-sensitive Ca^{2+} channels). The kinetics of Li^+ influx was analysed using the following equation:

$$([Li^+]_{iT})_t = ([Li^+]_{iT})_\infty [1 - \exp(-k_i t)] \quad (1)$$

where k_i is the rate constant for Li^+ influx, $([Li^+]_{iT})_t$ and $([Li^+]_{iT})_\infty$ are the total intracellular Li^+ concentrations at the different time points t and when the steady state has been reached, respectively.

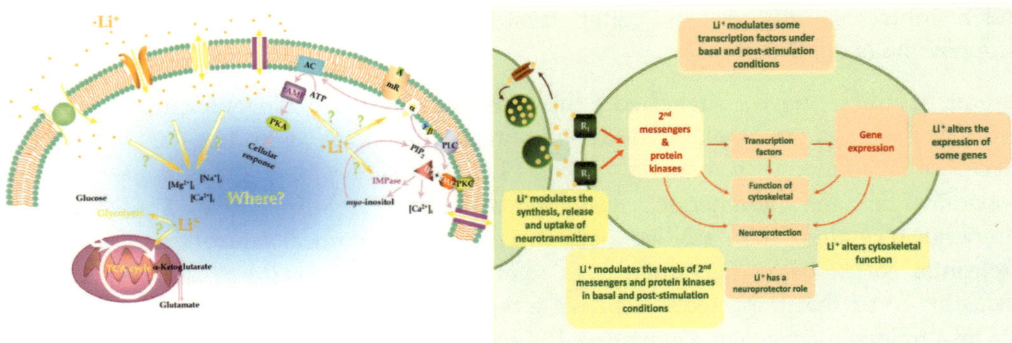

(A) (B)

Fig. 1. Some potential targets for Li^+ action inside the cells. **A)** Li^+ is transported across the cell membrane, through specific transport systems already present in the membrane or by passive diffusion. Li^+ has been proposed to bind the negatively charged groups of several membrane phospholipids modulating the activity of membrane proteins and possibly ion transport processes. Li^+ can also affect the intracellular Ca^{2+} concentration ($[Ca^{2+}]_i$) and compete with mono- and di-cations, such as Na^+ and Mg^{2+}, respectively, for their intracellular binding sites. Other possible targets for Li^+ action are guanine-nucleotide binding proteins (G-proteins), adenylate cyclase (AC) and inositol monophosphatase (IMPase), affecting in this way the levels of second messenger molecules such as 3´,5'-cyclic adenosine monophosphate (cAMP), diacylglycerol (DAG) or inositol 1,4,5-triphosphate (IP_3), and the metabolism of phosphatidylinositols. Li^+ has also been proposed to modulate the activity of some glycolytic and tricarboxylic acid (TCA) cycle enzymes and the levels of some metabolic intermediates. α, β and γ - subunits of G-proteins; A - agonist; ATP - adenosine triphosphate; mR - metabotropic receptor; $[Mg^{2+}]_i$ and $[Na^+]_i$ - intracellular Mg^{2+} and Na^+ concentrations, respectively; PIP_2 - phosphatidylinositol 4,5-bisphosphate; PKA - protein kinase A; PKC - protein kinase C; PLC - phospholipase C. **B)** It is known that multiple inter-related neurotransmitter systems are involved in mood regulation. Thus, Li^+ can affect the functional equilibrium between several interacting systems. This figure shows the processes which seem to have an important role in the mechanisms of mood stabilization. R_1, R_2= post-synaptic receptors; ● = neurotransmitters (Jope, 1999a; Manji et al., 1995; Manji et al., 2001)

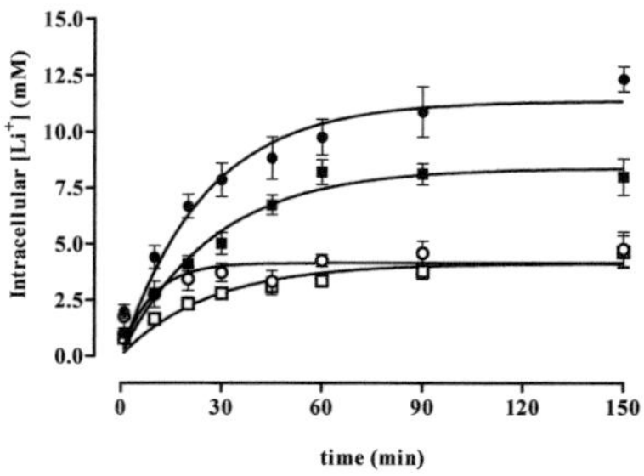

(A)

Experimental conditions	k_i (min^{-1})	$[Li^+]_{iT}$ (mmol (L cells)$^{-1}$)
Control	0.040 ± 0.006 (n=14)	11.39 ± 0.56
+ 50 µmol L^{-1} ouabain	0.036 ± 0.005 (n=7)	8.38 ± 0.38*
+ 45 mmol L^{-1} KCl	0.110 ± 0.038 (n=5)**	4.16 ± 0.27***
45 mmol L^{-1} KCl + 10 µmol L^{-1} nitrendipine	0.038 ± 0.007 (n=8)#	4.14 ± 0.28

(B)

Fig. 2. **A)** Graphical representation of the $[Li^+]_{iT}$ (determined by AA spectrophotometry) *versus* loading time during Li^+ influx experiments in bovine chromaffin cells in suspension (cytocrit of 2 to 3%) subject to loading with 15 mmol L^{-1} Li^+, at 37 °C. Data are for control (●), in the presence of 50 µmol L^{-1} ouabain (■), in the presence of 45 mmol L^{-1} KCl, a depolarising agent (O), and in the presence of 45 mmol L^{-1} KCl and 10 µmol L^{-1} nitrendipine (□). The lines correspond to the best exponential fits of the data to equation 1. In all cases, the intracellular Li^+ concentration increases up to 60–90 min and then reaches a steady state, except for the experiment done under depolarising conditions without nitrendipine, where the steady state is reached at 30 min; **B)** Li^+ influx rate constants (k_i) and the steady state total intracellular Li^+ concentrations ($[Li^+]_{iT}$), obtained from the curves presented in A), using equation 1. The values are average ± SEM for the number (n) of trials indicated in parenthesis. *, ** and ***: $p < 0.05$, $p < 0.01$ and $p < 0.001$ relative to control, respectively; # $p < 0.05$ relative to the KCl condition. (Fonseca et al., 2004)

In these experiments, the kinetics of Li^+ influx is not affected by the presence of ouabain, which suggests the non-involvement of (Na^+,K^+)-ATPase in Li^+ uptake by chromaffin cells, under resting conditions. However, when the cells are depolarised with KCl, a significant increase in the k_i value is observed, an effect that is completely suppressed in the presence of nitrendipine. This indicates that, under increased cellular excitability conditions, a new contribution to Li^+ influx appears which results from the activation of L-type voltage-sensitive Ca^{2+} channels. When the cells are depolarised, the intracellular Ca^{2+} concentration largely increases, increasing the activity of the Na^+/Ca^{2+} exchanger, known to be a high-capacity, low-affinity mechanism of Ca^{2+} efflux (Kao & Cheung, 1990; Powis et al., 1991). In our experiments, we propose that the Na^+/Ca^{2+} exchanger uses the external Na^+ and Li^+ to remove intracellular Ca^{2+} accumulated during cell depolarisation. Blocking of the L-type voltage-sensitive Ca^{2+} channels by nitrendipine prevents the depolarisation-dependent Ca^{2+} entry through these channels and therefore depresses the activity of the Na^+/Ca^{2+} exchanger, suppressing this new Li^+ entry pathway.

In the absence of active transport pathways for Li^+ influx, the $[Li^+]_{iT}$ obtained by AA spectrophotometry reflect the capacity of the cells to accumulate Li^+, which is controlled by the plasma membrane potential (-55mV in these cells, under resting conditions (Friedman et al.,1985)). When depolarisation occurs, the membrane potential becomes less negative, and the total amount of Li^+ that can be accumulated by the cells is lowered due to charge effects (Figure 2). This explains why the amount of Li^+ accumulated by the cells is significantly lowered when they are directly depolarised by KCl. The $[Li^+]_{iT}$ values observed under direct depolarising conditions in the presence and in the absence of nitrendipine are not significantly different, as expected, since under these conditions the membrane potential is kept constant by the high extracellular K^+ concentrations, even if nitrendipine affects the Li^+ uptake kinetics. The observation that ouabain lowers the steady state $[Li^+]_{iT}$ also shows its depolarising effect on these cells (Kitayama et al.,1990).

2.2 Li^+ degree of immobilisation inside cells by ^{7}Li NMR

The degree of immobilisation of Li^+ within different types of cells was investigated using ^{7}Li NMR relaxation measurements when the intracellular Li^+ concentration has reached a steady-state. Therefore, Li^+ uptake was first followed by ^{7}Li NMR spectroscopy, along with the shift reagent $[Tm(HDOTP)]^{4-}$ in cell suspensions and in agarose-embedded and perfused cells. The shift reagent was used to separate ^{7}Li NMR signals corresponding to Li^+ inside and outside the cells (Nikolakopoulos et al., 1998; Rong et al., 1993). Figure 3 shows ^{7}Li NMR spectra from an influx experiment performed in agarose gel-embedded bovine chromaffin cells, under the experimental conditions indicated in figure legend. A graphical representation of the percent of intracellular ^{7}Li resonance area, A_i, relative to the total area of intra- and extracellular ^{7}Li resonances, $A_i + A_e$, over time is shown in Figure 3 C. The kinetics of Li^+ influx was defined by equation 2:

$$[(A_i)_t / (A_i + A_e)_t] = [(A_i)_\infty / (A_i + A_e)_\infty] [1 - \exp(-k_i t)] \quad (2)$$

where k_i is the rate constant for Li^+ influx, $(A_i)_t$, $(A_e)_t$ and $(A_i)_\infty$, $(A_e)_\infty$ are the areas of the intracellular and extracellular $^7Li^+$ NMR signals at the different times t and when the intracellular Li^+ concentration has reached a steady state, respectively.

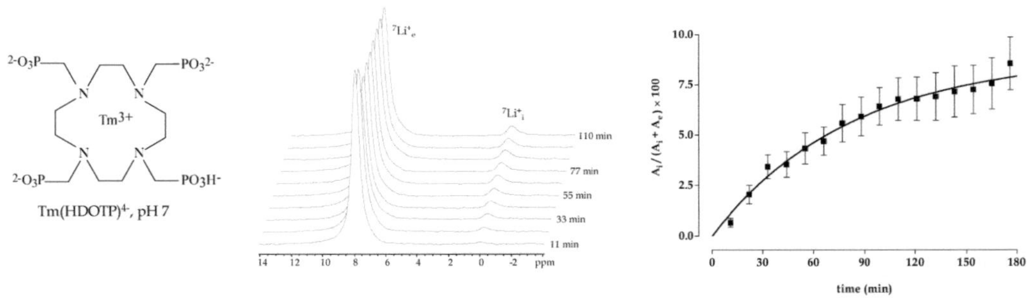

(A) (B) (C)

Fig. 3. **A)** Chemical structure of $[Tm(HDOTP)]^{4-}$; **B)** ^{7}Li NMR spectra (194.3 MHz) obtained over time in a Li^+ influx experiment performed with agarose gel-embedded bovine chromaffin cells (50 to 75 × 10^6) continuously perfused, at 1 mL min^{-1}, with culture medium containing 15 mmol L^{-1} Li^+ and 7 mmol L^{-1} $[Tm(HDOTP)]^{4-}$ (37 °C). Each spectrum represents the average of the total accumulation time of 11 min. $^7Li^+_e$ and $^7Li^+_i$ are the extra- and intracellular ^{7}Li NMR resonances, respectively. NMR experiments were performed on a Varian Unity-500 NMR spectrometer equipped with a multinuclear 10 mm broadband probe and a controlled temperature unit, using the following parameters: 64 transients for each spectrum, spectral width of 5.6 kHz, pulse width of 15 μs, acquisition time of 0.360 s and recycling time of 10.36 s. The signal-to-noise ratio was enhanced by exponential multiplication with a line broadening of 30 Hz. **C)** Time dependence of the percentage of intracellular $^7Li^+$ NMR signal area, normalised to the total area of intra- and extracellular ^{7}Li NMR signals $[(A_i)/(A_i + A_e)]$. The experimental data was fitted using equation 2 and the line corresponds to the best exponential fit of the data (Fonseca et., 2004).

For chromaffin cells immobilised in agarose gel threads, the influx rate constant has a contribution from the diffusion process of Li^+ across the gel before reaching the cell membrane (Nikolakopoulos et al., 1996). Therefore, under these conditions, the value obtained from equation 2 for Li^+ influx is an apparent k_i value, k_{iapp}. The average value obtained for k_{iapp} was 0.012 ± 0.003 min^{-1}, much lower than the value determined for cell suspensions using AA spectrophotometry (0.040 ± 0.006 min^{-1}, Figure 2 B, (Fonseca et al., 2004)) or ^{7}Li NMR spectroscopy (0.040 ± 0.003 min^{-1}, (Fonseca et al., 2000)), as expected.

Three hours after the beginning of the $^7Li^+$ influx NMR experiments, when the steady state intracellular Li^+ concentration was reached, the degree of immobilisation of Li^+ inside different types of cells was investigated using ^{7}Li NMR relaxation measurements by determining intracellular $^7Li^+$ T_1 and T_2 values and the respective T_1/T_2 ratio. This ratio is a sensitive measure of the rotational correlation time, τ_c, of the Li^+ ion, and hence of Li^+ immobilisation, independently of the fraction of bound Li^+ and of its binding affinity (Layden et al., 2003; Nikolakopoulos et al., 1998; Rong et al., 1993) (Figure 4).

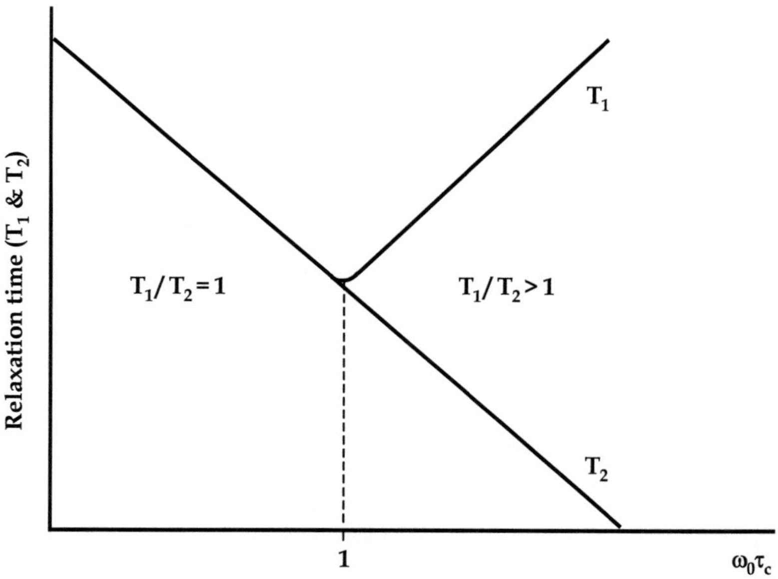

Fig. 4. T_1 and T_2 dependence on the rotational correlation time (τ_c) of a molecule. ω_0 stands for the Larmor angular frequency. When $\omega_0\tau_c << 1$ (for small molecules or freely moving ions) T_1 equals T_2 and the T_1/T_2 ratio equals 1. When $\omega_0\tau_c = 1$, T_1 reaches its minimum value and for $\omega_0\tau_c >> 1$ (large molecules or immobilised ions) T_1 increases proportionally to τ_c while T_2 decreases and the T_1/T_2 ratio becomes higher than 1 (Adapted from Gil & Geraldes, 1987).

Table 1 compares the results obtained for different types of cells:

Sample	$[Li^+]$ [b]	T_1 (s)	T_2 (s)	T_1/T_2
Bovine chromaffin cells suspensions (n=5) [c]	1.7	6.1 ± 0.2	0.02 ± 0.002	305 ± 32
Bovine chromaffin cells perfused (n=4) [d]	11.4	5.4 ± 1.3	0.05 ± 0.006	106 ± 28
Neuroblastoma cells (n=3) [e]	2.9	5.1 ± 0.8	0.05 ± 0.02	102 [h]
Lymphoblastoma cells (n=3) [f]	3.1	2.6 ± 0.4	0.06 ± 0.01	43 ± 4
Human red blood cells (n=3) [g]	3.5	6.5 ± 0.2	0.46 ± 0.01	14 [h]
Viscosity adjusted LiCl solution [g,i]	4.0	3.9 ± 0.4	3.6 ± 0.6	1.1± 0.2

[a] Each T_1 and T_2 value is an average ± SEM for the number (n) of trials indicated in parenthesis. [b] For the cell samples, this is the steady state total intracellular Li^+ concentration of Li^+-loaded cells, $[Li^+]_{iT}$, expressed as mmol (L cells)$^{-1}$, with errors less than 10%. [c] Data from Fonseca et al., 2000. [d] Data from Fonseca et al., 2004. [e] Data from Nikolakopoulos at al., 1998. [f] Data from Layden at al., 2003. [g] Data from Rong et al., 1993. [h] Errors are less than 10%. [i] Sample viscosity was adjusted to 5 centipoise (cP) with glycerol.

Table 1. ^{7}Li NMR relaxation time[a] and T_1/T_2 ratio values for intracellular Li^+ obtained for different types of cells.

The T_1/T_2 ratio obtained for bovine chromaffin cells, under the perfusion experimental conditions, is considerably lower than for the same cells in suspension at similar extracellular Li^+ concentrations (15 mmol L^{-1}) , indicating an increased degree of immobilization of Li^+ in the latter case. This difference could reflect some loss of viability of the cells during the NMR experiments in cell suspensions without perfusion. The disruption of the cell membrane and the probable nonintegrity of the cytoplasm may contribute to a higher immobilization of this ion through binding to the cytoplasmatic membrane and intracellular structures. Comparing the T_1/T_2 ratios of the various perfused cell systems at similar Li^+ loading levels (Table 1), the degree of Li^+ immobilization follows the following order: chromaffin ≈ neuroblastoma > lymphoblastoma > red blood cells (RBCs), reflecting the relative local mobility of the intracellular Li^+ binding sites of the different systems.

2.3 Monitorisation of agarose-embedded cells viability using ^{31}P NMR spectroscopy

Cell viability of the perfused agarose-embedded cells was monitored by obtaining ^{31}P NMR spectra during the time course of the ^{7}Li NMR experiments. Figure 5 shows representative ^{1}H-decoupled ^{31}P NMR spectra obtained after Li^+ NMR experiments in bovine chromaffin cells and neuroblastoma SH-SY5Y cells.

The criteria used to evaluate cell viability during the NMR experiments were, in the case of SH-SY5Y cells, the intracellular levels of ATP and PCr over the course of the perfusion experiments (by monitoring the areas of the PCr and of the α-, β-, and γ-ATP ^{31}P NMR resonances and the ratio of the PCr/β-ATP areas) and the chemical shift of inorganic phosphate (P_i) which is related to the intracellular pH of the sample. The areas of the ^{31}P NMR resonances of PCr and of ATP, as well as the chemical shift of the P_i resonance did not change appreciably during the course of NMR experiments, showing that there were no significant changes in energy metabolism or intracellular pH (Nikolakopoulos et al., 1998).

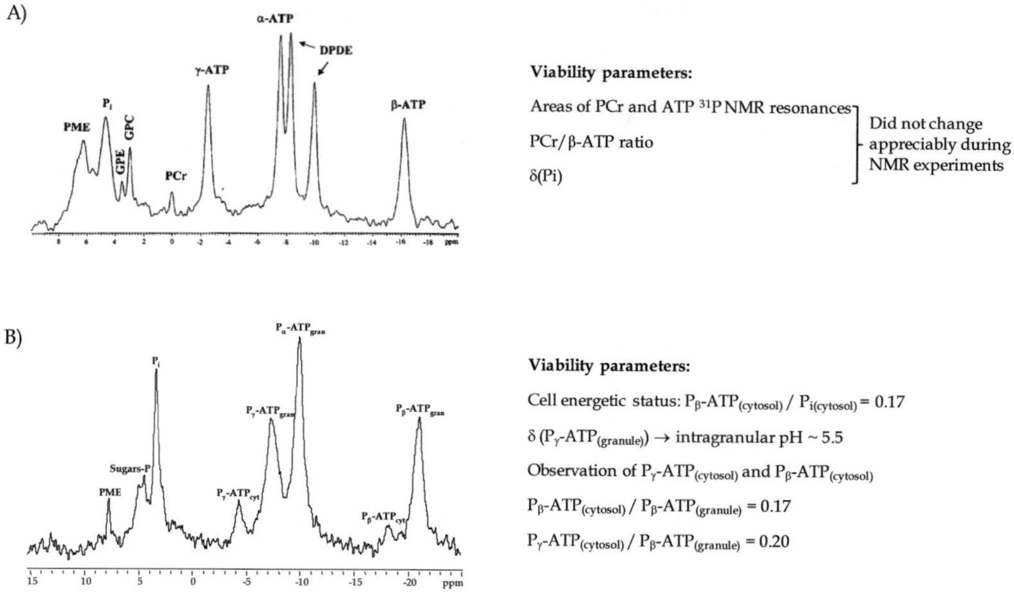

Fig. 5. ^{1}H-decoupled ^{31}P NMR spectrum (202.3 MHz) of Li^+-loaded agarose gel-embedded SH-SY5Y cells, after 5 h perfusion (Nikolakopoulos et al., 1998) (**A**) or bovine chromaffin cells, after 7 h 30 min perfusion (Fonseca et al., 2004) (**B**). For SH-SY5Y cells, ^{31}P chemical shifts are reported relative to phosphocreatine (PCr) referenced at 0 ppm. For chromaffin cells, H_3PO_4 85% was used as an external reference (0 ppm). PME: phosphomonoesters; P_i: inorganic phosphate; GPE: glycerophosphorylethanolamine; GPC: glycerophosphorylcholine; PCr: phosphocreatine; DPDE: diphosphodiesters; Sugars-P: sugars phosphate groups; α- β- and γ-ATP: α, β and γ phosphate groups of ATP; P_β– and P_γ–ATP_{cyt}: β and γ phosphate groups, respectively, of cytosolic ATP; P_α–, P_β– and P_γ–ATP_{gran}: α, β and γ phosphate groups of granular ATP, respectively.

In contrast to the SH-SY5Y cells, the ^{31}P NMR spectrum of perfused bovine chromaffin cells shows compartmentation of ATP in the cytosol and inside the granules. The ^{31}P NMR signal of cytosolic P_α-ATP is not observable, as it is part of the composite peak at - 10.8 ppm with the resonance from intragranular P_α-ATP, as well as from the vesicular P_α-ADP and the bisphosphate moiety of NAD^+ and NADH (Painter et al., 1989). In the particular case of chromaffin cells, the criteria used to evaluate cell viability were the ratio of the areas of the P_β-$ATP_{(cytosol)}$ and $P_{i(cytosol)}$ ^{31}P NMR resonances (P_β-$ATP_{(cytosol)}$/$P_{i(cytosol)}$) (which reflects the energetic state of the cells (Kaplan et al., 1989)), the chemical shift of the P_γ-$ATP_{(granule)}$ resonance (which is related to the intragranular pH (Njus et al., 1978)), the observation of the cytosolic P_γ-ATP and P_β-ATP signals (under good perfusion conditions, these resonances appear at approximately –4.5 and –18.4 ppm, respectively) and the area ratio P_β-$ATP_{(cytosol)}$/ P_β-$ATP_{(granule)}$. The analysis of these parameters, as shown in Figure 5B), showed that the energetic state and the viability of perfused chromaffin cells were kept throughout the NMR experiments (Fonseca et al., 2004).

2.4 Li^+/Mg^{2+} competition studies by ^{31}P NMR and fluorescence spectroscopy

It has been suggested that Li^+ may compete with Mg^{2+} (a very well-known protein cofactor) for Mg^{2+} binding sites in several biomolecules, due to their similar chemical properties. Li^+/Mg^{2+} competition have been studied in Mg^{2+}-dependent biomolecules (*e.g.* ATP, ADP, GTP, GDP, IP_3, G-proteins and phosphate groups of RBCs membrane phospholipids) and in cellular systems (such as Li^+-loaded human RBCs and lymphoblastoma cells) using fluorescence spectroscopy with the Mg^{2+} indicator furaptra, as well as ^7Li and ^{31}P NMR spectroscopy (Amari et al., 1999b; Layden et al., 1999, 2003; Mota de Freitas et al., 1994, 2006; Ramasamy et al., 1989; Rong et al., 1994; Srinivasan et al., 1999).

^{31}P NMR spectroscopic method was used to examine the competition between Li^+ and Mg^{2+} ions within intact human neuroblastoma SH-SY5Y cells. The ^{31}P NMR method is based on the measurement of chemical shift difference changes between the ^{31}P NMR resonances of the P_β and P_α phosphate groups ($\Delta\delta_{\alpha\beta}$) of ATP due to Mg^{2+} binding (Gupta et al., 1978). Upon Mg^{2+} binding to ATP, the β phosphate resonance is shifted downfield and the chemical shift difference between the β and α phosphates decreases (Amari et al., 1999a), being a parameter indicative of Mg^{2+} binding to ATP and used to measure Li^+/Mg^{2+} competition. The values of $\Delta\delta_{\alpha\beta}$ were taken from ^{31}P NMR spectra of Li^+-loaded (Figure 5A) and Li^+-free SH-SY5Y cells, and used to calculate intracellular free Mg^{2+} concentrations, $[Mg^{2+}]_f$, in both situations, as described in (Amari at al.,1999a). We found that the αβ chemical shift in Li^+-free cells was 8.67 ± 0.02 ppm (n=3), whereas in Li^+-loaded cells this difference increased significantly ($p<0.0005$) to 8.87 ± 0.02 ppm (n=3), corresponding to a $[Mg^{2+}]_f$ of 0.35 ± 0.03 mmol L^{-1} and 0.80 ± 0.04 mmol L^{-1}, respectively (Amari et al.,1999a). The increase in $[Mg^{2+}]_f$ after Li^+ loading suggests that Li^+ may compete with Mg^{2+} for its binding sites within the neuroblastoma cells.

The ^{31}P NMR method could not be used to study Li^+/Mg^{2+} competition in bovine chromaffin cells due to some particular characteristics of these cells. In fact, ATP compartmentation in chromaffin cells causes the overlap of the P_α ^{31}P NMR signals of cytosolic and granular ATP (Figure 5), preventing the use of this technique to determine $[Mg^{2+}]_f$ in the cytosol of these cells. Concerning the granule, no effect of Li^+ loading on the granular $\Delta\delta_{\alpha\beta}$ value was observed, indicating no significant Li^+/Mg^{2+} competition inside this organelle. Therefore, fluorescence spectroscopy using the Mg^{2+}-specific fluorescent probe furaptra was the method of choice to study Li^+/Mg^{2+} competition in these cells. According to established data (Amari et al., 1999a, 1999b), an increase in the ratio of fluorescence intensities at 335 nm and 370 nm, $R = (F_{335}/F_{370})$, during Li^+ cell loading is indicative of the displacement of Mg^{2+} by Li^+ from its binding sites, increasing the amount of intracellular free Mg^{2+} available to bind to furaptra (salt form) inside the cells. The R values can be converted into $[Mg^{2+}]_f$ using equation 3, which corrects for Li^+ binding to furaptra:

$$[Mg^{2+}]_f = (K_d\, S_{min}\, (R - R_{min}) \,/\, S_{max}\, (R_{max} - R)) + (K_d\, S'_{max}\, (R - R'_{max})[Li^+]_{if} \,/\, K'_d\, S_{max}\, (R_{max} - R)) \quad (3)$$

where R_{min}, R_{max} and R'_{max} are the ratios of the fluorescence intensities at 335 and 370 nm in the absence of metal ions and in the presence of saturating amounts of Mg^{2+} or Li^+, respectively; S_{min}, S_{max} and S'_{max} are the fluorescence intensities at 370 nm, respectively, in the absence of metal ions and in the presence of saturating amounts of Mg^{2+} or Li^+; K_d and K'_d are the dissociation constants of the furaptra-Mg^{2+} (1.5 mmol L^{-1} (Raju et al.,1989)) and furaptra-Li^+ (237 mmol L^{-1} (Amari et al., 1999b)) complexes, respectively. $[Li^+]_{if}$ is the intracellular free Li^+ concentration, corresponding to the Li^+ ions capable of competing with Mg^{2+} for furaptra (Amari et al., 1999b).

Figure 6 shows the fluorescence excitation spectra of furaptra in the presence of increasing concentrations of Mg^{2+} and a graphical representation of the time dependence of the R values for a control situation (in the absence of Li^+) and for a 90 min Li^+-loading experiment using a total Li^+ concentration in the medium of 15 mmol L^{-1}.

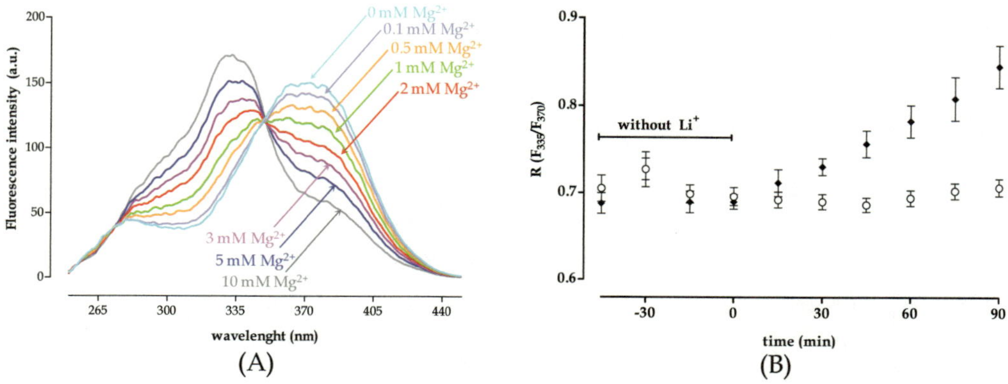

Fig. 6. **A)** Fluorescence excitation spectra of furaptra (0.2 μmol L^{-1}) in a modified Krebs medium (in mmol L^{-1}: NaCl 140, KCl 5, glucose 10, HEPES 20, EGTA 0.5, pH 7.35) containing 0-10 mmol L^{-1} free Mg^{2+}. The emission wavelength was fixed to 500 nm. The binding of Mg^{2+} to furaptra results in a blue shift in the excitation spectrum of this indicator from 370 to 335 nm with increasing amounts of Mg^{2+}. **B)** Time dependence of fluorescence intensity ratio R = (F_{335}/F_{370}) in bovine chromaffin cells previously loaded with furaptra, under control (Li^+-free) conditions (○) and when the cells were incubated for 90 min with 15 mmol L^{-1} Li^+ (◆). Bovine chromaffin cells, adherent to a 1 cm^2 square poly-L-lysine coated coverslip (0.8 × 10^6 cells per cm^2) were placed in a fluorescence cuvette containing Krebs medium (in mmol L^{-1}: NaCl 140, KCl 5, $CaCl_2$ 2, $MgCl_2$ 1, glucose 10, HEPES 20, pH 7.35). In the Li^+ experiment, this medium was then replaced by a modified Krebs medium containing 15 mmol L^{-1} Li^+ (NaCl was partially replaced in order to maintain the osmolarity of the medium). During the experiments, the medium was changed every 15 min to remove any fluorescent probe that might have been released from the cells to the incubation medium, preventing the binding of the probe to the extracellular Mg^{2+}, which would contribute to an overestimated value of $[Mg^{2+}]_f$ (Fonseca et al., 2004).

In the absence of Li^+, the basal R value, which corresponds to $[Mg^{2+}]_f$ = 0.54 ± 0.01 mmol (L cells)$^{-1}$ (n = 12), was maintained over time for 135 min. In the presence of Li^+, the $[Mg^{2+}]_f$ value was not significantly different from the control one ($[Mg^{2+}]_f$ = 0.52 ± 0.02 mmol (L cells)$^{-1}$, n=6) at time zero of Li^+ loading, but it significantly increased by 52% during the 90 min of the Li^+ loading process to a value of 0.79 ± 0.05 mmol (L cells)$^{-1}$ ($\Delta[Mg^{2+}]_f$ = 0.27 ± 0.05 mmol (L cells)$^{-1}$). In $[Mg^{2+}]_f$ calculations it was considered the $[Li^+]_{if}$ = 9.42 ± 0.01 mmol (L cells)$^{-1}$ obtained by ^{7}Li NMR for immobilised, perfused cells at steady state, which was shown to be the most accurate value (Fonseca et al., 2004). The $[Mg^{2+}]_f$ increase when Li^+ enters the viable cells, observed by fluorescence spectroscopy, confirms the capacity of Li^+ to displace Mg^{2+} from its intracellular binding sites.

The relative extent of Li^+/Mg^{2+} competition, which is cell type and $[Li^+]_{iT}$ dependent, may expressed by the percent $[Mg^{2+}]_f$ increase divided by $[Li^+]_{iT}$ $(\%(\Delta[Mg^{2+}]_f/[Mg^{2+}]_f)/[Li^+]_{iT}$ ratio). This extent is higher for neuroblastoma cells, similar for chromaffin and lymphoblastoma cells, and lower for RBCs (Fonseca et al., 2004). Based on the relative percent $[Mg^{2+}]_f$ increase/$[Li^+]_{iT}$ ratio values reported above, we suggest that the extent of Li^+/Mg^{2+} competition under pharmacological conditions is cell-type dependent, being affected by differences in Li^+ transport and immobilisation properties.

Li^+/Mg^{2+} competition has been shown to occur at therapeutic intracellular Li^+ levels (0.6–3.1 mmol (L cells)$^{-1}$) in human neuroblastoma SH-SY5Y cells (Layden et al., 2000). Changes in $[Mg^{2+}]_f$ of the order of 10%, observed for these cells at $[Li^+]_i$ = 0.6 mmol (L cells)$^{-1}$, are expected to have a large impact on the many biochemical and cell signalling pathways involving Mg^{2+}-dependent enzymes (Layden et al., 2003). Based on the experimentally observed proportional relationships in $\%(\Delta[Mg^{2+}]_f/[Mg^{2+}]_f)/[Li^+]_i$, much smaller percentage (3%) effects in $[Mg^{2+}]_f$ are to be expected in chromaffin cells at $[Li^+]_i$ = 0.6 mmol (L cells)$^{-1}$, similar to those proposed for lymphoblastoma cells (3%) but still higher than for RBCs (1.5%) (Layden et al., 2003), which possibly will have an undetectable cell impact.

In summary, ^{7}Li NMR spectroscopy proved to be a useful tool to investigate Li^+ transport, along with AA spectrophotometry, and intracellular binding in cellular models, whereas ^{31}P NMR and fluorescence spectroscopy, using the Mg^{2+}-specific fluorescent probe furaptra, allowed to quantify intracellular competition between Li^+ and Mg^{2+} ions. These studies provide further evidence for the generality of the ionic competition mechanism, contributing to the understanding of the pharmacological action of Li^+ at the molecular level.

3. Metabolic effects of Li^+ on neuronal and glial cells and rat brain

3.1 Li^+ effects on cell energetic metabolism

Altered intracellular signalling systems are thought to play an important role in the pathophysiology of bipolar disorder (Jope, 1999b; Lenox & Hahn, 2000; Manji et al., 1995). Since these processes, as well as brain activity in general, closely depend on energy metabolism and ATP availability, changes in brain energy metabolism may also be involved in the pathogenesis of this disease. Neuroimaging techniques (mainly positron emission tomography and blood oxygen level-dependent (BOLD)-based functional magnetic resonance imaging (fMRI)) suggest that both mania and depression are associated with alterations in the rates of substrate oxidation by the brain (Caligiuri et al., 2003; Goodwin et al., 1997; Kennedy et al., 1997). Other methods such as ^{31}P magnetic resonance spectroscopy (MRS) have provided evidence for dysfunction at the level of brain cells intermediary metabolism (Deicken et al., 1995; Kato et al., 1993, 1998; Kato & Kato, 2000). Since cerebral oxygen consumption and ATP synthesis are both tightly coupled to TCA cycle activity, such changes likely represent significant changes in cellular TCA cycle flux.

The therapeutic effects of Li^+ may be related to modifications of cerebral intermediary metabolic rates. Therefore, we studied the effect of Li^+ on TCA cycle flux from exogenous glucose in human neuroblastoma SH-SY5Y cells, a neuronal model (Biedler et al, 1973). Also, in light of studies suggesting that lactate may be an important neuronal oxidative substrate (Bouzier-Sore et al., 2003; Pellerin & Magistretti, 2004), the effect of Li^+ on competition between exogenous lactate and glucose for TCA cycle oxidation in these cells was also investigated. Modifications in the contribution of glucose and lactate to pyruvate and acetyl-CoA production could significantly alter the energy status of neuronal cells. Finally, the effect of Li^+ on the metabolism of glucose in primary cultures of cortical neurons and glial cells was also addressed.

3.1.1 Effects of Li^+ on the intermediary metabolism of SH-SY5Y cells

The effects of Li^+ on glucose and lactate metabolism in human neuroblastoma SH-SY5Y cells were evaluated by ^{13}C NMR isotopomer analysis, a powerful technique to study metabolic intermediary metabolism. Briefly, ^{13}C-labelled substrates ([U-^{13}C]glucose alone or a mixture of [U-^{13}C]glucose and [3-^{13}C]lactate) were used and allowed to be metabolised by the cells. Their fate was then deduced by ^{13}C NMR analysis of metabolite isotopomer distributions. Relative pathways feeding the TCA cycle were estimated from the relative areas of glutamate C2, C3 and C4 multiplets in the ^{13}C NMR spectra using the computer program tcaCALC (Malloy et al., 1988; Sherry et al., 2004). Figure 7 shows representative ^1H-decoupled ^{13}C NMR spectra of SH-SY5Y cell extracts obtained under control conditions (absence of Li^+) after incubation for 24 h with [U-^{13}C]glucose (Figure 7A) or a mixture of [U-^{13}C]glucose plus [3-^{13}C]lactate (Figure 7B).

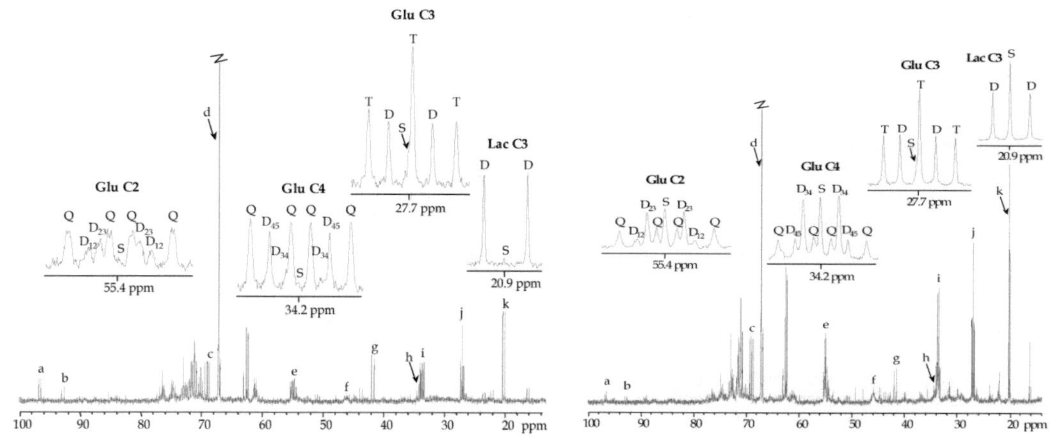

(A) (B)

Fig. 7. Representative ^1H-decoupled ^{13}C NMR spectra (125.7 MHz) from SH-SY5Y cell extracts obtained after incubating the cells for 24 h in Krebs-Ringer Bicarbonate (KRB) medium (in mmol L^{-1}: NaCl 119, KCl 4.7, $MgSO_4$ 1.2, KH_2PO_4 1.2, $CaCl_2$ 1.3, $NaHCO_3$ 24, pH 7.3) supplemented with 5 mmol L^{-1} [U-^{13}C]glucose (**A**) or a mixture of 5 mmol L^{-1} [U-^{13}C]glucose and 5 mmol L^{-1} [3-^{13}C]lactate (**B**), in the absence of Li^+ (control) (Fonseca et al., 2005). Carbon 2, 3 and 4 resonances of glutamate (Glu) and carbon 3 resonance of lactate (Lac) are expanded and assigned as follows: Q - quartet; T - triplet; D - doublet; D_{12} - doublet due to the coupling constant between carbons 1 and 2 (J_{12}) ~ 52 Hz; D_{23} - doublet due to J_{23} ~ 34 Hz; D_{34} - doublet due to J_{34} ~ 34 Hz; D_{45} - doublet due to J_{45} ~ 52 Hz; S - singlet. Resonance assignments: a – glucose C1 β; b – glucose C1 α; c – lactate C2; d – dioxane (reference, 67.4 ppm); e – glutamate C2; f – citrate C2,C4; g – malate C3; h – succinate C2,C2′; i – glutamate C4; j – glutamate C3; k – lactate C3.

^{13}C-incorporation was mainly observed on glutamate and lactate resonances. As for many other tissues, glutamate was the most abundant ^{13}C-enriched metabolite at the level of TCA cycle intermediates. The complex ^{13}C-^{13}C splitting patterns, expanded in Figure 7 for C4, C3 and C2 glutamate resonances, reveal the presence of different groups of this metabolite isotopomers as a result of glutamate labelling in different positions. The areas of the glutamate C2, C3 and C4 resonances in the spectra were calculated and the results introduced in tcaCALC program in order to obtain the following relative metabolic parameters: lac_{123} (the fraction of acetyl-CoA derived from the oxidation of [U-^{13}C]pyruvate), lac_3 (fraction of acetyl-CoA derived from the oxidation of [3-^{13}C]pyruvate, *via* pyruvate dehydrogenase (PDH), when applicable), fat_0 (the fraction of acetyl-CoA derived from unlabelled acyl sources) and y (anaplerotic flux from all sources). Further details about this method can be found in Fonseca et al., 2005. For the neuroblastoma SH-SY5Y cells, the metabolic model shown in Figure 8 was the one that provided the best fit between the glutamate isotopomer information and metabolic flux parameters as defined by the Monte Carlo and other statistical analyses of tcaCALC.

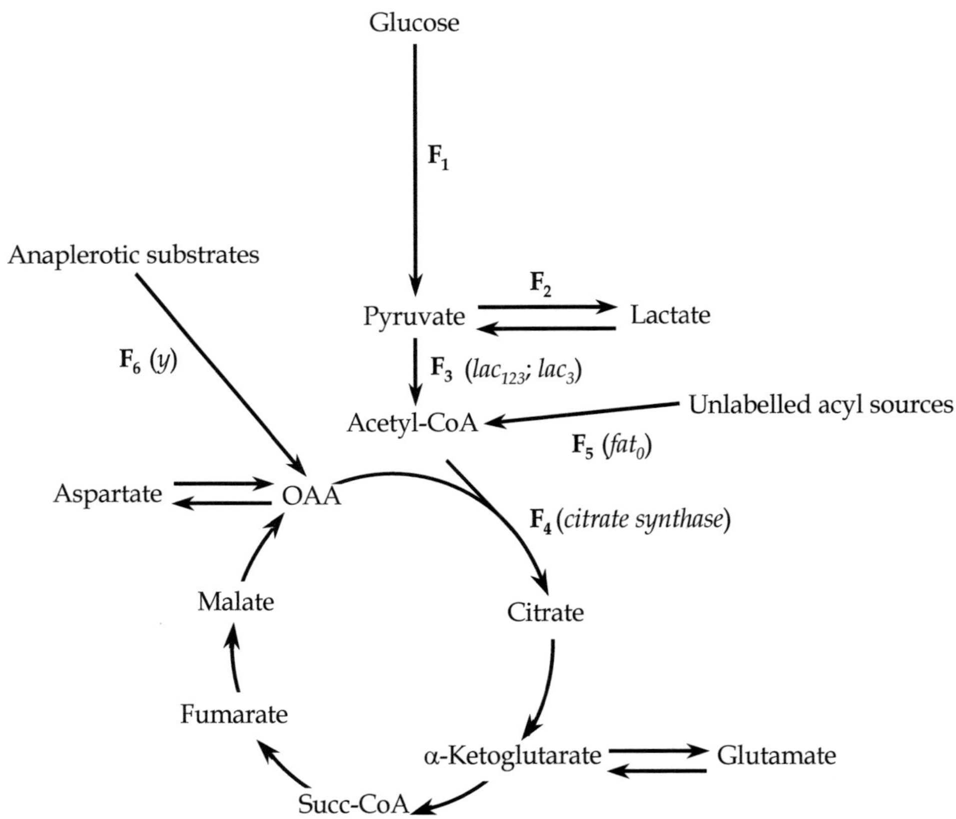

Fig. 8. Metabolic model for glucose and lactate metabolism in SH-SY5Y cells, based on tcaCALC best fitting solutions (Fonseca et al., 2005). The best fit was achieved with the parameter set of *lac123*, *y* and *fat0*. Note that *y* represents total flux from all anaplerotic pathways. Inclusion of carboxylation of labelled pyruvate and/or pyruvate cycling fluxes to the model did not significantly improve the fit; hence, these parameters were excluded .

Absolute TCA cycle fluxes were calculated using the following equations:

$$\text{Absolute PDH flux } (\mathbf{F_3}) = \text{Total glucose utilisation } (\mathbf{F_1}) - \text{Net lactate production } (\mathbf{F_2}) \quad (4)$$

$$\text{Absolute citrate synthase flux } (\mathbf{F_4}) = \mathbf{F_3}/lac_{123} \quad (5)$$

$$\text{Absolute endogenous acyl oxidation flux } (\mathbf{F_5}) = fat_O \times \mathbf{F_4} \quad (6)$$

$$\text{Absolute endogenous anaplerotic flux } (\mathbf{F_6}) = y \times F_4 \quad (7)$$

Extracellular glucose and lactate concentrations along the experimental period and, hence, glucose consumption (**F₁**) and lactate production (**F₂**) rates were calculated using an enzymatic method coupled to the increase in NADH absorption at 340 nm. Table 2 summarises the absolute metabolic fluxes for SH-SY5Y cells incubated in the presence of 5 mmol L^{-1} [U-^{13}C]glucose and the effects of 1 and 15 mmol L^{-1} Li^+ on these fluxes.

Absolute metabolic fluxes (μmol h^{-1} mg^{-1} protein)	**Control**	**1 mmol L^{-1} Li^+**	**15 mmol L^{-1} Li^+**
Glucose consumption (**F_1**)	0.877 ± 0.362 (n=6)	0.540 ± 0.170 (n=6)	0.411 ± 0.107 * (n=5)
Lactate production (**F_2**)	0.268 ± 0.086 (n=6)	0.275 ± 0.047 (n=6)	0.267 ± 0.064 (n=6)
Pyruvate dehydrogenase (**F_3**)	0.609 ± 0.302 (n=6)	0.216 ± 0.076 * (n=5)	0.098 ± 0.043 ** (n=4)
Citrate synthase (**F_4**)	0.666 ± 0.331 (n=3)	0.246 ± 0.087 ** (n=3)	0.108 ± 0.047 ** (n=3)
Endogenous acyl oxidation (**F_5**)	0.057 ± 0.029 (n=3)	0.030 ± 0.011 * (n=3)	0.009 ± 0.004 ** (n=3)
Endogenous anaplerosis (**F_6**)	0.091 ± 0.047 (n=3)	0.025 ± 0.012 ** (n=3)	0.011 ± 0.006 ** (n=3)

Table 2. Absolute fluxes in SH-SY5Y cells incubated for 24 h at 37 °C in KRB medium containing 5 mmol L^{-1} [U-^{13}C]glucose, in the absence (control) and presence of 1 or 15 mmol L^{-1} Li^+ (Fonseca et al., 2005). Values are means ± SD for the number (n) of experiments indicated in parenthesis. Glucose consumption rates were multiplied by 2 to express consumption in triose units. * and **: $p < 0.05$ and $p < 0.01$ relative to control, respectively.

SH-SY5Y cells have a relatively simple metabolic network featuring a single pyruvate pool and no cycling between pyruvate and oxaloacetate (Figure 8). The presence of 1 or 15 mmol L^{-1} Li^+ did not alter this optimal set of flux parameters. Under control conditions, TCA cycle oxidation accounted for about two-thirds of glucose consumption while the remaining third was converted to lactate. With Li^+, glucose conversion into pyruvate decreased, which is consistent with an inhibition of glycolytic flux, as already reported in literature (Kajda et al., 1979; Kajda & Birch, 1981; Nordenberg et al., 1982; Zager & Ames III, 1988). This study indicates that despite the decrease in glucose consumption by Li^+, lactate production rates were constant, which is consistent with an unchanged intracellular redox state. However, the fraction of glucose consumed by TCA cycle oxidation (given by the absolute PDH flux, **F_3**) was significantly reduced by Li^+, and this was coupled with reductions in citrate synthase and endogenous anaplerotic absolute fluxes, although no significant changes in the relative anaplerotic flux (*y*) were detected in the presence of Li^+. This suggests a direct inhibitory effect of Li^+ on TCA cycle flux. Possible inhibition sites of the TCA cycle by Li^+ could include aconitase (Abreu & Abreu, 1973). ^{13}C NMR spectra revealed a tendency for higher levels of ^{13}C-enriched citrate in the presence of Li^+, which is consistent with accumulation of citrate as a result of decreased aconitase activity. Moreover, increased citrate levels would inhibit key regulatory glycolytic enzymes such as phosphofructokinase thus contributing to the reduction in glucose conversion into pyruvate.

To determine if Li^+ has an effect on the competition between glucose and lactate oxidation when both substrates are available, the contribution of exogenous 5 mmol L^{-1} [U-^{13}C]glucose and 5 mmol L^{-1} [3-^{13}C]lactate to the TCA cycle acetyl-CoA pool was quantified by ^{13}C-isotopomer analysis in the absence and presence of Li^+ (Figure 7B). Both initial glucose and lactate concentrations were higher than the apparent K_m for glucose and lactate transporters, 1.16 mmol L^{-1} (Lust et al., 1975) and 1 mmol L^{-1} (Dringen et al., 1995), respectively; therefore, transport was not expected to be rate limiting for either substrate. Relative fluxes corresponding to the fraction of acetyl-CoA derived from [U-^{13}C]glucose (lac_{123}), [3-^{13}C]lactate (lac_3), and unlabelled endogenous sources (fat_0) are shown in Table 3.

Relative Flux Parameter	Control	1 mmol L^{-1} Li^+	15 mmol L^{-1} Li^+
Acetyl-CoA from [U-^{13}C]glucose (lac_{123})	0.26 ± 0.03 (n=3)	0.26 ± 0.03 (n=3)	0.26 ± 0.03 (n=3)
Acetyl-CoA from [3-^{13}C]lactate (lac_3)	0.62 ± 0.03 (n=3)	0.63 ± 0.03 (n=3)	0.65 ± 0.03 (n=3)
Acetyl-CoA from endogenous sources (fat_0)	0.12 ± 0.01 (n=3)	0.12 ± 0.03 (n=3)	0.10 ± 0.03 (n=3)
Endogenous anaplerotic flux (*y*)	0.20 ± 0.02 (n=3)	0.19 ± 0.02 (n=3)	0.15 ± 0.02 * (n=3)

Table 3. Relative metabolic fluxes for SH-SY5Y cells incubated for 24 h at 37 °C in KRB medium containing a mixture of 5 mmol L^{-1} [U-^{13}C]glucose and 5 mmol L^{-1} [3-^{13}C]lactate, in the absence (control) and presence of 1 or 15 mmol L^{-1} Li^+ (Fonseca et al., 2005). Values are average ± SD for the number (n) of experiments indicated in parentheses. All fluxes are relative to a citrate synthase flux of 1.0 . * $p < 0.05$, relative to control.

In control cells, over 60% of the acetyl-CoA was derived from [3-^{13}C]lactate, more than twice the contribution from [U-^{13}C]glucose, demonstrating that exogenous lactate is highly preferred over glucose for TCA cycle oxidation under these conditions. The presence of either 1 or 15 mmol L^{-1} Li^+ did not alter the relative utilisation of exogenous lactate and glucose. However, 15 mmol L^{-1} Li^+ resulted in a small but significant reduction in anaplerotic flux from endogenous sources.

3.1.2 Li^+ effects on the metabolic balances in cortical astrocytes and neurons

The metabolic balance between glucose or acetate consumption and lactate production by primary cultures of rat cortical astrocytes and neurons incubated with a mixture of 5 mmol L^{-1} glucose and 5 mmol L^{-1} acetate or with 1 mmol L^{-1} glucose, respectively, in the absence (control) and presence of 1 or 15 mmol L^{-1} Li^+, was evaluated by classical enzymatic assays. Table 4 shows the glucose or acetate consumption as well as lactate production rates calculated for cortical astrocytes and neurons, in the absence (control) and presence of two Li^+ concentrations.

Cell type	Metabolic rates (mmol L^{-1} h^{-1} mg^{-1})	Control	1 mmol L^{-1} Li^+	15 mmol L^{-1} Li^+
Astrocytes	Glucose consumption	3.19 ± 0.19 (n=5)	2.83 ± 0.27 (n=7)	3.46 ± 0.33 (n=5)
	Acetate consumption	1.37 ± 0.67 (n=5)	1.11 ± 0.11 (n=5)	1.24 ± 0.52 (n=4)
	Lactate production	4.02 ± 0.12 (n=5)	3.94 ± 0.14 (n=7)	4.48 ± 0.35 (n=6)
Neurons	Glucose consumption	2.50 ± 0.07 (n=6)	2.34 ± 0.11 (n=6)	2.22 ± 0.10 * (n=6)
	Lactate production	0.43 ± 0.08 (n=5)	0.48 ± 0.10 (n=5)	0.37 ± 0.04 (n=5)

Table 4. Glucose consumption and lactate production rates by cortical astrocytes or neurons incubated in the absence (control) and presence of 1 or 15 mmol L^{-1} Li^+. Acetate consumption rates by astrocytes under control and Li^+ conditions are also shown. Astrocytes were incubated in KRB medium containing a mixture of 5 mmol L^{-1} glucose and 5 mmol L^{-1} acetate, while neurons were incubated with 1 mmol L^{-1} glucose. Glucose and acetate consumption as well as lactate production were defined as the difference between the extracellular concentrations at the different time points (t) and the initial concentrations (at t = 0 h), i.e., ([glucose]t - [glucose]t=0h), ([acetate]t - [acetate]t=0h) and ([lactate]t - [lactate]t=0h), respectively (Fonseca et al., 2009). Values are means ± SEM for the number (n) of experiments indicated in parentheses. * $p < 0.05$ relative to control.

As expected, neurons showed a preferential oxidative metabolism with a relatively modest lactate production, since only 9 % of the glucose consumed was converted into lactate, while 91 % was oxidised in the neuronal TCA cycle. In contrast, astrocytes were found to be more glycolytic as approximately 63 % of the glucose consumed by astrocytes was converted into lactate, while only 37 % was oxidised in the astrocytic TCA cycle (Hassel et al., 1995). Our results revealed that 15 mmol L^{-1} Li^+ caused a statistically significant decrease in glucose uptake by neurons but no apparent effects in glucose or acetate uptake by astrocytes. The decrease in neuronal glucose uptake in the presence of Li^+, consistent with an inhibition of the glycolytic flux, was not paralleled by a concomitant significant change in lactate production, indicating that the decreased glucose consumption reflects a net decrease in glucose oxidation and TCA cycle activity (Fonseca et al., 2009). The ability of Li^+ to decrease glycolytic and TCA cycle fluxes is in agreement with the results obtained for SH-SY5Y cells (Fonseca et al., 2005) and other data in the literature (Abreu & Abreu, 1973; Kajda & Birch, 1981; Nordenberg et al., 1982).

In summary, the present study provides evidence that both TCA cycle and glycolysis are targets for Li^+ action in the neuroblastoma SH-SY5Y cell line and the inhibition of the TCA cycle, and hence of cell energy production, observed for therapeutic concentration of Li^+ (1 mmol L^{-1}) may constitute one hypothesis for the mechanism by which Li^+ exerts its antimanic effect. However, in cortical neurons only the highest Li^+ concentration (15 mmol

L^{-1}) was able to decrease neuronal glucose consumption and TCA cycle activity. Although Li^+ has a narrow therapeutic range—it is toxic *in vivo* for plasma concentrations higher than 2 mmol L^{-1} —the 15 mmol L^{-1} Li^+ concentration was found to be nontoxic for both neurons and astrocytes under our experimental conditions (data not shown). Furthermore, Li^+ concentrations achieved in the intercellular space are uncertain, but are expected to reach much higher levels than in plasma. Despite these limitations, the results presented here strengthen the importance of cell energetic metabolism as a target for Li^+ action, an area of study that has been underestimated and poorly reported in the literature.

3.2 Li^+ effects on glutamatergic and GABAergic neurotransmissions in adult rat brain and cultured brain cells, as revealed by ^{13}C NMR

Besides its metabolic effects, Li^+ was proposed to alter the balance between excitatory and inhibitory neurotransmitter systems, thus modulating glutamatergic and GABAergic neurotransmission (Antonelli et al., 2000; Brambilla et al., 2003; Gottesfeld, 1976; Jope et al. 1989; Marcus et al., 1986; O'Donnell et al., 2003; Otero Losada & Rubio, 1986; Petty, 1995; Rubio & Otero Losada, 1986; Shiah & Yatham, 1998) and facilitating in this way, mood recovery and stabilisation in Li^+-treated bipolar patients. However, the effects of Li^+ have been studied using different experimental models and protocols that have often yielded different or even contradictory results. In this study, we have used the ^{13}C NMR technique to investigate the effects of Li^+ on the metabolism of glutamate, glutamine and GABA in the adult intact rodent brain and in primary cultures of cortical neurons and astrocytes. This technique has been used successfully to study cerebral metabolic compartmentation and the glutamate-glutamine cycle as the basis of glutamatergic and GABAergic neurotransmissions (Rodrigues & Cerdán, 2005).

Adult male rats receiving a single dose of Li^+ intraperitoneally (7 mmol kg^{-1}), or saline (control), were infused with [1-^{13}C]glucose or [2-^{13}C]acetate. Glucose is considered a universal substrate for neurons and glial cells, although it is thought to be metabolized more in the neuronal TCA cycle; in contrast, acetate is considered a glial substrate because it is selectively taken up by astrocytes by a specialized transport system, which is absent or less active in neurons (Waniewski & Martin, 1998; Sonnewald & Kondziella, 2003). Brain extracts were prepared 3 h after Li^+ injection and analysed by ^{13}C NMR. The mean Li^+ concentrations in brain and plasma achieved 3 h after i.p. Li^+ injection were 1.5 ± 0.4 mmol kg^{-1} tissue wet-weight (n=5) and 1.3 ± 0.5 mmol L^{-1} (n=7), respectively. Figure 9 summarizes the results obtained from the quantitative analysis of ^{13}C incorporation in the aliphatic carbons of glutamate, glutamine and GABA from the brain of control and Li^+-treated animals.

With [1-^{13}C]glucose as substrate, Li^+ decreased the incorporation of ^{13}C in the observable carbons of glutamate, glutamine and GABA, with statistical significance for the C4 and C3 carbons of glutamate and GABA, respectively. Apparently Li^+ administration did not change the ^{13}C labelling of these metabolites after [2-^{13}C]acetate infusions. Because [1-^{13}C]glucose is believed to be a universal substrate for neurons and glial cells and [2-^{13}C]acetate is known to be mainly a glial substrate, our results suggest that the inhibition observed *in vivo* must occur primarily in the neuronal compartment, at an upstream level of the glutamate-glutamine-GABA cycles. Glucose consumption through glycolysis and the TCA cycle could be a possible target for this upstream Li^+ action (Fonseca et al., 2005, 2009;

Nordenberg et al., 1982; Plenge, 1976; Zager & Ames, 1988;). Li^+ significantly increased the glutamate C3/GABA C3 labelling ratio ($p < 0.01$), suggesting that Li^+ may affect the synthesis of GABA from its direct precursor glutamate in the neuronal compartment, *in vivo*, possibly through the inhibition of glutamate decarboxylase activity (Fonseca et al., 2009).

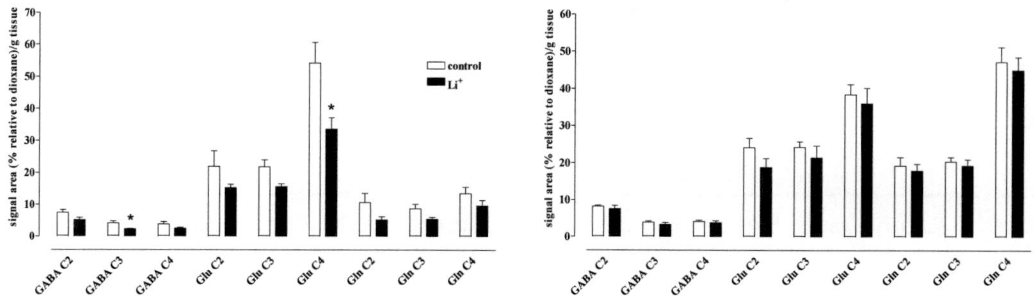

(A) (B)

Fig. 9. Graphical representation of ^{13}C NMR signal areas of the metabolites glutamate, glutamine and GABA, as observed in the ^{13}C NMR spectra obtained form rat brain extracts prepared after saline (control) and Li^+ administration, and infusion with [1-^{13}C]glucose (n=3 for both saline- and Li^+-treated rats) (**A**) or [2-^{13}C]acetate (n=4 for saline- and n=5 for Li^+-treated rats) (**B**) (Fonseca et al., 2009). Tissue wet weight and appropriate correction factor for nuclear Overhauser enhancement and signal saturation were taken into account. Values are means ± SEM for the indicated number of experiments. * $p < 0.05$ relative to control. Glu - glutamate; Gln – glutamine.

To investigate this mechanism in more detail, we used primary cultures of cortical neurons and astrocytes. The ^{13}C NMR spectra of extracts obtained from the cellular layers of astrocyte cultures after incubation with 5 mmol L^{-1} [2-^{13}C]acetate, or neuron cultures incubated with 1 mmol L^{-1} [U-^{13}C]glucose, showed an extensive incorporation of ^{13}C from the ^{13}C-labeled substrates in the C2, C3, and C4 carbons of glutamate and glutamine, C2 and C3 of aspartate (and C2, C3, and C4 of GABA and C3 of lactate for neurons). Table 5 shows the relative ratios between the areas of the C4 and C3 carbon resonances of glutamine and glutamate obtained from the ^{13}C NMR spectra of astrocyte extracts, as well as the ratios between the areas of the C2, C3, and C4 carbon resonances of glutamate and the C4, C3 and C2 carbon resonances of GABA, respectively, obtained from the ^{13}C NMR spectra of neuron extracts. The reason for selecting and comparing the areas of these specific carbon resonances of glutamate and GABA was based on the well-known metabolic fate of ^{13}C labelling from ^{13}C-labeled substrates because the order of ^{13}C-labeled carbons in glutamate is reversed in GABA (e.g., the labels in (2(4)-^{13}C) glutamate end up subsequently in (4(2)-^{13}C) GABA) (Sonnewald & Kondziella, 2003).

When 15 mmol L^{-1} was used, an increase in glutamate C2/GABA C4 ($p<0.05$), glutamate C3/GABA C3 and glutamate C4/GABA C2 ^{13}C labelling ratios was observed, suggesting that Li^+ may decrease GABA synthesis from glutamate, its direct precursor in cortical GABAergic neurons. These findings reflect a direct inhibitory effect of Li^+ on glutamate decarboxylase activity and are in agreement with the in vivo studies, where a more pronounced reduction in ^{13}C labelling of GABA relative to glutamate was observed. Indeed, previous studies have reported Li^+-induced inhibitory effects on glutamate decarboxylase (Otero Losada & Rubio, 1986; Rubio & Otero Losada, 1986).

Cell type	Area ratio	Control	1 mM Li^+	15 mM Li^+
Astrocytes	Gln C4/Glu C4	0.76 ± 0.37	0.58 ± 0.29	0.32 ± 0.17
	Gln C3/Glu C3	1.12 ± 0.54	1.13 ± 0.43	0.64 ± 0.29
Neurons	Glu C2/GABA C4	1.60 ± 0.15	1.64 ± 0.21	2.95 ± 0.45 *
	Glu C3/GABA C3	2.12 ± 0.26	1.69 ± 0.29	3.64 ± 1.06
	Glu C4/GABA C2	2.04 ± 0.19	1.99 ± 0.35	3.12 ± 0.79

Table 5. Relative ratios between the areas of the C4 and C3 carbon resonances of glutamine (Gln) and glutamate (Glu), as well as between the areas of the C2, C3 and C4 carbon resonances of GABA and the C4, C3 and C2 carbons of Glu, respectively, calculated from the ^{13}C NMR spectra of astrocyte or neuron extracts. Cells were incubated for 24 h at 37 °C in the absence (control) and presence of 1 and 15 mmol L^{-1} Li^+, in a modified KRB medium containing 5 mM [2-^{13}C]acetate (astrocytes) or 1 mM [U-^{13}C]glucose (neurons) (Fonseca et al., 2009). Values are means ± SEM of 3 independent experiments. * $p < 0.05$, relative to the control.

In summary, Li^+ was found to decrease the incorporation of ^{13}C labelling into GABA carbons from its precursor glutamate in neurons, both ex vivo and in vivo, probably through the inhibition of glutamate decarboxylase. These inhibitory effects, together with those detected in glucose consumption and TCA cycle activity of neurons, configure a complex mechanism of Li^+ action involving inhibitory actions at multiple sites. The results presented here may provide a new insight into the basis of the metabolic effects of Li^+ on brain metabolism involving the modulation of the main excitatory and inhibitory systems, which may facilitate mood recovery and stabilization in bipolar patients.

4. Effect of mood stabilizers on the regulation of cyclic AMP levels by dopaminergic and β-adrenergic receptors

4.1 Intracellular lithium and cyclic AMP levels are mutually regulated in neuronal cells

Recent research has been focused on how Li^+ changes the activity of cellular signal transduction systems, in particular those involving AC. It has been suggested that cAMP levels are abnormal in bipolar patients and are regulated by mood stabilizing agents. So, it is important to know whether this second messenger regulates Li^+ transport into neuronal cells and determine the effect of Li^+ on the homeostasis of intracellular cAMP levels, which depends on the AC activity. Thus, the effect of intracellular cAMP on Li^+ uptake, at therapeutic plasma concentrations, in SH-SY5Y human neuroblastoma cells and in primary cultures of rat cortical and hippocampal neurons was studied. The cells were stimulated with forskolin, a direct activator of the catalytic subunit of AC, or with the cAMP analogue dibutyryl-cAMP, to increase intracellular cAMP levels. Intracellular Li^+ was quantified by AA spectrophotometry and cAMP levels were determined under basal and forskolin stimulated-conditions, and in the presence and absence of Li^+, through a radioactive assay using [8-^{3}H]cAMP.

The kinetics of Li$^+$ influx and Li$^+$ uptake by SH-SY5Y cells and cortical and hippocampal neurons, in the presence and absence of forskolin, was studied and the results presented in Figure 10 details of experimental procedure in figure legend).

It was observed (Figure 10 A,B) that under forskolin stimulation both the Li$^+$ influx rate constant k_i, [0.028 ± 0.005 min^{-1} (ctrl) and 0.041 ± 0.005 min^{-1} (fsk); $p < 0.05$] and the Li$^+$ accumulation in SH-SY5Y cells were increased (27.6 ± 1.8 nmoL mL^{-1} *vs* the control value 17.9 ± 1.7 nmoL mL^{-1}; $p < 0.01$). Dibutyryl-cAMP also increased Li$^+$ uptake confirming that these effects were due to an increase in intracellular cAMP and not to a non-specific effect of forskolin [126.9 ± 11.6 % for forskolin ($p < 0.01$ relative to ctr) and 142.6 ± 13.9 % for db-cAMP ($p < 0.001$, relative to ctr) and not significantly different between them]. Identical results were obtained with cortical [133.5 ± 5.9 % for forskolin ($p < 0.01$) and 154.0 ± 9.3 % for db-cAMP ($p < 0.001$)] and hippocampal neurons [133.3 ± 11.9 % for forskolin ($p < 0.05$) and 141.4 ± 6.9 %; for db-cAMP ($p < 0.05$)] (Figure 10 C,D). To obtain information about the transport pathways responsible for Li$^+$ uptake under resting and forskolin stimulated conditions, experiments were carried out using inhibitors of specific transporters at SH-SY5Y cells membrane (details are in the legend of the figure) and the intracellular Ca^{2+} chelator BAPTA. The graphs of Figure 11 show the data obtained.

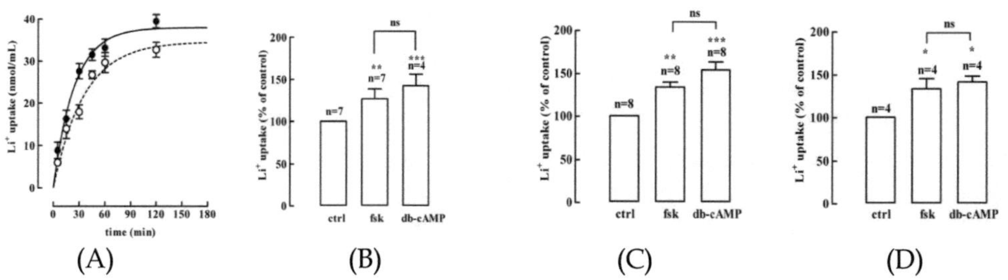

Fig. 10. Kinetics of Li$^+$ influx in SH-SY5Y cells, in the absence (○) (ctrl) and in the presence (●) of forskolin (fsk) **(A)**. The cells were pre-incubated with RO-201724, the cAMP phosphodiesterase inhibitor (Miles et al., 1987), (25 μmol L^{-1}), for 15 min, and then with fsk (10 μmol L^{-1}), for 15 min. LiCl at 1 mmol L^{-1} concentration was then added to the medium and at 5, 15, 30, 45, 60 or 120 min; the amount of Li$^+$ taken up by cells was measured by AA spectrophotometry. The rate constants obtained for Li$^+$ influx were 0.028 ± 0.005 min^{-1} (ctrl) and 0.041 ± 0.005 min^{-1} (fsk) ($p < 0.05$). Values are means ± SEM of 4-20 independent experiments. Li$^+$ uptake by SH-SY 5Y cells **(B)**, by cortical **(C)** or hippocampal neurons **(D)**, pre-treated or not (ctrl) with fsk (10 μmol L^{-1}; 15 min) or db-cAMP (500 μmol L^{-1}; 30 min) after pre-incubating the cells with RO-201724 (25 μmol L^{-1}, 15 min), and before incubation with 1 mmol L^{-1} Li$^+$, for 30 min. The total amount of intracellular Li$^+$ was measured by AA spectrophotometry, as described in (Montezinho et al., 2004). Data are presented as a percentage of intracellular Li$^+$ content relative to the control. Values are means ± SEM., for the indicated number of independent experiments. *, $p < 0.05$; **, $p < 0.01$; ***, $p < 0.001$, significantly different from control; ns= not significant (Montezinho et al., 2004).

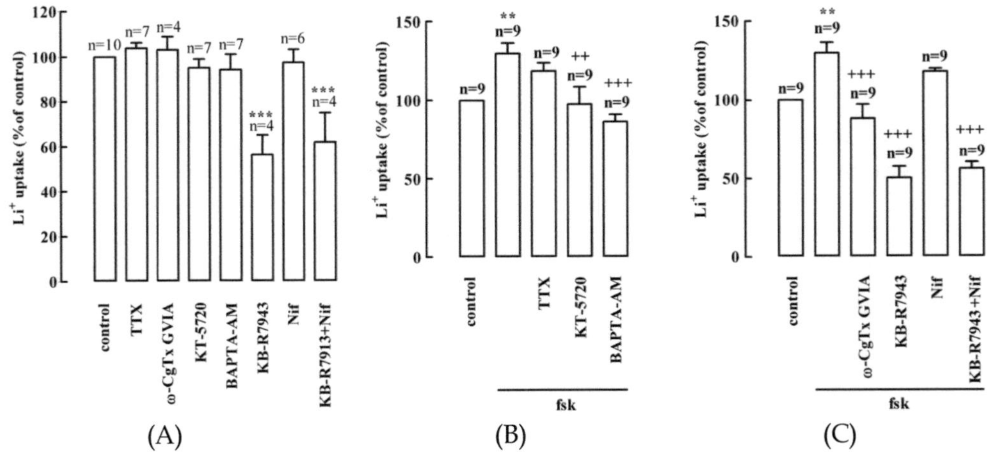

Fig. 11. Pharmacological characterization of Li^+ uptake by SH-SY5Y cells under resting conditions (**A**) and after fsk stimulation (**B**) and (**C**). The cells were pre-treated with RO-201274 (25 μmol L^{-1}, 15 min), and then with the following drugs, at different concentrations and pre-incubation times: TTX (1 μmol L^{-1}; 5 min), ω-Conotoxin GVIA (ω-CgTx GVIA;0.5 μmol L^{-1}; 30 min), KT-5720 (10 μmol L^{-1}; 10 min), BAPTA-AM (10 μmol L^{-1}; 30 min), KB-R7943 (20 μmol L^{-1}; 5 min) and nifendipine (Nif) (1 μmol L^{-1}; 5 min), to study the contribution of voltage-sensitive sodium channels (VSSC), N-type voltage-sensitive calcium channels (VSCC), PKA, $[Ca^{2+}]_i$, Na^+/Ca^{2+} exchanger and L-type VSCC, respectively, for Li^+ uptake by the cells under resting conditions (**A**). To study the contribution of all these transport pathways to the fsk-induced Li^+ uptake, after exposure to the drugs, in the same concentrations and pre-incubation times, the cells were incubated with fsk (10 μmol L^{-1}), for 15 min (B, C). Then, in all cases, 1 mmol L^{-1} LiCl was added to the medium and after 30 min the amount of Li^+ taken up by the cells was measured by AA spectrophotometry. Total intracellular Li^+ content is presented as a percentage relative to the control. Values are means ± SEM, for the indicated number of independent experiments. ***, $p< 0.01$ and **, $p< 0.01$, significantly different from control, respectively, under resting and fsk stimulated conditions; $^+$, $p< 0.05$; $^{++}$, $p< 0.01$; $^{+++}$, $p< 0.001$, significantly different from fsk stimulation in the absence of any of these drugs (Montezinho et al., 2004).

Under resting conditions, the inhibitor of the Na^+/Ca^{2+} exchanger, KB-R7943, reduced the influx of Li^+ and completely inhibited the effect of forskolin on the increase of Li^+ uptake, although this effect was attenuated also by KT-5720 (protein kinase A (PKA) inhibitor), ω-Conotoxin GVIA (N-type voltage-sensitive calcium channels (VSCC) inhibitor) or BAPTA (Ca^{2+} chelator). This indicates that the Na^+/Ca^{2+} exchanger is the principal responsible for Li^+ influx into SH-SY5Y cells in these conditions.

The involvement of intracellular Ca^{2+} concentration ($[Ca^{2+}]_i$) in Li^+ uptake, under resting and after forskolin stimulation, was investigated in the presence of different inhibitors of pathways which contribute to $[Ca^{2+}]_i$ homeostasis by measuring intracellular free Ca^{2+} levels using fluorescence spectroscopy (Figure 12).

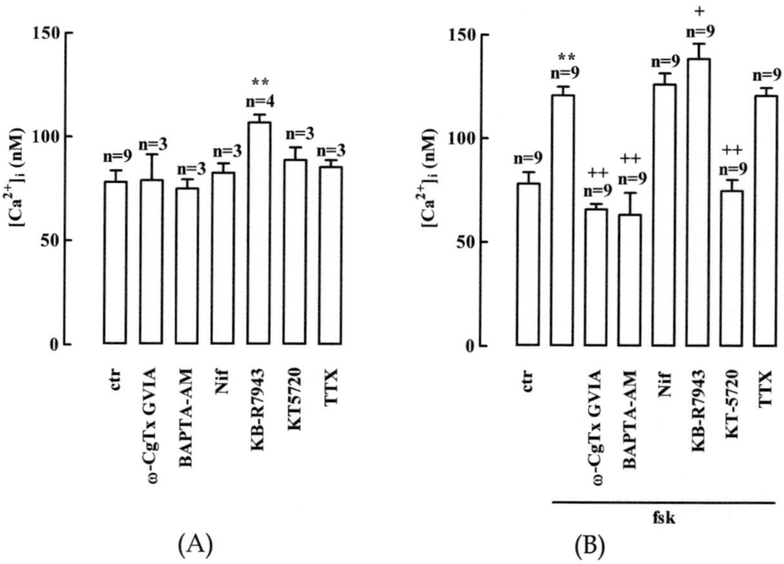

Fig. 12. Pharmacological characterization of the pathways for [Ca2+]i homeostasis in SH-SY5Y cells under resting conditions **(A)** and after stimulation with fsk **(B)**. Cells were pre-treated with RO-201724 (25 μmol L^{-1},15 min), and then with ω-CgTx GVIA (0.5 μmol L^{-1}; 30 min), Nif (1 μmol L^{-1}; 5 min), KB-R7943 (20 μmol L^{-1}; 5 min), KT-5720 (10 μmol L^{-1}; 10 min) or TTX (1 μmol L^{-1}; 5 min), to test the contribution of N- and L-type VSCC, Na^+/Ca^{2+} exchanger, PKA and VSSC, respectively, to the maintenance of the $[Ca^{2+}]_i$. The effect of the Ca^{2+} chelator, BAPTA, was also determined by pre-incubating the cells with 10 μmol L^{-1} BAPTA-AM, for 30 min. The cells were stimulated or not with 10 μmol L^{-1} fsk, for 15 min, and the $[Ca^{2+}]_i$ was determined by fluorescence spectroscopy using fura-2. Values are means ± SEM, for the indicated number of independent experiments. **, $p< 0.01$, significantly different from control; +, $p< 0.05$; ++, $p< 0.01$, significantly different from fsk stimulation in the absence of any drug (Montezinho et al., 2004).

It was observed that under resting conditions only KB-R7943 increased $[Ca^{2+}]_i$ in SH-SY5Y cells, confirming the role of the Na^+/Ca^{2+} exchanger on intracellular homeostasis (Figure 12A). Intracellular Ca^{2+} chelation (with BAPTA), or inhibition of N-type VSCC (ω-CgTx GVIA), or inhibition of cAMP-dependent protein kinase A (KT-5720) abolished the effect of forskolin on Li^+ uptake (Figure 12B). All these conditions decrease free $[Ca^{2+}]_i$, as demonstrated by the quantification of Ca^{2+} levels, demonstrating the involvement of Ca^{2+} on forskolin-induced Li^+ uptake.

SH-SY5Y cells were exposed to 1 mmol L^{-1} Li^+ for different periods of time to check its effect on cAMP levels. Figure 13A shows that after 24 h to Li^+ exposure, basal cAMP levels increased, which can be explained by the inhibition of G_i, the G proteins most active in basal conditions. Pre-incubation of the cells with 1 mmol L^{-1} Li^+, during 1, 24 or 48 h, decreased cAMP production in response to forskolin (Figure 13B) which may be due to inhibition of AC activity as a result from Li^+/Mg^{2+} competition.

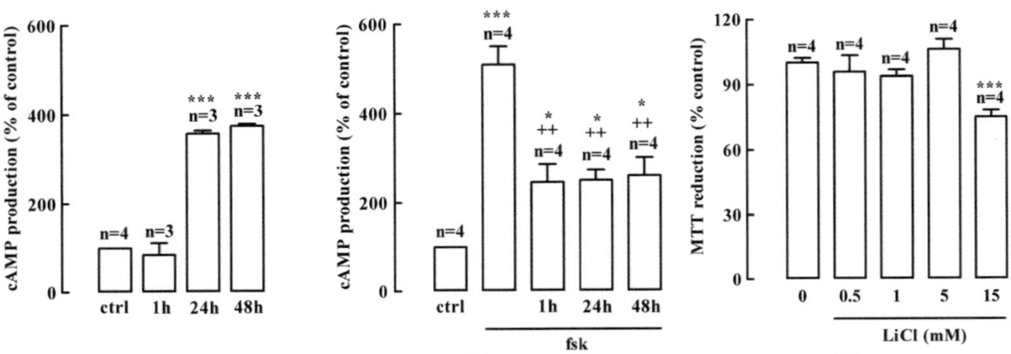

(A) (B) (C)

Fig. 13. Effects of Li^+ on cAMP production by SH-SY5Y cells. **(A)** Cells were pre-treated with RO-201724 (25 μmol L^{-1}, 15 min), and then incubated or not (ctrl) with 1 mmol L^{-1} Li^+, for the indicated periods of time. **(B)** Where indicated, the cells were stimulated with 10 μmol L^{-1} fsk, for 15 min, after treatment with Li^+. The cAMP levels were measured as described (Montezinho et al., 2004). **(C)** MTT biochemical assay with SH-SY5Y cells treated with LiCl (0, 0.5, 1, 5 and 15 mmol L^{-1}), during a period of 48h. Data are presented as a percentage relative to the control. Values are means ± SEM, for the indicated number of independent experiments. *, $p< 0.05$; ***, $p< 0.001$, significantly different from the control. ++, $p< 0.01$, significantly different from fsk stimulation in the absence of Li^+ (Montezinho et al., 2004).

These data indicate that forskolin AC activation increases Li^+ influx in the three cellular models, an effect mediated by PKA and due to changes in $[Ca^{2+}]_i$. The intracellular cAMP accumulation increases $[Ca^{2+}]_i$, due to its entry through N-type VSCC, thus activating Na^+/Ca^{2+} exchanger allowing the increase of Li^+ influx, by substituting Na^+, in exchange with Ca^{2+} extrusion (Deval et al., 2002; Fonseca et al., 2004; Montezinho et al., 2004).

Overall, these results demonstrate that intracellular cAMP levels regulate Li^+ uptake in a Ca^{2+}-dependent manner, and Li^+ plays an important role in the homeostasis of this second messenger in neuronal cells. This is relevant data to understand the mechanism of action of Li^+, taking into account several aspects already mentioned in the literature such as: bipolar disorder is associated with deregulation in AC mediated signal transduction processes, which are affected by Li^+ (Berns & Nemeroff, 2003; Brunello & Tascedda, 2003; Manji & Lenox, 2000a, 2000b; Manji et al., 2001), and with increased $[Ca^{2+}]_i$ (Hough et al., 1999) pointing to an interaction between intracellular signalling mechanisms mediated by Ca^{2+} and cAMP (Cooper et al., 1995); bipolar patients with a $[Ca^{2+}]_i$ higher than normal respond better to Li^+ treatment, indicating a possible correlation between Ca^{2+} levels and Li^+ response (Ikeda & Kato, 2003).

4.2 Effect of mood stabilizers on dopamine D_2-like receptor-mediated inhibition of adenylate cyclase

Second messenger-mediated pathways represent targets for Li^+ action; thus, it is important to investigate whether other mood stabilizing agents exert similar effects on the same signalling pathways. Bipolar disorder seems to be associated with an enhanced signalling activity of the cAMP cascade and most of its events have been implicated in the action of mood stabilizing drugs, which somehow modulate brain cAMP levels. Therefore, the effects of Li^+, carbamazepine and valproate on basal and forskolin-evoked cAMP accumulation were studied both ex vivo, in cultured cortical neurons, and in vivo in the rat prefrontal cortex using microdialysis in freely moving animals, a technique which detects extracellular cAMP levels, thus accurately reflecting intracellular changes in AC activity (Masana et al., 1991, 1992). It has been demonstrated that a fraction of intracellular cAMP generated by activation of AC is extruded into the extracelular fluid in proportion with its accumulation in cells. Therefore, the efflux of cAMP can be used to study the cAMP second messenger system in intact brains, using in vivo microdialysis (Mørk & Geisler, 1994). Moreover, the capacity of dopamine D_2-like receptors stimulation, with quinpirole, to block the increase of forskolin-stimulated cAMP levels was measured under control conditions and after treatment with the mood stabilizing drugs. The cAMP was quantified using the [8-^{3}H] and [^{125}I] radioimmunoassay kits, respectively for the ex vivo and in vivo experiments.

Figure 14 shows the determination of intracellular cAMP levels in cultured cortical neurons as a response to different experimental conditions (protocols are detailed in the figure legend).

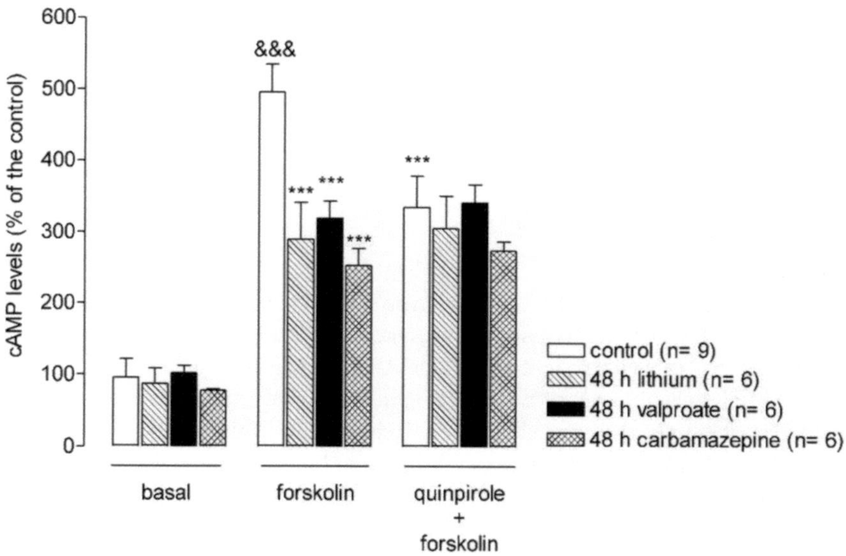

Fig. 14. Effects of lithium, valproate and carbamazepine on the intracellular cAMP levels in cultured cortical neurons under basal conditions and after stimulation with forskolin or quinpirole plus forskolin. Neurons were pre-exposed or not (control) to 1 mmol L^{-1} lithium, 0.05 mmol L^{-1} carbamazepine or 0.5 mmol L^{-1} valproate, for 48 h. After this period, neurons were incubated with Ro-201724 (25 µmol L^{-1}; 15 min) and were then treated or not (basal) with forskolin (10 µmol L^{-1}; 15 min), or with forskolin together with quinpirole (10 µmol L^{-1}; 15 min). In the latter experimental conditions the cells were pre-treated with 10 µmol L^{-1} quinpirole for 5 min. The cAMP levels were measured as described (Montezinho et al., 2007). The levels of cAMP are presented as percentage relative to the control. Data are means ± SEM, for the indicated number of independent experiments, performed in duplicate. Data were analysed by one-way ANOVA, followed by post-hoc Bonferroni´s test. &&&, $p < 0.001$ compared to control untreated cells; ***, $p < 0.01$ compared to control forskolin-stimulated cells (Montezinho et al., 2007).

Lithium, carbamazepine or valproate had no effect on the basal cAMP production, but partially inhibited forskolin-induced cAMP accumulation (control: 495.8 ± 39.0 %, $p < 0.001$ (5× basal value); lithium: 290.1 ± 51.8 %, $p< 0.001$; carbamazepine: 273.9 ± 13.1 %, $p< 0.001$; valproate: 319.6 ± 23.4 %, $p< 0.001$). In the presence of quinpirole, the agonist of dopamine D_2-like receptors, no effect on the basal cAMP accumulation was observed (data not shown), but inhibition of forskolin-enhanced cAMP levels occurred in untreated neurons (335.6 ± 42.8 %, $p< 0.001$); the activation of dopamine D_2-like receptors had no additional effect in cortical neurons treated with the mood stabilizers (lithium: 305.2 ± 46.2 %; carbamazepine: 273.9 ± 13.1 %; valproate: 341.9 ± 24.7 %, $p< 0.001$).

Similar experiments were carried out *in vivo*. Extracellular cAMP levels were determined in prefrontal cortex of freely moving rats by microdialysis and the obtained results are shown in Figure 15 (detailed experimental conditions are in the legend of this figure).

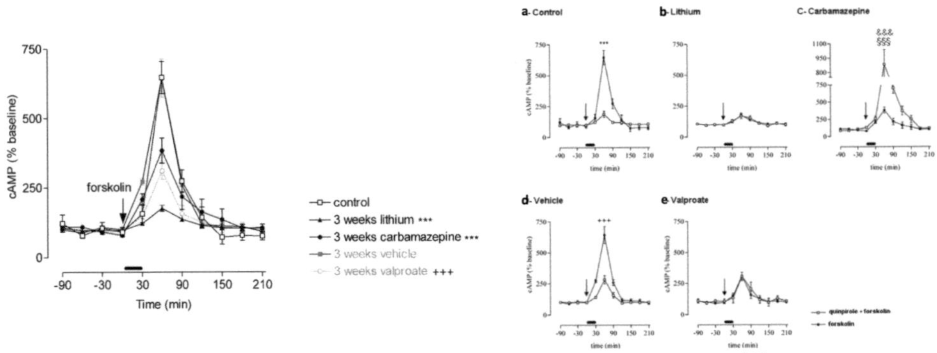

(A) (B)

Fig. 15. Effect of local infusion of forskolin (100 μmol L^{-1}) **(A)** and forskolin (100 μmol L^{-1}) or quinpirole (100 μmol L^{-1}) simultaneously with forskolin (100 μmol L^{-1}) **(B)** in the prefrontal cortex of freely moving rats (starting at arrow and maintained during 30 min) on the extracellular cAMP levels of **(A)** control-, vehicle-, lithium-, valproate- or carbamazepine-treated rats; **(B)** control rats **a)** or rats chronically treated with lithium **b)**, carbamazepine **c)**, vehicle **d)** or valproate **e)**. Baseline levels of cAMP were taken as the average cAMP content in the four consecutive samples collected 3 h after the insertion of the probe. Thereafter, the agent of interest (100 μmol L^{-1} forskolin or 100 μmol L^{-1} quinpirole plus 100 μmol L^{-1} forskolin) dissolved in Ringer solution was infused through the probe, for 60 min, and seven samples were then collected. Extracellular cAMP levels in the dialysates were measured by radioimmunoassay analysis and expressed as percentage of the basal value. Results are expressed as means ± SEM, from 4 to 6 independent experiments. The rats were treated or not, during three weeks, with therapeutic doses of lithium (n= 6) or carbamazepine (n=6) in the diet, whereas others were treated with therapeutic doses of valproate (n= 6) by intraperitoneal injections once daily, for three weeks. Control rats (n= 6) received standard diet, whereas vehicle rats (n= 4) received intraperitoneal injections with saline solution (NaCl 0.9%) once daily during three weeks (Montezinho et al., 2007). Data are analyzed by one-way ANOVA, followed by post-hoc Bonferroni´s test. ***, $p< 0.001$; compared to the cAMP produced 60 min after the infusion of forskolin in control rats. $^{+++}$, $p< 0.001$; compared to the cAMP produced 60 min after the infusion of forskolin in vehicle-treated rats. §§§, $p< 0.001$; compared to the cAMP produced 60 min after the infusion of quinpirole plus forskolin in control rats. $^{\&\&\&}$, $p< 0.001$, compared to the cAMP produced 60 min after the infusion of forskolin in carbamazepine-treated rats.

After forskolin infusion in the prefrontal cortex of the rats, for 30 min, the average basal cAMP levels (control: 9.68 ± 2.10 fmoL/20 µL; lithium-treated animals: 13.17 ± 1.21 fmoL/20 µL ($p< 0.05$ relative to control); carbamazepine-treated rats, 7.82 ± 0.34 fmoL/20 µL ($p< 0.01$ relative to control); valproate-treated rats: 10.42 ± 1.06 fmoL/20 µL) increased to approximately 647.0 ± 58.3 % and 644.5 ± 68.9 % of the basal values (control and vehicle, respectively) whereas in lithium-, carbamazepine- or valproate-treated rats there was a statistically significant decrease in the forskolin-stimulated cAMP levels determined after 60 min of forskolin infusion (lithium: 176.0 ± 11.5 %, $p< 0.001$; carbamazepine: 383.2 ± 44.9 %, $p< 0.001$; valproate: 310.3 ± 29.1 %, $p< 0.001$), when compared to the values from control or vehicle-treated rats (Figure 15A)

Forskolin-induced increase in extracellular cAMP was significantly inhibited in the presence of quinpirole, in control and vehicle-treated rats (respectively 187.6 ± 24.6 %, $p< 0.001$ and 282.7 ± 34.4, $p< 0.001$) (Figure 15B a) and d)), as measured 60 min after the infusion of the dopamine D_2-like receptors agonist. However, in lithium- or valproate-treated rats no difference occurred in the forskolin-induced increase in extracellular cAMP in the absence or presence of the agonist (Figure 15B b) and e)) whereas in carbamazepine-treated animals an unexpected increase was observed (885.5 ± 152.7 %, $p< 0.001$) (Figure 15B c)) after infusion of quinpirole plus forskolin, being the obtained value even higher than that observed in control animals under the same conditions (187.6 ± 24.6 %) (Figure 15B a)) or in carbamazepine-treated rats infused only with forskolin (383.2 ± 44.9 %) (Figure 15B c)).

No statistical significant effect on basal cAMP levels neither under control conditions (control rats: 100.4 ± 18.4 %; vehicle-treated rats: 104.0 ± 11.3 %) nor in carbamazepine-treated animals (114.4 ± 8.9 %) was observed after activation of dopamine D_2-like receptors (data not shown).

Taken together, these results indicate that lithium, carbamazepine and valproate, at therapeutically relevant concentrations, modulate basal and forskolin evoked cAMP production. These drugs had no effect on basal cAMP levels in vitro, but differential effects were observed in vivo, in agreement with literature data (Jope,1999a, 1999b; Masana et al., 1991, 1992; Montezinho et al., 2004; Mørk & Jensen, 2000). The different behaviour with cells and in vivo may be due to the lack of afferent projections releasing catecholamines, which could be regulated by the *chronic in vivo* drug infusion (Dziedzicka-Wasylewska et al., 1996; Ichikawa & Meltzer, 1999). Direct stimulation of AC with forskolin increased the levels of cAMP both *in vitro* and *in vivo*, and this effect was significantly inhibited by all three mood stabilizers. Activation of dopamine D_2-like receptors partially inhibited the forskolin-induced increase in cAMP in untreated cell cultures, likely due to AC inhibition (Memo et al., 1992), but no further attenuation in cAMP levels was observed in cultured cortical neurons treated with the three mood-stabilizing drugs, suggesting that, in this case, there are no additive effects, although these drugs can act through different mechanisms (Chen et al., 1996a, 1996b; Gallagher et al., 2004; Montezinho et al., 2004; Mørk & Jensen, 2000; Post et al., 1992).

Similar results were obtained upon chronic *in vivo* treatment with Li^+ and valproate in the rat prefrontal cortex. However, in carbamazepine-treated animals, the activation of dopamine D_2-like receptors, surprisingly, enhanced the responsiveness of AC to activation by forskolin, possibly due to what is described as super-sensitization of AC, a neuroadaptative mechanism that occurs under prolonged inhibition of this enzyme (Johnston CA & Watts, 2003). This may have an important role in tolerance and dependency effects generally observed in chronic abuse drugs administration, which are known to induce AC sensitization due to the persistent activation of $G\alpha_i$-coupled receptors. The molecular mechanisms involved in super-sensitization are not completely clarified, however, some studies show that chronic inhibition of PKA induces AC sensitization in some cellular systems and that activation of this kinase has the opposite effect (Johnston et al., 2002).

The down-regulation of the basal cAMP levels by chronic carbamazepine treatment could therefore underlie the sensitization of the dopamine D_2-like receptors in prefrontal cortex *in vivo*. Inhibition of cAMP formation decreases cAMP-dependent PKA activity and inhibits subsequent PKA-mediated phosphorylation events. Recent results demonstrated that chronic inhibition of PKA induced supersensitization of AC in neuronal cell line and that activators of PKA attenuated sensitization. Although inhibition of cAMP and PKA is not generally required for the development of super-sensitization of AC, inhibition of PKA may contribute to the development of sensitization in select cellular models (Johnston et al., 2002).

From these data, it can be speculated that mechanisms such as the sensitization of dopamine D_2-like receptors due to the chronic inhibition of AC by carbamazepine, contribute to the increased relapsed rate observed in patients under monotherapy with this drug for 3-4 years (Post et al., 1990), although it is an effective drug in the acute treatment of bipolar disorder.

In conclusion, abnormal function of brain AC and dopamine systems may be implicated in bipolar disorder. Chronic treatment with lithium, carbamazepine and valproate affects the basal and evoked cAMP levels in a distinct way, both ex vivo and in vivo, resulting in differential responses to dopamine D_2-like receptors activation, and suggesting that additional signalling systems are involved as well.

4.3 The interaction between dopamine D_2-like and β-adrenergic receptors in the prefrontal cortex is altered by mood stabilizing agents

Biogenic monoaminergic neurotransmission has been suggested to be involved in bipolar disorder as well as in the therapy for this disease. The dopamine D_2 receptor is the predominant member of the family of D_2-like dopamine receptors in the brain (Emilien et al., 1999) and the β_1-adrenergic receptors are the most abundant members of β-adrenergic receptors in the cerebral cortex (Rainbow et al., 1984). The effects of the mood stabilizing drugs lithium, carbamazepine or valproate on the intracellular signalling mediated by receptors coupled to activation and inhibition of AC were investigated. Thus, experiments with the dopaminergic and adrenergic systems, particularly on D_2-like and β-adrenergic receptors, respectively positive and negatively coupled to AC, were performed both in cultured rat cortical neurons and in prefrontal cortex of freely moving rats, using microdialysis (Masana et al., 1991, 1992).

Isoproterenol and quinpirole were used to stimulate β-adrenergic and dopamine D_2-like receptors, respectively, and cAMP levels inside neurons and in the dialysates were determined by a radioimmunoassay using the [8-^{3}H-] and [^{125}I], respectively for the *ex vivo* and *in vivo* experiments. Moreover, the cAMP levels produced by the stimulation of β-adrenergic receptors with isoproterenol as well as the ability of the activation of dopamine D_2-like receptors to block the increase of isoproterenol-evoked cAMP levels were measured under control conditions and after the treatment with mood stabilizing drugs. Immunohistochemistry and western blot techniques were used to investigate if β_1-adrenergic and D_2 dopaminergic receptors are co-localized in the rat prefrontal cortex and if the three drugs have any effect on the levels of these receptors on the membranes of cortical neurons.

The cAMP accumulation in cultured cortical cells is presented (Figure 16), using the experimental conditions described in the legend of the figure.

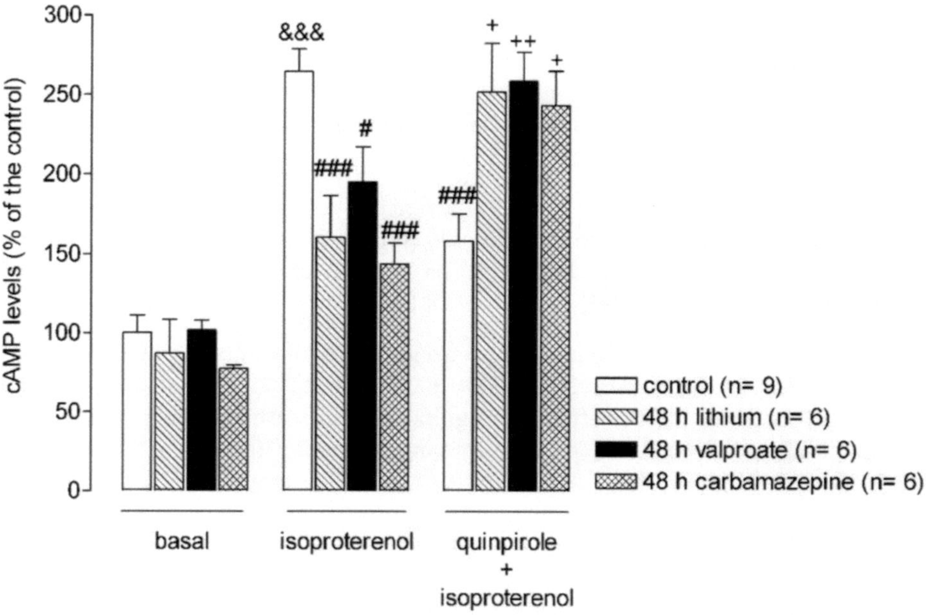

Fig. 16. Effects of lithium, valproate and carbamazepine on the intracellular cAMP levels in cortical neurons under basal conditions and after stimulation with isoproterenol or quinpirole plus isoproterenol. Neurons were pre-exposed or not (control) to 1 mmol L^{-1} lithium, 0.05 mmol L^{-1} carbamazepine or 0.5 mmol L^{-1} valproate, for 48 h. After this period, neurons were incubated with RO-201724 (25 µmol L^{-1}, 15 min) and were then treated or not (basal) with isoproterenol (10 µmol L^{-1}; 15 min), or with isoproterenol together with quinpirole (10 µmol L^{-1}; 15 min). In the latter experimental conditions the cells were pre-treated with 10 µmol L^{-1} quinpirole for 5 min. The cAMP levels were measured as described (Montezinho et al., 2006). The average value of basal cAMP level for all cells tested (101.4 ± 21.0 nmoL µg^{-1} protein) was set to 100 %. Data are means ± SEM, for the indicated number of independent experiments, performed in duplicate. Data were analyzed by one-way ANOVA, followed by post-hoc Bonferroni's test for multiple comparisons. $^{###}$, $p< 0.001$; $^{\#}$, $p< 0.05$ compared to control isoproterenol-treated cells; $^{+}$, $p< 0.05$; $^{++}$, $p< 0.01$ compared to control quinpirole plus isoproterenol-stimulated cells; $^{\&\&\&}$, $p< 0.001$ compared to control of untreated cells; *, $p< 0.05$; **, $p< 0.01$ compared to neurons pre-exposed to LiCl and valproate, respectively (Montezinho et al., 2006).

In the *ex vivo* studies, the average value of 101.4 ± 21.0 nmol μg^{-1} protein for basal cAMP level was set to 100 %. Isoproterenol stimulation produced a cAMP increase to 264.6 ± 14.3 %, $p < 0.001$ (2x above basal levels) which was partially inhibited by lithium, carbamazepine or valproate (lithium: 160.7 ± 25.6 %, $p < 0.001$; carbamazepine: 143.9 ± 13.1 %, $p < 0.001$; valproate: 195.4 ± 21.8 %, $p < 0.05$) although no effect was observed by these drugs on the basal cAMP production. Quinpirole, had no effect on the basal cAMP accumulation (data not shown); however, this dopamine D_2-receptor agonist inhibited isoproterenol-enhanced cAMP levels in untreated neurons (158.4 ± 16.8 %, $p < 0.001$) and this effect was decreased in neurons pre-exposed to lithium, carbamazepine or valproate (lithium, 249.1 ± 28.4 %, $p < 0.05$; carbamazepine, 243.8 ± 21.5 %, $p < 0.05$; valproate, 259.0 ± 18.4 %, $p < 0.01$) (Figure 16).

To complement these results, in vivo studies were also performed. Preliminary experiments showed that 180 min after probe insertion cAMP levels were stable and were maintained throughout the experiment being the basal cAMP levels for control animals 9.68 ± 2.10 fmol/20 μL, not different in valproate-treated rats, but significantly increased in lithium-treated, 13.17 ± 1.21 fmol/20 μL ($p < 0.05$) and decreased in carbamazepine-treated rats, 7.82 ± 0.34 fmol/20 μL ($p < 0.01$).

The concentrations of 2.5 mmol L^{-1} and 100 μmol L^{-1}, respectively, for isoproterenol and quinpirole were used in all in vivo experiments (Figure 17), as they were found to be the minimal concentrations that produced an observable effect under our experimental conditions (Mørk and Geisler, 1994).

After 30 min of isoproterenol infusion in the prefrontal cortex of control and vehicle-treated rats an increase in cAMP levels to, respectively, 234.6 ± 12.2 % and 272.1 ± 12.0 % (Figure 17 A,D) of the basal values occurred, and this effect was significantly decreased in rats pre-treated with the three drugs (lithium: 131.9 ± 5.8 %, $p < 0.01$; carbamazepine: 159.0 ± 8.5 %, $p < 0.01$; valproate: 137.9 ± 8.7 %, $p < 0.01$ (Figure 17 B,C and E). A decrease was also observed in the total isoproterenol-evoked increase in the extracellular cAMP (lithium: 85.1 ± 3.7 %, $p < 0.01$; carbamazepine: 87. 5 ± 1.6 %, $p < 0.05$; valproate: 69.5 ± 2.4 %, $p < 0.001$), when compared to control or vehicle-treated rats values (Figure 17 A,D), similarly to what happened in the *ex vivo* studies. When both isoproterenol and quinpirole were locally infused, the effect of the D_2-like receptors agonist significantly inhibited the effect of isoproterenol-induced increase of extracellular cAMP (control rats: 132.5 ± 5.8 %, $p < 0.01$; vehicle-treated rats: 141.0 ± 9.1, $p < 0.01$) measured 30 min after the agonist infusion (control rats: 77.0 ± 1.6 %, $p < 0.001$; vehicle-treated rats: 74.9 ± 2.7, $p < 0.001$) (Figure 17 A, D) and the total increase in extracellular cAMP evoked by isoproterenol within 150 min (Figure 17 F)). In lithium-treated rats the activation of D_2-like receptors did not inhibit the isoproterenol-induced increase in extracellular cAMP as demonstrated by measurements 30 min after infusion (188.2 ± 19.1 %, $p < 0.01$ (Figure 17 B), vs control 132.5 ± 5.8 % (Figure 17 A) or lithium- treated animals infused only with isoproterenol (131.9 ± 5.8 %) (Figure 17 B) and the total extracellular cAMP produced within 150 min (77.0 ± 1.6 vs 94.7 ± 4.4 %, $p < 0.01$, (Figure 17 F), as it was observed in cortical neuron cultures. In contrast to the behaviour of the three drugs in the *ex vivo* studies and in lithium-treated rats, in carbamazepine- or valproate-treated animals, quinpirole did not significantly change the isoproterenol-induced increase in extracellular cAMP (176.9 ± 8.3 % and 158.0 ± 11.7 %, respectively (Figure 17 C ,E) *vs* control 132.5 ± 5.8 % , vehicle-treated rats 137.9 ± 8.7 %, (Figure 17 A,D), or carbamazepine- 159.0 ± 8.5 % or valproate - treated rats 137.9 ± 8.7 %, and infused only with isoproterenol (Fig. 3C and E) produced 30 min post-infusion, or the total evoked extracellular accumulation of cAMP measured during 150 min following infusion with quinpirole and isoproterenol (84.7 ± 3.5 % and 85.9 ± 2.2 %, respectively vs control 77.0 ± 1.6 % and vehicle-treated rats 74.9 ± 2.7 %, (Figure 17 F). As a control, it was observed that the activation of D_2-like receptors had no effect on basal cAMP levels under control conditions (control rats: 100.4 ± 18.4 %; vehicle-treated rats: 104.0 ± 11.3 %) but in lithium-treated rats the activation with quinpirole, significantly increased cAMP levels to 155.1 ± 9.3 %, $p < 0.05$ and in carbamazepine- and valproate-treated rats the basal cAMP levels didn't significantly change (114.4 ± 8.9 % and 98.7 ± 13.6 %, respectively *vs* 100.4 ± 18.4 % and 104.0 ± 11.3 %).

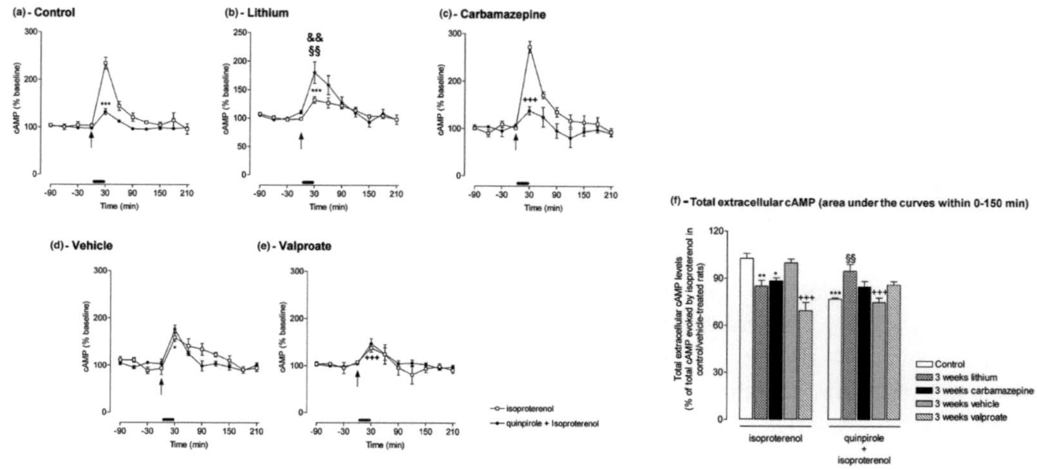

Fig. 17. Effect of the local infusion of isoproterenol (2.5 mmol L^{-1}) or quinpirole (100 µmol L^{-1}) simultaneously with isoproterenol (2.5 mmol L^{-1}) (starting at arrow) on the maximal extracellular cAMP levels produced 30 min post-infusions (a) control rats; b) lithium-treated rats; c) carbamazepine-treated rats; d) vehicle-treated rats; e) valproate-treated rats) and on total cAMP levels (correspondent to the area under the curves), measured within 150 min post-infusions (f). Basal cAMP level was taken as the average cAMP concentration in the four consecutive samples collected from 3 h after the insertion of the probe, prior to isoproterenol (8.9 ± 2.6 fmol/20 µL, n= 6) and quinpirole plus isoproterenol infusions (9.4 ± 1.1 fmol/20 µL, n= 4). Thereafter, the agent of interest (2.5 mmol L^{-1} isoproterenol or 100 µmol L^{-1} quinpirole plus 2.5 mmol L^{-1} isoproterenol) dissolved in Ringer solution was infused through the probe, for 30 min, and seven samples were then collected. Extracellular cAMP levels in the dialysates were measured by radioimmunoassay analysis and expressed as percentages of the basal value (a)-e)) or expressed as a percentage of the percentage of the total amount of cAMP produced during 150 min post-infusion with isoproterenol in control or vehicle-treated rats (f). Results are means ± SEM, from 4 to 6 independent experiments. The rats were treated or not with doses yielding therapeutic plasma levels of lithium, carbamazepine or valproate, during three weeks, as indicated in (Montezinho et al., 2006). Data are analyzed by one-way ANOVA, followed by post-hoc Bonferroni's test for multiple comparisons. ***, $p < 0.001$; compared to the cAMP produced 30 min after the infusion of isoproterenol in control rats. $^{+++}$, $p < 0.001$; compared to the cAMP produced 30 min after the infusion of isoproterenol in vehicle-treated rats. §§, $p < 0.01$; compared to the cAMP produced 30 min after the infusion of quinpirole plus isoproterenol in control rats. $^{\&\&}$, $p < 0.001$; compared to the cAMP produced 30 min after the infusion of quinpirole plus isoproterenol in lithium-treated rats (Montezinho et al., 2006).

Taken together, *ex vivo* and *in vivo* data showed that stimulation of β-adrenergic receptors with isoproterenol increased cAMP levels and this effect was significantly inhibited by lithium, carbamazepine or valproate. The activation of dopamine D_2-like receptors with quinpirole decreased the isoproterenol-induced raise in cAMP in control conditions. This inhibition was observed *in vivo* after chronic treatment of the rats with carbamazepine or valproate, but not after treatment with lithium or in cultured rat cortical neurons after 48 h exposure to the three mood stabilizers.

Immunohistochemistry studies (data not shown) (Montezinho et al., 2006) confirmed the co-existence of dopamine D_2 and β_1-adrenergic receptors in the majority of the cells of the rat prefrontal cortex, allowing the interactions at the second messenger level. The results obtained from prefrontal cortex sections of lithium- carbamazepine or valproate-treated suggested a loss of dopamine D_2 receptors, whereas in valproate-, carbamazepine-, and lithium-treated rats the β_1-adrenergic receptors seemed to be, respectively, down-regulated, upregulated or not changed (Montezinho et al., 2006).To confirm these data, the levels of the dopamine D_2 and β_1-adrenergic receptor proteins were determined in membranes prepared from cultured rat cortical neurons and from rat prefrontal cortex, pre-treated or not with the mood stabilizing agents, by western blot analysis, using the same anti-D_2 and anti-β_1 specific antibodies used for immunohistochemistry. An evident immunoreactivity was observed at ≈50 kDa and ≈64 kDa, the predicted molecular weights for the dopamine D_2 and β_1 adrenergic receptor proteins, respectively, in protein extracts from cultured rat cortical neurons and from cortex of rats treated with the three drugs (data not shown) (Montezinho et al., 2006). The quantitative analysis of these bands, by densitometry, confirmed that lithium, valproate and carbamazepine treatments significantly decreased dopamine D_2 receptor protein levels in membranes from treated cultured rat cortical neurons and in prefrontal cortex of treated animals when compared with control values. Concerning β_1-adrenergic receptors only valproate decreased the expression of the receptor present in cultured cortical neurons; however, *in vivo* treatment with carbamazepine, valproate or lithium, up-regulated, down-regulated or had no effect on β_1-adrenergic receptor levels, respectively, when compared with control and vehicle-treated rats (data not shown) (Montezinho et al., 2006).

In conclusion, this study shows that there is a cross-talk between dopamine D_2-like and β-adrenergic receptors activities in the rat brain cortical region, which is differentially affected by therapeutic concentrations of the mood stabilizing drugs lithium, carbamazepine and valproate. Indeed, *in vivo* and *in vitro* data showed that activation of dopamine D_2-like receptors inhibits β_1-adrenergic receptor stimulated cAMP production. *Ex vivo* this inhibition was attenuated by lithium, carbamazepine and valproate, whereas *in vivo* only lithium had such effect. Consistent with this regulatory role on AC activity, dopamine D_2 and β_1-adrenergic receptors are co-localized in the rat prefrontal cortex and their protein levels are changed by mood stabilizers as determined by immunohistochemistry and immunoblotting, respectively.

These data show that the three mood stabilizers affect dopamine D_2 -like receptor-mediated regulation of β-adrenergic signalling, both *ex vivo* and *in vivo* although in a different way. The discrepancy between *ex vivo* and *in vivo* results can be explained through at least three factors: the cellular interactions in the 3D structure of intact rat brain are different from those in a 2D typical organization of neuronal cultures (Dziedzicka-Wasylewska et al. 1996; Ichikawa & Meltzer 1999); the absence of glial cells, which contain β-adrenergic and D_2 -like receptors (Stone & John 1990; Stone et al. 1990), in the cortical neuron cultures used; cortical neuron cultures were exposed to the mood stabilizing agents for 48 h whereas the animals were treated with the same drugs during three weeks and, according to literature data, the mechanisms involved in acute and chronic Li$^+$ treatment are different (Mørk & Geisler, 1987b, 1989c; Newman & Belmaker, 1987). This highlights the importance of the *in vivo* studies. Each of these drugs acts by a unique mechanism *in vivo*: chronic treatment with lithium increases cAMP levels which may be attributed to the inhibition of G_i, (Jope, 1999a; Masana et al., 1991, 1992; Montezinho et al., 2004); however, taking into account data here presented, the loss of dopamine D_2-like receptors, which are coupled to the inhibition of AC, may also account for the increase in the basal cAMP levels, thus mediating lithium antimanic action (Schatzberg et al., 2004; Silverstone and Silverstone, 2004; Yatham *et al.*, 2002),. In contrast with lithium, *in vivo* carbamazepine treatment decreases cAMP levels despite an observed increase in the β_1 -receptor expression. This effect may be exerted primarily at the level of cAMP production, by the direct inhibition of AC (Chen et al., 1996a), which can also explain the attenuated isoproterenol-evoked cAMP production and the lack of a further effect of D_2 receptor activation. The changes in receptor levels could represent a compensatory mechanism in the *in vivo* conditions. *In vivo* administration of valproate significantly reduced both the β_1-adrenergic and dopamine D_2 receptor levels, which can explain the decrease in isoproterenol-evoked receptor-mediated cAMP production, and no further attenuation of cAMP production after dopamine D_2 receptors stimulation. The β-adrenergic down regulation by valproate was also showed *in vitro* (Chen et al., 1996b).

These findings suggest that additional mechanisms are operative *in vivo* as compared to *in vitro*. It can be speculated that dopaminergic and noradrenergic neurotransmitter levels released by afferent projections participate in the regulation of these monoaminergic systems. In agreement with this hypothesis, a decrease in the dopamine concentration in prefrontal cortex and an increase in striatum (Dziedzicka-Wasylewska et al., 1996) were reported after the *in vivo* intragastrical administration of lithium. In contrast, it was described that therapeutic concentrations of carbamazepine or valproate enhanced basal releases of dopamine in the prefrontal cortex (Ichikawa & Meltzer, 1999). In addition, carbamazepine and valproate are metabolized *in vivo*, what might contribute to the differences observed in the effects of these drugs *in vitro* and *in vivo*. Moreover, the presence of glia cells expressing β_1-adrenergic (Stone *et al.*, 1990; Stone & John, 1990) and D_2 receptors (Khan *et al.*, 2001) in rat brain cortex increases the complexity of the regulation of the cAMP levels *in vivo* when compared with the *in vitro* data. Lastly, the other D_2-like receptor subtypes, namely the D_3 and D_4 receptors, and the β_2 and β_3-adrenergic receptors may also play a role in the regulation of cAMP levels and the expression pattern of these receptors *in vitro* may be different from the one observed *in vivo*.

It has been described that the function of the β-adrenergic receptor signalling in the striatum and prefrontal cortex depends upon dopaminergic activity (Herve et al., 1990). Thus, treatments affecting dopaminergic neurotransmission can influence the success and the rapidity of action of mood stabilizing drug therapy. These results support the hypothesis that therapeutic intervention in bipolar disorder may be improved by affecting β-adrenergic receptor signalling *via* effects on dopamine D_2 receptor pathways.

5. Conclusions

The work presented in this chapter focused on Li^+ effects on several cellular systems and *in vivo*, in particular how it affects intracellular Mg^{2+} binding, cell energy metabolism, GABAergic and glutamatergic neurotransmitter systems, intracellular cAMP levels and their modulating systems, contributing to a better understanding of Li^+ action in bipolar disorder. The results showed that:

- Under depolarising conditions, the Na^+/Ca^{2+} exchanger is proposed to be the new contribution to Li^+ influx, resulting from the activation of L-type voltage-sensitive Ca^{2+} channels, where Li^+ replaces Na^+.
- Li^+ intracellular immobilization is cell type dependent.
- Li^+ is able to displace Mg^{2+} from its intracellular binding sites, providing further evidence for the generality of the ionic Li^+/Mg^{2+} competition mechanism. The extent of Li^+/Mg^{2+} competition was suggested to be cell-type dependent, being affected by differences in Li^+ transport and immobilization properties.
- Li^+ had a remarkable inhibitory effect on the energetic metabolism of glucose in SH-SY5Y cells. The results were consistent with an inhibition of glycolytic and TCA cycle fluxes, with an unchanged intracellular redox state. Li^+ did not interfere with the competitive metabolism of glucose and lactate, or the residual contribution of unlabeled endogenous sources for the acetyl-CoA pool.
  Similarly to SH-SY5Y cells, but to a much lesser extent, neuronal glycolytic flux was also found to be decreased in cultured rat cortical neurons after incubation with 15 mmol L^{-1} Li^+.
- Li^+ decreased the incorporation of ^{13}C labeling into GABA carbons from its precursor glutamate in neurons, both *in vivo* and *ex vivo*, suggesting an inhibitory effect on glutamate decarboxylase.
- Li^+ plays an important role in the homeostasis of the second messenger cAMP in neuronal cells (Fig. 18 B) a.).
- In cultured cortical neurons the mood stabilizing drugs Li^+, valproate and carbamazepine do not affect basal cAMP levels, although *in vivo* different effects are observed. The increase in β-adrenergic- and forskolin-mediated production of cAMP is attenuated by the three drugs both *ex vivo* and *in vivo* (Fig. 18 B) b.).
- The three mood stabilizers act on the dopamine D_2-like and β-adrenergic receptor-mediated regulation of cAMP levels both *in vivo* and *ex vivo* (Fig. 18 B) b.), but by different mechanisms. This may explain the differences in their efficacy in the treatment of manic episodes in clinical cases.

Fig.18 summarizes some of the results obtained concerning targets and effects of Li^+, carbamazepine and valproate.

This work is a contribution to clarify some of the hypotheses that have been advanced for Li^+ action namely the ionic competition model, the effects on energetic metabolism and on signal transduction pathways. Li^+ acts on multiple targets and interferes with many biological processes. Although biological effects cannot be studied isolated, the data presented are some pieces to join the puzzled information in the literature, which hopefully will help to make "the whole story" - a step forward to establish the general mechanism of action of this ion in bipolar disorder.

Multinuclear NMR spectroscopy was found to be a powerful tool in the study of Li^+ transport, Li^+ immobilization and Li^+ effects on different metabolic pathways involved in cell energy production and signalling. Application of mathematical models to ^{13}C NMR data obtained from cultured brain cells and rat brain *in vivo* seems to be a promising strategy to identify, more precisely, the metabolic pathways specifically affected by Li^+. Microdialysis in freely moving animals showed to be very useful for the *in vivo* quantification of neurotransmitters, providing information about signal transduction processes.

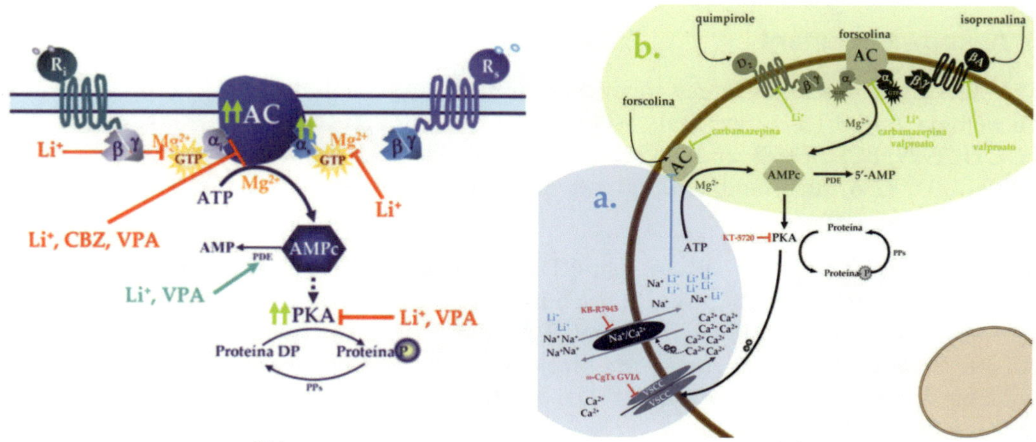

(A) (B)

Fig. 18. **A**) Signal transduction cascades involved in the pathophysiology of bipolar disorder (alterations represented by ↑↑) and in the therapeutic action of mood-stabilizing agents (Li^+), valproate (VPA) e carbamazepine (CBZ) (represented by the red lines (**inhibition**) and blue arrow (**activation**). Li^+ and VPA activate phosphodiesterases (PDE) which hydrolyze 3'5'-cyclic-adenosine monophosphate (cAMP), and the three mood stabilizing agents reduce the activity of adenylate cyclase (AC). Li^+ inhibits the activity of stimulatory and inhibitory G proteins ($G_{\alpha s}$ e $G_{\alpha i}$), which is explained through Li^+/Mg^{2+}competition. These alterations contribute to the reduction of protein kinase A (PKA) activity decreasing its phosphorylation capacity and originating long term perturbations. Phosphatases (PPs) convert phosphoproteins (P-proteins) in the dephosphorylated form (DP-protein). cAMP is hydrolyzed to 5'-AMP by phosphodiesterases (PDE) (Jope 1999a; Manji *et al.*, 1995, 2001). **B**) **a)** The activation of adenylate cyclase (AC) with forskolin increases Li^+ uptake by SH-SY5Y cells and by cultured hippocampal and cortical neurons, at therapeutic concentrations of this cation. The inhibitory of the Na^+/Ca^{2+} exchanger, KB-R7943, reduces Li^+ influx under resting conditions and completely inhibits the effect of forskolin on the accumulation of this cation. Inhibition of N-type voltage-sensitive Ca^{2+} channels (VSCC), with ω-CgTx GVIA, or inhibition of protein kinase A (PKA) with KT-5720, also abolishes the effect of forskolin on Li^+ uptake. The effect of cAMP in SH-SY5Y cells is mediated by PKA and occurs *via* changes in the intracellular free concentration ($[Ca^{2+}]_i$). Accordingly, intracellular accumulation of cAMP seems to increase the $[Ca^{2+}]_i$ due to Ca^{2+} entre through N-VSCC, which activates the Na^+/Ca^{2+} (Li^+/Ca^{2+} exchanger in order to extrude the Ca^{2+} taken up by the cells. **b)** Direct stimulation of AC with forskolin increase cAMP levels both *in vitro* and *in vivo*, and this effect was significantly inhibited by Li^+, valproate and carbamazepine. In carbamazepine-treated animals, the activation of dopamine D_2-like receptors enhances the responsiveness of

AC to activation by forskolin, possibly as a consequence of chronic inhibition of the activity of this enzyme. An increase in the activation of AC and cAMP production is described as a super-sensitization of AC. In vivo, each of the mood stabilizing drugs modulates dopamine D_2-like (D_2) and β-adrenergic (βA) receptor-regulated cAMP levels by a distinct mechanism. The effects of carbamazepine are most likely due to direct inhibition of AC, whereas Li^+ may act by affecting dopamine D_2 receptor mediated signalling and valproate by down-regulating β-adrenergic transmission.

6. Acknowledgement

M.M.C.A.Castro acknowledges financial support from Fundação para a Ciência e Tecnologia (F.C.T.), Portugal (Project PECS/P/SAU166/95 and Project POCTI/1999/BCI/36160) and FEDER. Liliana P. Montezinho was supported by F.C.T. (SFRH/BD/3286/2000) and FEBS grants and Carla P. Fonseca by a F.C.T. grant (Praxis XXI/BD/21462/99).

In a very special way the authors thank Duarte Mota de Freitas, of Loyola University of Chicago, USA, who initiated M.M. Castro in this research area, C. F. Geraldes of Dept. of Life Sciences, Univ. of Coimbra, Portugal, S. Cerdán of Instituto Investigaciones Biomedicas Alberto Sols, Madrid, Spain, and Arne Mørk of Lundbeck, Copenhagen, Denmark, for their valuable contribution to these studies. The contribution of R. Ramasamy, C.B. Duarte and J. Jones with important discussions is also acknowledged. The authors are also thankful to J. Nikolakopoulos, B. Layden, Z. Zachariah and A. Sierra for the active participation in the experimental work, to Margarida Figueiredo and Helena Freitas (University of Coimbra) for using the AA spectrophotometer, to M. João Bastos and Helena Castro for helping in the AA measurements, and to Sandra Almeida and Inês Araújo for their assistance in the preparation of cortical and hippocampal cultures, respectively. The authors would also like to thank Anette Frederiksen and Kirsten Jørgensen (Lundbeck) for their skillful technical assistance on rat microdialysis surgeries and immunohistochemistry studies, respectively.

7. References

Abreu, L.A. & Abreu, R.R. (1973). Effect of lithium on brain aconitase activity. *Experientia.* 29, 446-447.

Amari, L., Layden, B., Nikolakopoulos, J., Rong, Q., Mota de Freitas, D., Baltazar, G., Castro, M.M.C.A. & Geraldes C.F.G.C. (1999a). Competition between Li^+ and Mg^{2+} in neuroblastoma SH-SY5Y cells: a fluorescence and ^{31}P NMR study. *Biophys J.*, 76, 2934-2942.

Amari, L., Layden, B., Rong, Q., Geraldes, C.F.G.C. & Mota de Freitas, D. (1999b). Comparison of fluorescence, ^{31}P NMR, and ^{7}Li NMR spectroscopic methods for investigating Li^+/Mg^{2+} competition for biomolecules. *Anal. Biochem.*, 272, 1-7.

Antonelli, T., Ferioli, V., Lo, G.G., Tomasini, M.C., Fernandez, M., O'Connor, W.T., Glennon, J.C., Tanganelli, S. & Ferraro, L. (2000). Differential effects of acute and short-term lithium administration on dialysate glutamate and GABA levels in the frontal cortex of the conscious rat. *Synapse,* 38, 355-362.

Berns, G. S. & Nemeroff, C. B. (2003). The neurobiology of bipolar disorder. *Am. J. Med. Genet. C. Semin. Med. Genet.*, 123, 76-84.

Biedler, J.L., Helson, L. & Spengler, B.A. (1973). Morphology and growth, tumorigenicity, and cytogenetics of human neuroblastoma cells in continuous culture. *Cancer Res.,* 33, 2643-2652.

Bouzier-Sore, A.K., Serres, S., Canioni, P. & Merle, M. (2003). Lactate involvement in neuron-glia metabolic interaction: ^{13}C-NMR spectroscopy contribution. *Biochimie,* 85, 841-848.

Brambilla, P., Perez, J., Barale, F., Schettini, G. & Soares J.C. (2003). GABAergic dysfunction in mood disorders. *Mol. Psychiatry,* 8, 721-37, 715.

Brunello ,N. & Tascedda, F. (2003). Cellular mechanisms and second messengers: relevance to the psychopharmacology of bipolar disorders. *Int. J. Neuropsychopharmacol.,* 6, 181-189.

Caligiuri, M.P., Brown, G.G., Meloy, M.J., Eberson, S.C., Kindermann, S.S., Frank, L.R., Zorrilla, L.E. & Lohr, J.B. (2003). An fMRI study of affective state and medication on cortical and subcortical brain regions during motor performance in bipolar disorder. *Psychiatry Res.,* 123, 171-182.

Castro, M.M.C.A., Nikolakopoulos, J., Zachariah, C., Freitas, D.M., Stubbs, E.B.Jr., Geraldes, C.F.G.C. & Ramasamy, R. (1996). 7Li NMR Study of Lithium Ion Transport in Perfused Human Neuroblastoma Cells. *Metal Ions in Biology and Medicine,* 4, 192-194.

Chen, G., Pan, B., Hawver, D.B., Wright, C.B., Potter, W,Z. & Manji, H.K, (1996a). Attenuation of cyclic AMP production by carbamazepine. *J.Neurochem.,* 67, 2079-2086.

Chen, G., Manji, H.K., Wrigh,t C.B., Hawver, D.B. & Potter, W.Z .(1996b). Effects of valproic acid on beta-adrenergic receptors, G-proteins, and adenylyl cyclase in rat C6 glioma cells. *Neuropsychopharmacology* , 15, 271-280.

Cooper, D. M., Mons, N. & Karpen, J. W. (1995). Adenylyl cyclases and the interaction between calcium and cAMP signaling. *Nature,* 374, 421-424.

Deicken, R.F., Weiner, M.W. & Fein, G. (1995). Decreased temporal lobe phosphomonoesters in bipolar disorder. *J. Affect. Disord* .,33, 195-199.

Deval, E., Raymond, G. & Cognard, C. (2002.) Na^+-Ca^{2+} exchange activity in rat skeletal myotubes: effect of lithium ions. *Cell Calcium,* 31, 37-44.

Dringen, R., Peters, H., Wiesinger, H. & Hamprecht, B. (1995). Lactate transport in cultured glial cells. *Dev. Neurosci.,* 17, 63-69.

Dziedzicka-Wasylewska, M., Mackowiak, M., Fijat, K., Wedzony, K. (1996). Adaptive changes in the rat dopaminergic transmission following repeated lithium administration. *J. Neural. Transm.,* 103, 765-776.

Emilien, G., Maloteaux, J. M., Geurts ,M., Hoogenberg, K. *et al.* (1999.) Dopamine receptors-physiological understanding to therapeutic intervention potential. *Pharmacol. Ther.,* 84, 133-156.

Fonseca, C.P., Montezinho, L.P., Baltazar, G., Layden, B., Mota de Freitas, D., Geraldes, C.F.G.C. & Castro, M.M.C.A. (2000). Li^+ influx and binding, and Li^+/Mg^{2+} competition in bovine chromaffin cell suspensions as studied by ^7Li NMR and fluorescence spectroscopy. *Metal Based Drugs,* 7, 357-364.

Fonseca, C.P., Montezinho, L.P., Nabais, C., Tomé, A.R., Freitas, H., Geraldes, C.F.G.C. & Castro, M.M.C.A. (2004). Effects of Li^+ transport and intracellular binding on Li^+/Mg^{2+} competition in bovine chromaffin cells. *Biochim Biophys Acta* (*Mol.Cell Res.*), 1691, 79-90.

Fonseca, C.P., Jones, J.G., Carvalho, R.A., Jeffrey, F.M.H., Montezinho, L.P., Geraldes, C.F.G.C. & Castro, M.M.C.A. (2005). Tricarboxylic acid cycle inhibition by Li^+ in the human neuroblastoma SH-SY5Y cell line: a ^{13}C NMR isotopomer analysis. *Neurochemistry Int.*, 47, 385-393.

Fonseca, C.P., Sierra, A, Geraldes, C.F.G.C., Cerdán, S. & Castro, M.M.C.A. (2009). Mechanisms underlying Li^+ effects in glutamatergic and GABAergic neurotransmissions in the adult rat brain and in primary cultures of neural cells as revealed by ^{13}C NMR. *J. Neurosci. Res.*, 87, 1046-1055.

Friedman, J.E., Lelkes, P.I., Lavie, E., Rosenheck, K., Schneeweiss, F. & Schneider, A.S. (1985). Membrane potential and catecholamine secretion by bovine adrenal chromaffin cells: use of tetraphenylphosphonium distribution and carbocyanine dye fluorescence. *J. Neurochem.*, 44, 1391-1402.

Gallagher, H.C., Bacon, C.L., Odumeru, O.A., Gallagher, K.F., Fitzpatrick, T., Regan, C.M. (2004). Valproate activates phosphodiesterase-mediated cAMP degradation: relevance to C6 glioma G1 phase progression. *Neurotoxicol Teratol.*, 26, 73-81.

Gil, V.M.S. & Geraldes, C.F.G.C. (1987). Relaxação de spins nucleares, In: *Ressonância Magnética Nuclear*, Fundação Calouste Gulbenkian, Chap. 6, pp. 401-497.

Goodwin, G.M., Cavanagh, J.T., Glabus, M.F., Kehoe, R.F., O'Carroll, R.E. & Ebmeier K.P. (1997). Uptake of 99mTc-exametazime shown by single photon emission computed tomography before and after lithium withdrawal in bipolar patients: associations with mania. *Br. J. Psychiatry*, 170, 426-430.

Gottesfeld, Z. (1976). Effect of lithium and other alkali metals on brain chemistry and behavior. I. Glutamic acid and GABA in brain regions. *Psychopharmacologia*, 45, 239-242.

Gupta, R.K., Benovic, J.L. & Rose, Z.B. (1978). The determination of the free magnesium level in the human red blood cell by ^{31}P NMR. *J. Biol .Chem .*, 253, 6172-6176.

Hassel, B, Sonnewald, U. & Fonnum, F. (1995). Glial-neuronal interactions as studied by cerebral metabolism of [2-13C]acetate and [1-13C]glucose: an ex vivo 13C NMR spectroscopic study. *J. Neurochem.*, 64, 2773–2782.

Herve, D., Trovero, F., Blanc, G., Vezina, P., Glowinski, J., Tassin, J.P. (1990). Involvement of dopamine neurons in the regulation of beta-adrenergic receptor sensitivity in rat prefrontal cortex. *J. Neurochem.*, 54, 1864-1869.

Hough C., Lu S. J., Davis C. L., Chuang D. M. *et al.* (1999). Elevated basal and thapsigargin-stimulated intracellular calcium of platelets and lymphocytes from bipolar affective disorder patients measured by a fluorometric microassay. *Biol. Psychiatry*, 46, 247-255.

Ichikawa, J. & Meltzer, H.Y. (1999). Valproate and carbamazepine increase prefrontal dopamine release by $5\text{-}HT_{1A}$ receptor activation. *Eur. J. Pharmacol.*, 380, R1-R3.

Ikeda A. & Kato T. (2003). Biological predictors of lithium response in bipolar disorder. *Psychiatry Clin. Neurosci.*, 57, 243-250.

Johnston, C. A., Beazely, M. A., Vancura, A. F., Wang, J. K. *et al.* (2002.) Heterologous sensitization of adenylate cyclase is protein kinase A-dependent in Cath.a differentiated (CAD)-D2L cells. *J. Neurochem.* ,82, 1087-1096.

Johnston, CA & Watts, VJ. (2003). Sensitization of adenylate cyclase: a general mechanism of neuroadaptation to persistent activation of Galpha(i/o)-coupled receptors? *Life Sci.*, 73, 2913-2925.

Jope, R.S., Miller, J.M., Ferraro, T.N. & Hare, T.A. (1989). Chronic lithium treatment and status epilepticus induced by lithium and pilocarpine cause selective changes of amino acid concentrations in rat brain regions. *Neurochem. Res.*, 14, 829-834.

Jope, R.S. (1999a). A bimodal model of the mechanism of action of lithium. *Mol Psychiatry*, 4, 21-25.

Jope, R.S. (1999b). Anti-bipolar therapy: mechanism of action of lithium. *Mol. Psychiatry*, 4, 117-128.

Kajda, P.K., Birch, N.J., O'Brien, M.J. & Hullin, R.P. (1979). Rat-brain pyruvate kinase: purification and effects of lithium. *J. Inorg. Biochem.*, 11, 361-366.

Kajda, P.K. & Birch, N.J. (1981). Lithium inhibition of phosphofructokinase. *J. Inorg. Biochem.*, 14, 275-278.

Kao, L.S. & Cheung, N.S. (1990). Mechanism of calcium transport across the plasma membrane of bovine chromaffin cells. *J. Neurochem.*, 54, 1972-1979.

Kaplan, O., Aebersold, P. & Cohen, J.S. (1989). Metabolism of peripheral lymphocytes, interleukin-2-activated lymphocytes and tumor-infiltrating lymphocytes from ^{31}P NMR studies. *FEBS Lett* ., 258, 55-58.

Kato, T., Takahashi, S., Shioiri, T. & Inubushi, T. (1993). Alterations in brain phosphorous metabolism in bipolar disorder detected by in vivo ^{31}P and ^{7}Li magnetic resonance spectroscopy. *J. Affect. Disord* .,27, 53-59.

Kato, T., Inubushi, T. & Kato, N. (1998). Magnetic resonance spectroscopy in affective disorders. *J. Neuropsychiatry Clin. Neurosci.*, 10, 133-147.

Kato, T. & Kato, N. (2000). Mitochondrial dysfunction in bipolar disorder. *Bipolar Disord.*, 2, 180-190.

Kennedy, S.H., Javanmard, M. & Vaccarino, F.J. (1997). A review of functional neuroimaging in mood disorders: positron emission tomography and depression. *Can. J .Psychiatry* ,42, 467-475.

Khan, Z.U., Koulen, P., Rubinstein, M., Grandy, D.K., Goldman-Rakic, P.S. (2001). An astroglia-linked dopamine D_2-receptor action in prefrontal cortex. *Proc. Natl. Acad. Sci. U S A.* 98, 1964-1969.

Kitayama, S., Ohtsuki, H., Morita, K., Dohi, T. & Tsujimoto, A. (1990). Bis-oxonol experiment on plasma membrane potentials of bovine adrenal chromaffin cells: depolarizing stimuli and their possible interaction. *Neurosci. Lett.*, 116, 275-279.

Layden, B., Fonseca, C.P., Minadeo, N., Abdullahi, H., Castro, M.M.C.A., Geraldes, C.F.G.C. & Mota de Freitas, D. (1999). Comparison of fluorescence, ^{31}P NMR and ^{7}Li NMR spectroscopic methods for the investigation of Li^+/Mg^{2+} competition in a model system and applications to cellular systems, In:: *Lithium - 50 years: recent advances in biology and medicine*, Lukas K.C., Becker R.W. & Gallichio V.S. (Eds.), pp. 45-62, Weidner Publishing, Cheshire, Connecticut, USA.

Layden, B., Diven, C., Minadeo, N., Bryant, F.B. & Mota de Freitas, D. (2000). Li^+/Mg^{2+} competition at therapeutic intracellular Li^+ levels in human neuroblastoma SH-SY5Y cells. *Bipolar Disorder*, 2, 200-204.

Layden, B.T., Abukhdeir, A.M., Williams, N., Fonseca, C.P., Carroll, L., Castro, M.M.C.A., Geraldes, C.F.G.C., Bryant, F.B. & Mota de Freitas, D. (2003). Effects of Li^+ transport and Li^+ immobilisation on Li^+/Mg^{2+} competition in cells: implications for bipolar disorder. *Biochem. Pharmacol* ., 66, 1915-1924.

Lenox, R.H. & Hahn, C.G. (2000). Overview of the mechanism of action of lithium in the brain: fifty-year update. *J. Clin. Psychiatry*, 61 Suppl 9, 5-15.

Lust, W.D., Schwartz, J.P. & Passonneau, J.V. (1975). Glycolytic metabolism in cultured cells of the nervous system. I. Glucose transport and metabolism in the C-6 glioma cell line. *Mol. Cell. Biochem.*, 8, 169-176.

Malloy, C.R., Sherry, A.D. & Jeffrey, F.M. (1988). Evaluation of carbon flux and substrate selection through alternate pathways involving the citric acid cycle of the heart by ^{13}C NMR spectroscopy. *J. Biol. Chem.*, 263, 6964-6971.

Manji, H.K., Potter, W.Z. & Lenox, R.H. (1995). Signal transduction pathways. Molecular targets for lithium's actions. *Arch. Gen. Psychiatry*, 52, 531-543.

Manji, H.K. & Lenox R. H. (2000a). The nature of bipolar disorder. *J. Clin. Psychiatry* 61 Supp 13, 42-57.

Manji, H.K. & Lenox R. H. (2000b). Signaling: cellular insights into the pathophysiology of bipolar disorder. *Biol. Psychiatry*, 48, 518-530.

Manji, H. K., Moore G. J. & Chen G. (2001). Bipolar disorder: leads from the molecular and cellular mechanisms of action of mood stabilisers. *Br. J. Psychiatry*, 178, S107-S119.

Marcus, S.R., Nadiger, H.A., Chandrakala, M.V., Rao T.I. & Sadasivudu, B. (1986). Acute and short-term effects of lithium on glutamate metabolism in rat brain. *Biochem. Pharmacol.*, 35, 365-369.

Masana, M.I., Bitran, J.A., Hsiao, J.K., Mefford, I.N., Potter, W.Z. (1991). Lithium effects on noradrenergic-linked adenylate cyclase activity in intact rat brain: an *in vivo* microdialysis study. *Brain Res.*, 538, 333-336.

Masana, M.I., Bitran, J.A., Hsiao, J.K., Potter, W.Z. (1992). *In vivo* evidence that lithium inactivates G_i modulation of adenylate cyclase in brain. *J Neurochem.,*, 59, 200-205.

Memo, M., Pizzi, M., Belloni, M., Benarese, M., Spano, P. (1992). Activation of dopamine D_2 receptors linked to voltage-sensitive potassium channels reduces forskolin-induced cyclic AMP formation in rat pituitary cells. *J. Neurochem.*, 59, 1829-1835.

Miles, K., Anthony, D. T., Rubin, L. L., Greengard, P. & Huganir R. L. (1987). Regulation of nicotinic acetylcholine receptor phosphorylation in rat myotubes by forskolin and cAMP. *Proc Natl Acad Sci U S A*, 84, 6591-6595.

Montezinho, L.P., Fonseca, C.P., Geraldes, C.F.G.C., Castro,M.M.C.A. (2002). Quantification and localization of intracellular free Mg^{2+} in bovine chromaffin cells. *Metal Based Drugs*, 9, 69-80.

Montezinho, L.P., Duarte C.B., Fonseca, C.P., Glinka, Y., Layden, B., Mota de Freitas, D., Geraldes, C.F.G.C., Castro, M.M.C.A. (2004). Intracellular lithium and cyclic AMP levels are mutually regulated in neuronal cells. *J Neurochem*, 90, 920-930.

Montezinho, L.P., Castro, M.M.C.A., Duarte C.B., Penschuck,S., Geraldes, C.F.G.C., Mørk, A. (2006). The interaction between dopamine D_2-like and beta-adrenergic receptors in the prefrontal cortex is altered by mood stabilizing agents. *J. Neurochem.*, 96, 1336-1348.

Montezinho, L.P., Mørk, A., Duarte C.B., Penschuck,S., Geraldes, C.F.G.C., Castro, M.M.C.A. (2007). Effect of mood stabilizers on dopamine D2-like receptors-mediated inhibition of adenylate cyclase. *Bipolar Disorders*, 9, 290-297.

Mørk, A. & Geisler, A. (1987). Mode of action of lithium on the catalytic unit of adenylate cyclase from rat brain. *Pharmacol. Toxicol.* 60, 241-248.

Mørk, A. & Geisler, A. (1989). The effects of lithium *in vitro* and *ex vivo* on adenylate cyclase in brain are exerted by distinct mechanisms. *Neuropharmacology*, 28, 307-311.

Mørk, A. & Geisler, A. (1994). Lithium *in situ* decreases extracellular levels of cyclic AMP in the dorsal hippocampus of living rats. *Pharmacol. Toxicol.*, 74, 300-302.

Mørk, A. & Jensen, J. (2000). Effects of lithium and other mood-stabilizing agents on the cyclic adenosine monophosphate signaling system in the brain., In: Bipolar medications. Mechanisms of action. Manji H, Bowden C, Belmaker R (Ed). Washington: American pychiatric press, Inc.,: pp 109-128.

Mota de Freitas, D., Amari, L., Srinivasan, C., Rong, Q., Ramasamy, R., Abraha, A., Geraldes, C.F.G.C. & Boyd, M.K. (1994). Competition between Li^+ and Mg^{2+} for the phosphate groups in the human erythrocyte membrane and ATP: an NMR and fluorescence study. *Biochemistry* , 33, 4101-4110.

Mota de Freitas, D., Castro, M.M.C.A. & Geraldes, C.F.G.C. (2006). Is Competition Between Li^+ and Mg^{2+} the Underlying Theme in the Proposed Mechanisms for the Pharmacological Action of Lithium Salts in Bipolar Disorder? *Acc. Chem. Res.*, 39, 283-291.

Newman M. E. & Belmaker R. H. (1987) Effects of lithium *in vitro* and *ex vivo* on components of the adenylate cyclase system in membranes from the cerebral cortex of the rat. *Neuropharmacology* 26, 211-217.

Nikolakopoulos, J., Zachariah, C., Mota de Freitas, D. & Geraldes, C.F.G.C. (1996). Comparison of the use of gel threads and microcarrier beads in Li^+ transport studies of human neuroblastoma SH-SY5Y cells. *Inorg. Chim. Acta,* 251, 201-205.

Nikolakopoulos, J., Zachariah, C., Mota de Freitas, D., Stubbs, E.B., Ramasamy, R., Castro, M.M.C.A. & Geraldes, C.F.G.C. (1998). ^7Li nuclear magnetic resonance study for the determination of Li^+ properties in neuroblastoma SH-SY5Y cells. *J .Neurochem.,* 71, 1676-1684.

Njus, D., Sehr, P.A., Radda, G.K., Ritchie, G.A. & Seeley P.J. (1978). Phosphorus-31 nuclear magnetic resonance studies of active proton translocation in chromaffin granules. *Biochemistry* , 17, 4337-4343.

Nordenberg, J., Kaplansky, M., Beery, E., Klein, S. & Beitner, R. (1982). Effects of lithium on the activities of phosphofructokinase and phosphoglucomutase and on glucose-1,6-diphosphate levels in rat muscles, brain and liver. *Biochem. Pharmacol.,* 31, 1025-1031.

O'Donnell, T., Rotzinger, S., Ulrich, M., Hanstock, C.C., Nakashima, T.T. & Silverstone, P.H. (2003). Effects of chronic lithium and sodium valproate on concentrations of brain amino acids. *Eur Neuropsychopharmacol,* 13, 220-227.

Otero Losada, M.E. & Rubio, M.C. (1986). Acute and chronic effects of lithium chloride on GABA-ergic function in the rat corpus striatum and frontal cerebral cortex. *Naunyn Schmiedebergs Arch. Pharmacol.,* 332, 169-172.

Painter, G.R., Diliberto, E.J., Jr. & Knoth, J. (1989). ^{31}P nuclear magnetic resonance study of the metabolic pools of adenosine triphosphate in cultured bovine adrenal medullary chromaffin cells. *Proc. Natl .Acad. Sci .U S A,* 86, 2239-2242.

Pellerin, L. & Magistretti, P.J. (2004). Neuroscience. Let there be (NADH) light. *Science,* 305, 50-52.

Petty, F. (1995). GABA and mood disorders: a brief review and hypothesis. *J. Affect. Disord.,* 34, 275-281.

Plenge, P. (1976). Acute lithium effects on rat brain glucose metabolism - *in vivo. Int. Pharmacopsychiatry,* 11, 84-92.

Post, R.M., Leverich, G.S., Rosuf,f A.S. (1990). Carbamazepine prophylaxis in refractory affective disorders: a focus on long-term follow-up. *J. Clin. Psychopharmacol.,* 10, 318–327.

Post, R.M., Weiss, S.R., Chuang, D.M. (1992). Mechanisms of action of anticonvulsants in affective disorders: comparisons with lithium. *J. Clin. Psychopharmacol.,* 12, 23S-35S.

Powis, D.A., O'Brien, K.J. & Von Grafenstein, H.R. (1991). Calcium export by sodium-calcium exchange in bovine chromaffin cells. *Cell Calcium,* 12, 493-504.

Rainbow, T.C., Parsons, B. & Wolfe, B.B. (1984). Quantitative autoradiography of beta 1- and beta 2-adrenergic receptors in rat brain. *Proc.Natl.Acad.Sci.U.S.A*, 81, 1585-1589.

Raju B., Murphy E., Levy L.A., Hall R.D. and London R.E. (1989) A fluorescent indicator for measuring cytosolic free magnesium. *Am J Physiol* 256, C540-C548.

Ramasamy, R. & Mota de Freitas, D. (1989). Competition between Li^+ and Mg^{2+} for ATP in human erythrocytes. A ^{31}P NMR and optical spectroscopy study. *FEBS Lett.*, 244, 223-226.

Rodrigues, T.B. & Cerdán, S. (2005). ^{13}C MRS: an outstanding tool for metabolic studies. *Concepts Magnet Reson*, 27A, 1–16.

Rong, Q., Espanol, M., Mota de Freitas, D. & Geraldes, C.F.G.C. (1993). ^{7}Li NMR relaxation study of Li^+ binding in human erythrocytes. *Biochemistry*, 32, 13490-13498.

Rong, Q., Mota de Freitas, D. & Geraldes, C.F.G.C. (1994). Competition between lithium and magnesium ions for the substrates of second messenger systems: a nuclear magnetic resonance study. *Lithium*, 5, 147-156.

Rubio, M.C. & Otero Losada, M.E. (1986). GABAergic responses to lithium chloride: dependence on dose, treatment length and experimental condition. *Adv. Biochem. Psychopharmacol.*, 42, 69-77.

Schatzberg, A.F. (2004). Employing pharmacologic treatment of bipolar disorder to greatest effect. *J.Clin.Psychiatry* 65, 15-20.

Sherry, A.D., Jeffrey, F.M. & Malloy C.R. (2004). Analytical solutions for ^{13}C isotopomer analysis of complex metabolic conditions: substrate oxidation, multiple pyruvate cycles, and gluconeogenesis. *Metab. Eng.*, 6, 12-24.

Shiah, I.S. & Yatham, L.N. (1998). GABA function in mood disorders: an update and critical review. *Life Sc.i* 63, 1289-1303.

Silverstone, P.H. & Silverstone, T. (2004). A review of acute treatments for bipolar depression. *Int.Clin.Psychopharmacol.*, 19, 113-124.

Silverstone, P. H., McGrath, B. M., Wessels, P. H., Bell, E. C. *et al.* (2005). Current pathophysiological findings in bipolar disorder and in its subtypes. *Curr. Psychiatric Reviews*, 1, 75-101.

Sonnewald, U. & Kondziella, D. (2003). Neuronal glial interaction in different neurological diseases studied by *ex vivo* ^{13}C NMR spectroscopy. *NMR Biomed.*, 16, 424-429.

Srinivasan, C., Minadeo, N., Geraldes, C.F.G.C. & Mota de Freitas, D. (1999). Competition between Li^+ and Mg^{2+} for red blood cell membrane phospholipids: A ^{31}P, ^{7}Li, and ^{6}Li nuclear magnetic resonance study. *Lipids*, 34, 1211-1221.

Stone, E. A. & John, S. M. (1990). *In vivo* measurement of extracellular cyclic AMP in the brain: use in studies of beta-adrenoceptor function in nonanesthetized rats. *J. Neurochem.* 55, 1942-1949.

Stone, E. A., Sessler, F. M. & Liu, W. M. (1990). Glial localization of adenylate-cyclase-coupled beta-adrenoceptors in rat forebrain slices. *Brain Res.*, 530, 295-300.

Trifaró, J.M. (1982). The cultured chromaffin cell: a model for the study of biology and pharmacology of paraneurones. *Trends Pharmacol. Sci.*, 3, 389-392.

Waniewski, R.A. & Martin, D.L. (1998). Preferential utilization of acetate by astrocytes is attributable to transport. *J. Neurosci.*, 18, 5225-5233.

Yatham, L.N. (2002). The role of novel antipsychotics in bipolar disorders. *J.Clin.Psychiatry*, 63, 10-14.

Zager, E.L. & Ames, A., III (1988). Reduction of cellular energy requirements. Screening for agents that may protect against CNS ischemia. *J. Neurosurg.*, 69, 568-579.

5

Memantine: A New Mood Stabilizer for Treatment-Resistant Bipolar Disorders

Gino Serra[1], Giulia Serra[2], Alexia E. Koukopoulos[3], Francesca Demontis[1] and Athanasio Koukopoulos[3]
[1]*Dipartimento di Scienze del Farmaco, University of Sassari,*
[2]*Ospedale S. Andrea, Dipartimento NESMOS, La Sapienza University, Roma,*
[3]*Centro Lucio Bini, Roma,*
Italy

1. Introduction

Memantine is a non-competitve NMDA receptor antagonist, but, at variance with the most potent NMDA receptor blockers, such as Ketamine, Phencyclidine and MK-801, has a low affinity for the receptor and its action is voltage/use dependent (Gillin et al., 2009; Johnson & Kotermanski, 2006; Rammes et al., 2008) . Moreover it has been recently demonstrated that this compound selectively blocks the extrasynaptic (excitotoxic) receptor but preserves the normal synaptic function (La Spada, 2009).

These peculiar pharmacological properties explain the lack of psychotomimetic/psychedelic effect and of interference with the normal physiological functions [memory and learning, synaptic plasticity, etc. etc. (Van Dongen, Editor, 2009)].

The drug has been on the market in Germany as Akatinol Memantine since 1982 for the treatment of Parkinsonism, cerebral and peripheral spasticity, and organic brain syndrome, without apparent cause for concern before its approval in 2002 and 2004 by EMEA and FDA for the treatment of moderate to severe Alzheimer's Disease.

Although its actual efficacy on the AD patient's quality of life has proven to be moderate (Emre et al., 2008; Kaduszkiewicz & Hoffmann, 2008), several pre-marketing and post-marketing studies have demonstrated the excellent safety and tolerability profile of the drug (Farlow & Phillips, 2008; Jones, 2010).

Moreover, the drug has been used off-label in a number of neurological and psychiatric conditions, including depression, with conflicting and inconclusive results (Zdanis & Tampi, 2008).

We have recently suggested, on the basis of preclinical experimental evidence obtained in our laboratory (Serra 2009, 2010) the use of memantine, as having an antimanic and mood-stabilizing effect, in treatment-resistant bipolar disorder. This use is absolutely at variance with the prevalent hypotheses which take NMDA receptor antagonists to be potential antidepressant drugs [mainly on the basis of their effect on the forced swimming test (Rogoz et al., 2002, Reus et al., 2010), a widely used animal model of depression].

In keeping with our experimental pharmacological evidence we have observed in bipolar patients a significant clinical antimanic and sustained mood-stabilizing effect of memantine, as augmenting agent, in treatment-resistant bipolar disorders.

Lithium is the drug of choice for the treatment of mania and the prophylaxis (as a mood stabilizer) of manic and/or depressive recurrences of Bipolar Disorders (BD) and Unipolar Depression (Goodwin, 2002). In cases of severe mania with a strong component of psychomotor agitation and/or psychotic symptoms, antipsychotic drugs are generally combined.

Anticonvulsants (carbamazepine, valproic acid, lamotrigine) are used [in mania and/or as mood stabilizers (Rapoport, 2009)] in combination with lithium when the latter does not produce a satisfactory response or in monotherapy in patients for whom lithium is contraindicated. The effectiveness of these drugs, both as antimanics and mood stabilizers, is however modest, and therefore patients who do not respond adequately to lithium (Koukopoulos et al. 1995, Tohen et al., 2005) today constitute the real problem of long-term treatment of mood disorders.

The use of antipsychotics as mood stabilizers is not advisable both because of their dubious efficacy (Culpepper & Ghaemi, 2011) and their long-term safety and/or tolerability.

Traditional antipsychotics, as well as having undesirable effects of a psychiatric type (emotional blunting, etc.), may, if administered for a long time, provoke sometimes debilitating and irreversible neurological damage (Serra & Gessa, 1990). The so-called atypical antipsychotics, on which the jury is still out on whether they do not cause long-term neurological damage (Gardener et al., 2005), have the drawback (except for some) of being able to cause what is called "metabolic syndrome" (Gardner et al., 2005) which constitutes a serious cardiovascular risk factor and thus risk of early death (which is aggravated by the fact that BD patients *per se* have an increased cardiovascular risk). Finally, it should be stressed that most antipsychotics, whether traditional or atypical, may cause, albeit rarely, death by sudden cardiac arrest (Ray et al., 2009).

Concern has also been raised by a recent alarm (FDA Alert, 2008) over the possibility that anti-epileptics may bring a two-fold increase in suicide risk compared to placebo.

Along with the approved old anticonvulsants a plethora of "new" antiepileptics [Gabapentin, Oxcarbazepine, Clonazepam, Pregabaline, Tiagabine, Topiramate, etc. (Kaufman, 2011)] are usually used off-label in combination with lithium, anticonvulsants and antipsychotics, when these associations fail to produce a satisfactory clinical response. Thus, the overwhelming majority of bipolar patients, today, receive a combination of 4 or 5 drugs or even more, often with questionable pharmacological rationale and little additional relevant clinical benefit.

The problem of the therapy and prophylaxis of mood disorders in patients who do not respond to lithium, whose numbers appear to be continually rising, therefore remains unsolved (Koukopoulos et al., 1995; Ghaemi, 2008).

Indeed, while the acute phases of the manic/hypomanic and depressive episodes of the illness appear to be relatively easy to control, the failure on the stabilization of treatment-resistant bipolar disorders is becoming an emergengy of psychiatry today and represents an increased risk of suicide.

Unfortunately, the currently available lithium-alternative mood-stabilizers not only are of limited efficacy, but have a number of acute troublesome and sometimes long-term severe side effects along with clinically relevant drug-drug interaction (Mc Namara, 2011).

Hence, the need for new more effective and safe mood-stabilizing drugs.

2. Treatment-resistant bipolar disorders

Mood disorders are currently classified (DSM IV TR) as Depressive Disorders and Bipolar Disorders. Depressive Disorders include Major Depressive Disorder (single or recurrent) and Dysthymic Disorder.

Bipolar Disorders comprise: BD I (which presents with an alternation of episodes of major depression and recurrent episodes of mania); BD II (which is made up of episodes of major depression and recurrent hypomanias); Cyclothymic Disorder (for at least two years several hypomanic and depressive episodes which must not be major). Further, a Mixed Episode is when symptoms of major depression and mania are present in the same episode.

The depression that manifests in BD is also commonly defined as Bipolar Depression, while that of Major Depressive Disorder is often called Unipolar Depression (or Unipolar Depressive Disorder).

Since affective disorders are often highly recurrent prophylaxis is of primary importance. Many cases, however, do not respond to current mood stabilizing treatments.

Cole et al. (1993) identified four different patterns of treatment-resistant Bipolar Disorder: rapid cycling (37%), other form of cycling (32%), chronic depression (26%) and mixed states (6%). They identified some risk factors for treatment resistant Bipolar Disorder, including female gender (for rapid cycling), high prevalence of family history of affective disorders (72%) and electroencephalografic abnormalities (54% of recordings).

Particularly resistant to prophylactic treatments are the patients with rapid cycling course.

Dunner and Fieve (1974) defined as Rapid Cyclers those patients that have a course with four or more affective episodes in the 12 previous months, and this course specifier was adopted by DSM-IV.

Today, most clinicians waive the duration criterion of episodes and intervals because in many cases the episodes become progressively shorter and we see ultra-rapid (within the course of weeks or days) and ultradian cyclers (mood shifts within 24 hours). Some patients indeed show an alternation of phase within a few hours.

The general impression of clinicians today is that the course of recurrences of manic-depressive illness has substantially changed in the last 40 years. The recurrences of many patients have become more frequent. One sees more manias and hypomanias and therefore more bipolar cases than before, more rapid cyclers, and more chronic depressions (Koukopoulos et al 1983; Ghaemi, 2008). This phenomenon is called today mood destabilization.

In his monograph on *la folie a double forme,* Ritti (1883) presented only 17 cases with a rapid course; he collected them from various French and German authors. Kraepelin (1913) illustrated one case with a rapid cycling course in his graphs of type of course. This was case C, and he commented, "I had to seek for a long time in my cases until I finally found at least one course of the type that case C represents.' In a meta-analysis of 8 studies including 2054 bipolar patients, Kupka et al. (2003) found a percentage of 16.3% of rapid cyclers. Women and bipolar II patients were slightly but significantly more prevalent. Maj et. al (1994) found a prevalence of 19.5% of rapid cyclers among BPI patients.

Among the first 500 BP patients of the STEP-BD study 20% were rapid cyclers while in 2008 among 1,742 BD patients the percentage at entry was 32% (Schneck et al, 2008).

Among women, rapid cycling is more frequent than among men and the age of onset of the bipolar disorder is earlier than among non-rapid cycling patients (Yildiz et al, 2004).

Particularly frequent is rapid cycling among prepubertal and early adolescent bipolar patients. Geller at al. (1998) found a proportion of 83.3% rapid, ultra-rapid or ultradian cyclers among such young patients.

Rapid cycling is more frequent among BPII than among BPI and in patients with hyperthymic and cyclothymic temperament. They are very energetic, very emotional and reactive (Kukopoulos et al., 1983). One could posit the hypothesis that they are equally reactive to chemical stimulations. Because of this temperament they are diagnosed often as borderline personality disorders and in the past as hysterics. In patients with cyclothymic temperament the rapid-cycling course could be viewed as the accentuation of longstanding subclinical mood oscillations.

Patients that more likely remain rapid-cyclers for many years are those with a DMI (depression-mania-interval) or DmI (depression-hypomania-interval) cycle patterns, those with a switch process and/or agitated depression in their course and those who have not been stabilized after the first year of an adequate treatment.

The investigation of the course of manic-depressive illness is a difficult task in itself, and the investigation of possible factors that influence this course is extremely difficult given the variability of the spontaneous course. Yet the above-mentioned changes and the general increase of bipolar cases today make this investigation necessary. Of all the possible factors that unfavorably influence the course of the disease what most urgently needs to be studied are the treatments themselves, firstly because they certainly influence the disease, and second, because treatments are given by the doctor and therefore can be easily modified. This is certainly not the case of other factors like menopause, older age, life situations, and so forth.

The increase of substance abuse, including alcohol and cannabis among young people certainly plays an important role in the worsening of the course of affective disorder.

Antidepressant treatments should also be reconsidered. They have been largely employed in the treatment of affective disorders during the last five decades, although many authors observed an increase of the frequency of episodes compared to the course of the disease before the introduction of antidepressants (Arnold & Kryspin Exner, 1965; Freyhan, 1960; Heinz & Grunze, 2008; Hoheisel, 1966; Lauber, 1964; Till, 1970).

Particularly associated to the use of antidepressants is the rapid cycling course (Kukopulos et al., 1983). The stabilization of rapid cyclers is often very difficult. It requires prolonged effort: it is necessary to know the life of the patient, his premorbid temperament and personality, the history of his illness and its course, the treatments used in the past, and the events and treatments associated with the onset of rapid cycling.

Patients with rapid cycling course are usually of cyclothymic or hyperthymic temperament, therefore particularly excitable and emotional persons. It is important to advise them to avoid coffee, tea, alcohol and all psychostimulant substances, intense physical exercise and also to avoid, as much as possible, stressful situations and events. Sleep is particularly important and they should go to sleep early and try to sleep as long as possible.

The resistance of rapid cyclers to all mood-stabilizing treatments is well established. None of these treatments (carbamazepine, lithium, lamotrigine, topiramate and valproate) showed a clear advantage over the others (Baldessarini et al, 2000; Tondo et al, 2003). Also the efficacy of atypical antipsychotics as mood stabilazers in maintenance therapy has not been demonstrated, and current data do not support their use as maintenance agents (Zupancic , 2011). The authors, however, agree with Muzina (2009) about the fundamental importance of lithium in the treatment of rapid cyclers.

We apply the following strategy. Given that most cases of rapid cycling are induced by antidepressants, it is of primary importance to suspend the antidepressants and to continue the treatment with mood stabilizing agents. In most cases without antidepressants the depressive phase will last longer but the following mania/hypomania will be less intense. The anti-manic action of mood stabilizers will be more effective in the absence of antidepressants. Gradually, both the excitatory and the depressive phase will be attenuated and stabilization will eventually be achieved, over a variable period of time. This therapeutic strategy is based on the idea that the suppression of the excitatory phase prevents or attenuates the following depression (Koukopoulos & Ghaemi, 2009). Indeed, all mood stabilizers like lithium, anticonvulsants, calcium antagonists, old or new antipsychotics (Koukopoulos & Ghaemi, 2009), and also memantine (Koukopoulos et al., 2010) are essentially anti-excitatory agents .

In the presence of suicide risk, electroconvulsive therapy (ECT) could be used to end the depressive phase and immediately afterwards mood-stabilizing treatment should be administered. A study from Sainte-Anne hospital in Paris shows that maintenance ECT is effective against the RC course, with full or partial remission for 100% of RC patients (Vanelle et al, 1994). Other authors have shown that maintenance ETC is effective against rapid cycling course: Minnai et al. (2011) found a percentage of remission of 58% of patients, and other similar results were obtained by Fazzari (2009, *personal comunication*). Continuous treatment was more effective against mania/hypomania than against depression, yet in all persisting rapid cycler cases the mania/hypomania remitted only partially.

3. Memantine as antimanic and mood-stabilizer: Pharmacological rationale

3.1 Dopamine and bipolar disorders

The first evidence-based hypothesis of the neurobiology of depression was proposed by Schildkraut in 1965 (Schildkraut, 1965). Mainly on the basis of the supposed mechanism of therapeutic effect of tricyclic antidepressants and MAO inhibitors (blockade of serotonin and noradrenaline reuptake or inhibition of MAO, respectively) he suggested that depression may be associated with a decreased function of serotonin and noradrenaline transmission. Further support for this hypothesis is provided by the observation that reserpine, which induces depression in humans, causes a depletion of the neurotransmitters in monoaminergic neurons.

As a consequence of this hypothesis the role of dopamine (DA) in the pathophysiology of mood disorders has long been neglected although reserpine also depletes dopaminergic neurons.

In 1975 Randrup et al (1975) suggested that mania might be associated with increased dopaminergic transmission, while depression could be associated with decreased DA transmission.

In 1979 (Serra et al, 1979) we first reported that antidepressant drugs act not only on serotonin and noradrenaline, but also activate dopaminergic transmission. In fact we found that chronic treatment with antidepressants, by inducing a dopamine autoreceptor subsensitivity, potentiates dopaminergic transmission in rats, and suggested that this effect may play an important role both in the therapeutic action and in the capacity of antidepressants to induce mania/hypomania.

Since then, in the last three decades, an amount of preclinical and clinical evidence has been accumulated strongly suggesting a key role of dopamine in the pathophysiology of bipolar disorders (Berk et al, 2007; Cousins & Butts, 2009; Dihel & Gerson, 1992).

A detailed description of this clinical and preclinical evidence supporting a key role of dopamine in the pathophysiology of bipolar disorders is beyond the aim of this chapter: the reader may find such a description in a number of excellent recent reviews (Berk et al, 2007; Cousins & Butts, 2009; Dihel & Gerson, 1992; Dunlop & Nemeroff, 2007).

3.2 Antidepressants sensitize mesolimbic dopamine D2 receptors

In our first report (Serra et al, 1979) we observed that chronic treatment with antidepressants reduces the sedative effect of apomorphine and potentiates its stimulatory action on locomotor activity, thus potentiating dopamine transmission. We interpreted these effects as a consequence of a development of sedative dopamine autoreceptor subsensitivity, and suggested that the potentiation of dopamine transmission may play an important role in the capacity of these drugs both in therapeutic action and in inducing switching from depression to mania/hypomania.

Subsequent studies in the following 30 years have produced conflicting results (Chiodo & Antelman, 1980a, 1980b; Diggory & Buckett, 1984; Dziedzicka-Wasylewska, 1997; Holcomb et al, 1982; Muscat et al., 1988; Serra et al, 1980, 1981a, 1981b; Spyraki & Fibiger, 1981) on the ability of ADs to induce subsensitivity in DA autoreceptors.

However numerous studies have confirmed the capacity of virtually all AD treatments (TCAs, MAO-inhibitors, Mianserine, SSRI/NSRI, electroconvulsive shock, REM-sleep deprivation), to increase the motor-stimulant effect of DA receptor agonists (Collu et al., 1997a; D'Aquila et al., 2000a; Gershon et al., 2007; Spyraki & Fibiger, 1981; Serra et al., 1990, 1991, 1992; Willner, 1997). In particular, Spyraki and Fibiger in 1981 (Spyraki & Fibiger, 1981) observed that chronic Desipramine treatment potentiated dopaminergic transmission by inducing a supersensitivity of postsynaptic dopamine receptors in the mesolimbic system.

Strong evidence now exists suggesting a key role of mesolimbic DA in the mechanism of action of antidepressants (Collu et al., 1997b; D'Aquila et al, 2000a; Fibiger & Philips, 1981; Gershon et al, 2007; Spyraki & Fibiger, 1981; Serra et al, 1990, 1992; Willner, 1997)

Gerson et al. (2007) recently published an elegant review on the experimental and clinical evidence indicating the role of dopamine D2 receptors in the mechanism of action of antidepressants.

In 1990 the availability of selective agonists for DA receptor subtypes prompted us to re-evaluate the effect of chronic antidepressant treatments on pre- and post-synaptic dopamine receptor sensitivity (Serra et al, 1990). We found that chronic antidepressant treatment does not induce a subsensitivity of dopamine autoreceptors, but, confirming the results of Spiraki and Fibiger (1981), we observed that such a treatment sensitizes dopamine D2, but not D1, receptors selectively in the mesolimbic system The key role of mesolimbic DA D2 receptor sensitization in the mechanism of action of antidepressants is now widely accepted (Collu et al., 1997a; D'Aquila et al, 2000a; Fibiger & Philips, 1981; Gershon et al, 2007; Spyraki & Fibiger, 1981; Serra et al, 1990, 1991, 1992; Willner, 1997).

In keeping with these observations it may be suggested that the increased dopaminergic transmission in the mesolimbic system (the reward system) due to D2 receptor sensitization, induced by antidepressants, may contribute to their therapeutic effect, and in particular for such symptoms as anhedonia, loss of motivation, decreased libido and psychomotor retardation. (Collu, 1997b; D'Aquila, 2000a; Serra et al. 1990, 1992).

Moreover, the sensitization of mesolimbic dopamine D2 receptors induced by antidepressants may be responsible, in "vulnerable subjects" (bipolar disorder, the presence of previous mixed states, early age at onset, a cyclothymic or hyperthymic temperament, genetic factors?), for the switches from depression to mania/hypomania (Collu et al, 1997b; D'Aquila e al, 2000a; Gessa et al, 1995; Serra et al, 1990, 1992, Serra & D'Aquila, 2008; Serra, 2009, 2010) which, in turn, trigger a rapid-cycling course (Serra et al, 2008; Serra, 2009, 2010).

3.3 Antidepressants induce a "bipolar-like behaviour"

Koukopoulos et al. (1980) suggested that "The intensification of an underlying hypomanic process by antidepressants would precipitate another depression and establish continuous circularity".

In accordance with this hypothesis we recently found that chronic treatment with imipramine induced a 'bipolar-like behaviour' (i.e. a cycle of mania-depression) in rats (D'Aquila et al, 2003; D'Aquila et al, 2004). In fact, as expected, imipramine induces a sensitization of dopamine D2 receptors (mania/hypomania), which is followed after 12, 33 and 40 days of imipramine withdrawal by a progressive desensitization of dopamine D2 receptors (depression) (D'Aquila et al, 2003) associated with a depressive-like behaviour as assessed in the forced swimming test animal model of depression (D'Aquila et al, 2004).

This observation provides strong experimental support for the hypothesis that antidepressant-induced mania/hypomania is the trigger phenomenon of rapid-cycling course (Collu et al, 1997b; Serra & D'Aquila, 2008; Serra, 2009, 2010) and that the prevention of both spontaneous or antidepressant-induced mania/hypomania (Koukopoulos & Reginaldi, 1973; Koukopoulos & Ghaemi, 2009) (in neurobiological terms, dopamine D2 receptor sensitization) is essential to avoid the development of a rapid-cycling bipolar disorder.

The prevention of mania, whether induced by antidepressants or spontaneous, is the essential element in the therapy and prophylaxis of bipolar disorders (Koukopoulos & Reginaldi, 1973; Koukopoulos & Ghaemi, 2009). When, in fact, treatments currently in use do not achieve this aim the course of the disorder becomes 'malign', i.e refractory to treatment, as in the rapid-cycling course.

According with the clinical observations that demonstrate the ineffectiveness of currently used mood stabilizers in preventing antidepressant-induced switch from depression to mania (Leverich et al., 2006; Tondo et al., 1981, 2010), we found that the concomitant administration of lithium (D'Aquila et al, 2000b), carbamazepine (D'Aquila et al., 2001), valproate (D'Aquila et al 2006), lamotrigine (unpublished results) with imipramine fails to prevent the development of dopamine D2 receptor sensitization. Actually, carbamazepine seems to be effective, but its effect is due to the reduction of imipramine plasma levels due to the induction of the drug metabolism.

3.4 Memantine prevents the "bipolar-like behaviour" induced by antidepressants

These observations led us to further investigate the mechanism by which antidepressants induce dopamine receptor sensitization. In fact, this mechanism may represent a possible target to develop drugs effective in preventing this phenomenon as potential new antimanic and mood stabilizing agents.

There is ample experimental evidence showing that the NMDA glutamate receptor plays an essential role in the phenomenon of sensitization. Its stimulation is, in fact, necessary for the sensitization of amphetamine (Battisti et al., 2000; Groning et al., 2004; Ohmori et al, 1994; Pacchioni et al, 2002; Vezina & Quen, 2000; Wolf et al, 1994), methylphenidate (Gaytan et al, 2000), cocaine (Heusner & Palmiter 2005; Li et al, 2000; Kim et al, 1996; Rompré & Bauco 2006), apomorphine (Acerbo et al, 2004; Pacchioni et al, 2002; Voikar et al, 1999) and other dopamine mimetics (Kalivas, 1995; Rockhold, 1998), nicotine (Kelsey et al, 2002), morphine (Jeziorski et al, 1994; Trujillo, 2002) and ethanol (Broadbent & Weiemier, 1999; Camarini et al, 2000; Kotlinska et al, 2006), as well as several types of stress such as, for instance, 'restraint stress' (Pacchioni et al, 2002) and 'social defeat stress' (Yap et al, 2005).

The stimulation of NMDA receptors is required for the development of dopamine receptor sensitization induced by antidepressants. Indeed, we found that the administration of MK-801, a selective non-competitive NMDA receptor blocker, completely prevents the dopamine receptor sensitization induced by imipramine (D'Aquila et al, 1992) and by electroconvulsive shock (D'Aquila et al, 1997).

These observations strongly suggest that the non-competitive blockade of NMDA receptors should result in an anti-manic and mood stabilizing action, and that it should also be effective in the treatment of the disorders resistant to currently used antimanic and mood stabilizers.

Consistent with this hypothesis, we have recently found that memantine prevents both the up-regulation induced by chronic imipramine (Malesa & Serra, 2011) and the down-regulation (Demontis & Serra, 2011) of dopamine D2 receptors associated with a depressive-like behaviour (Cubeddu & Serra, 2011) observed after imipramine withdrawal.

These observations provide strong experimental evidence supporting the hypothesis of the antimanic and prophylactic effect of memantine in bipolar disorders resistant to conventional treatments.

Moreover, an antimanic-like activity of memantine has been observed in other animal models of mania by Gao et al.,(2011).

These observations prompted us to suggest the use of memantine, the only safe and well-tolerated non-competitive NMDA antagonist that may be proposed for long-term clinical use, as antimanic and mood-stabilizer in treatment-resistant bipolar disorder (Serra, 2009, 2010).

In addition, it may be interesting to recall that memantine, like the "gold standard" antimanic and mood-stabilizer, lithium, seems to posses a powerful neuroprotective activity (La Spada, 2009), an effect that may contribute to its mood-stabilizing properties. Indeed, in accordance with the neurotrophic hypothesis of mood-disorders (Dumas, 2004; Serra & Fratta, 2007), an excessive glutamatergic stimulation of NMDA receptors [that seems to be associated with mania (Ongur et al., 2009)]could result in a neurodegeneration that appears to be associated with depression (Dumas, 2004, Macqueen, 2003, Videbech & Ravnkilde, 2004) and can be reversed by an effective antidepressant treatment (Sheline et al., 2003; Malberg, 2004, Malberg & Blendy, 2005; Paizanis et al., 2007).

Thus, it may be suggested that memantine, by blocking NMDA receptors, prevents both the up (mania) and down (depression) regulation of dopamine D2 receptors, and the neurodegeneration that results from the excessive glutamatergic neurotransmission during mania, which might underlie the following depressive phase.

4. Memantine, a drug with excellent safety and tolerability

4.1 Pre-marketing data

"This module reflects the initial scientific discussion for the approval of Ebixa (EMEA, 2004). The product has been on the market for nearly twenty years in a European country without apparent cause for concern, which can be considered as giving some reassurance. In addition there has been clinical exposure in the older clinical trials.

The most frequent adverse events reported with memantine have been dizziness, followed by headache and fatigue. Agitation occurred less with memantine than with placebo. There is no suggestion of a psychedelic effect that could be feared as a result of activation of the NMDA receptors. Even if the target population would have had difficulties in reporting this kind of effects the fact that the levels of agitation were decreased is in favour of absence of such theoretical psychedelic effects. Despite the absence of studies formally addressing the question of withdrawal and dependence, there are no signals in the data available suggesting its existence. Taking into account the indication granted, the clinical evidence available gives reassurance of a sufficient safety profile."

4.2 Post-marketing data

A recent review (Jones, 2010) of the most recent safety/tolerability data for memantine (derived from meta-analyses, pooled analyses, European SPCs, and EMEA publications) confirmed that memantine has a favorable tolerability profile when used in monotherapy or in combination with other agents. Moreover, results of studies of a total treatment period of up to two years show that memantine is safe and tolerated, with an adverse event profile almost indistinguable from that of placebo. Side effects are usually mild to moderate in severity, and are commonly (1-10%) represented by dizziness, constipation, headache, hypertension and somnolence.

The incidence of serious adverse events (SAEs) was slightly lower for memantine than for placebo.

Warnings and precautions are few: caution is recommended in epileptic patients or with a former history of convulsions or predisposing factors for epilepsy. Close supervision is recommended in patients with myocardial infarction, uncompensated congestive heart failure or uncontrolled hypertension.

There are no contraindications for the use of memantine, apart from the sensitivity to tablet excipients. However, due to the lack of clinical experience, the drug should be avoided in patients with severe hepatic impairment.

Drug-drug interactions: memantine should not be administered alongside other compounds acting at NMDA receptors such as amantadine, ketamine and dextromethorphan, due to the risk of psychotic symptoms. Moreover memantine might enhance the effects of antiparkinson drugs such as levodopa, dopamine agonists and anticholinergic compounds. On the contrary, the effects of neuroleptics and barbiturates might be reduced. Finally, memantine may also influence the effect of baclofen and dantrolene (antispamodic agents).

5. Memantine as antimanic and mood stabilizer in treatment-resistant bipolar disorder: Clinical studies

We have recently carried out 3 naturalistic clinical studies in order to evaluate the antimanic and mood stabilizing effects of memantine, as augmenting agent, in treatment-resistant bipolar disorder.

In the first study (Koukopoulos et al, 2010) we administered memantine (10-30 mg/day), as augmenting agent, to 18 treatment-resistant bipolar patients monitored for 24 weeks. The severity of the patients' condition before memantine and the change after memantine augmentation was evaluated on the Clinical Global Impression-Bipolar (Spearing et al., 1997) Overall Bipolar Illness scale.

The patients had been ill for an average of 21 years and had been resistant to very intense standard treatments (lithium, anticonvulsants, typical and atypical antipsychotics, electroconvulsive therapy, and antidepressants). Of these 18 patients, 13 were bipolar I and 5 bipolar II, 10 were rapid cyclers, 5 were continuous circular with long cycles, and 3 had a course with free intervals. Thirteen patients exhibited psychotic symptoms. The 10 rapid cyclers had a mean duration of rapid cycling course of 11 years.

The average of CGI-BP score before memantine was 6.6, which indicates a very severe condition. After 24 weeks of memantine addition 72.2% of patients were very much or much improved. Among the rapid cyclers 60% reached stability. The mean time to improvement was 55 days.

The second study (Koukopoulos et al, 2011) encompassed 40 treatment-resistant bipolar patients monitored for 12 months. Of these 40 patients 21 were bipolar I and 19 bipolar II, 19 were rapid cyclers, 9 were continuous circular with long cycles, and 12 had a course with free intervals. Nineteen exhibited psychotic symptoms. All patients had been resistant for many years to very intense long-term standard treament. The mean duration of illness was 22 years, while the average duration of rapid cycling course of rapid cyclers was 8.6 years.

The average of CGI-BP Overall Bipolar Illness score before the memantine addition was 6.7, indicating a very severe condition.

After 6 months of memantine augmentation (10-30 mg/day added to ongoing treatment which was left unmodified) 72.5% of patients were very much or much improved and remained stabilized for 12 months. Among the rapid cyclers 68.5% reached stability after 6 months and remained free of recurrences for 12 months.

Finally, in order to evaluate the long-term effect of memantine, we studied the action of the drug on 22 treatment-resistant bipolar patients with a follow-up of 24 months (Serra, 2011). Of these 22 patients 16 were bipolar I, 6 bipolar II, 10 were rapid cyclers, 6 continuous circular with long cycles, and 6 had a course with free intervals, 15 exhibited psychotic symptoms. Almost all patients were very severely ill (average CGI-BP Overall Bipolar Illness score before memantine addition 6.6). All patients had been resistant for many years to very intense standard treatments (including lithium, anticonvulsants, typical and atypical antipsychotics, electroconvulsive therapy, antidepressants).

The mean duration of illness was 22.4 years, and the duration of rapid cycling course of rapid cyclers was 11 years. After 6 months of memantine augmentation 77.3% were very much or much improved and 60% of rapid cyclers reached stability. The mean time to improvement was 69 days. Similar results were obtained on the evaluation at 12, 18, and 24 months.

All patients who were very much or much improved at 6 months and continued on memantine remained free of recurrences at 12 and 18 months, and all but one at 24 months.

The side effects observed during our studies are the already described dizziness (one patient), constipation (one patient) and drowsiness (one patient), thus confirming the excellent safety and tolerability profile of the drug, also when used in combination with other drugs currently used in the treatment of bipolar disorders.

Our results strongly suggest that memantine, as augmenting agent, was associated with clinically substantial antimanic and sustained mood-stabilizing effects in treatment-resistant bipolar disorder patients with excellent safety and tolerability.

The significant clinical relevance of our observations is not diminished by the use of memantine as augmentation treatment, considering the long history of drug-resistance of the study patients.

As in all naturalistic studies, the limitation of our clinical observations is the lack of a placebo control. Hovewer, the prior history of treatment resistance of the study patients and the long-lasting effect of the drug argues against a placebo effect contributing in any substantial way to what we observed. Nevertheless, we are going to start an RCT to confirm our naturalistic observations.

Consistent with our results is the observation by Keck et al (2009) of an antimanic effect of memantine monotherapy in bipolar disorder patients.

In addition, our results are consistent with a recent analysis on the effect of memantine on the management of behavioural disorders in AD patients (Gauthier et al., 2010) which suggest an antimanic and mood-stabilizing effect of memantine.

In fact, the analysis demonstrates that memantine reduces "manic" symptoms such as agitation, aggression, irritability, lability, and even delusions and hallucinations in AD patients. Moreover, suggestive of a prophylactic/ mood-stabilizing effect, the observation that patients who do not exhibit such symptoms at baseline, showed a reduction of their emergence.

Taken together these clinical observations and considering the safety/tolerability and the drug-drug interaction of the drug, we are tempted to suggest, pending the results of our controlled clinical trial, the use of memantine, as augmenting agent, in treatment-resistant bipolar disorders, which have no therapeutic alternative

Discontinuation of long-term lithium treatment leads to early and severe affective recurrences (Baldessarini & Tondo, 1998), which are often resistant to other mood stabilizers. In order to evaluate the effect of memantine in this condition we administered the drug to three patients who are lithium responders, but discontunued lithium because of severe renal complications (two patients) or excessive tremor (one patients).

Case 1) A woman born in 1930 suffered of BD II with rapid cycling course. She was perfectly stable on Lithium and Valproic acid since 1980. In june 2009 Lithium was withdrawn because of renal impairment. She was put on 20 mg/day Memantine. In Dec.2009 she had a depressive recurrence treated with six ECT and in May 2010 she had a second depressive recurrence treated with thirteen ECT. Since then she is well and stable on Memantine 20 mg/day and Valproic acid 600mg.

Case 2) A woman born in 1934, suffering from BD II, was stable on lithium since 2001 In Nov. 2009 lithium was withdrawn because of renal impairment. A hypomanic and a depressive relapse followed. In june 2010 Memantine 20 mg/day was added to Valproic acid. She is still having mood oscillations but much milder than those she had before Lithium and before Memantine.

Case 3) A woman born in 1937 and suffering from BD II was started on Lithium and Valproic acid on June 2009. On May 2010 Memantine 20 mg/day were added and her mood oscillations became milder. On March 2011 Lithium was withdrawn because of tremor. A short depressive phase followed and for the moment she is well.

These observations suggest that memantine could stabilize the course of bipolar disorder in patients who discontinued long-term lihium treatment.

However, further studies in a large population sample are needed, before suggesting the use of memantine to prevent recurrences due to lithium discontinuation.

6. Conclusions

Memantine is a non-competitive NMDA receptor antagonist, with peculiar pharmacological properties, which explain its excellent safety and tolerability profile, at variance with the most potent NMDA receptor blockers such as Ketamine, Phencyclidine and MK-801.

The drug has been used in Germany since 1982, and approved by EMEA in 2002 and by FDA in 2004 for the treatment of moderate to severe Alzheimer's Disease.

Moreover it has been recently used off-label in a number of neurological and psychiatric disorders, including depression, with conflicting and inconclusive results.

At variance with the hypothesis that, mainly on the basis of the effect of these compounds in the forced swimming test animal model of depression, suggests that NMDA receptors antagonist might have an antidepressant activity, we have recently demonstrated that MK-801 and Memantine show an antimanic and mood-stabilizing-like effect in an animal model of bipolar disorder resistant to standard treatments.

Thus, we have suggested (Serra, 2009, 2010) the use of Memantine as antimanic and mood-stabilizing agent in the treatment of resistant bipolar disorders.

This hypothesis prompted us to carry out three naturalistic trials to test memantine as augmenting agent, in the management of treatment resistant bipolar disorders.

The results of these studies strongly suggest that memantine has a clinically relevant antimanic effect and a long-term prophylactic action in treatment of resistant and very severe bipolar patients, with very good safety and tolerability.

Although our naturalistic observations lack placebo control and need to be confirmed by an RCT (which we are going to start), we believe that they provide enough information to suggest the safe use of memantine augmentation in severely ill bipolar patients, who are resistant to conventional treatments and have not therapeutic alternative.

7. Acknowledgements

The authors would like to thank Denis Greenan for his precious contribution to the drafting of the paper. This work was funded by a grant of Fondazione Banco di Sardegna

8. References

Acerbo, M.J., Lee, J.M., Delius, J.D. (2004). Sensitization to apomorphine, effects of dizocilpine NMDA receptor blockades. *Behav Brain Res.* May 5; 151(1-2):201- 208.

Arnold, O.H., Kryspin-Exner, K. (1965). Zur Frage der Beeinflussung des Venlaufes des manish-depressiven Krankheitsgeschehens durch Antidepressiva. *Wiener Medizinische Wochenschrift* 45/46:929-934,1965.

Baldessarini R.J., Tondo L. (1998). Reccurrence risk in bipolar manic depressive disorders after discontinuig lithium maintenace treatment: an overview. Clin Drug Investig. 15 (4): 337-351.

Baldessarini, R.J., Tondo, L., Floris, G., Hennen, J. (2000). Effects of rapid cycling on response to Lithium maintenance treatment in 360 bipolar I and 11 disorder patients. *J. Affect. Disord.* 61, 13—22.

Baldessarini, R.J. in Goodman & Gilman's. *The Pharmacological Basis of Therapeutics 11 Edition 2005;* 429-500.

Battisti, J.J., Shreffler, C.B., Uretsky, N.J., Wallace, L.J. (2000). NMDA antagonists block expression of sensitization of amphetamine- and apomorphine-induced stereotypy. *Pharmacol Biochem Behav.* Oct; 67(2):241-246.

Berk, M., Dodd, S., Kauer-Sant'anna, M., Malhi, G.S., Bourin, M., Kapczinski, F., Norman, T. (2007). Dopamine dysregulation syndrome: implications for a dopamine hypothesis of bipolar disorder. *Acta Psychiatr Scand Suppl.* (434):41-49.

Broadbent, J., Weitemier, A.Z. (1999). Dizocilpine (MK-801) prevents the development of sensitization to ethanol in DBA/2J mice. *Alcohol Alcohol.* May-Jun; 34(3):283-288.

Camarini, R., Frussa-Filho, R., Monteiro, M.G., Calil, H.M. (2000). MK-801 blocks the development of behavioral sensitization to the ethanol. *Alcohol Clin Exp Res.* Mar; 24(3):285-290.

Chiodo, L.A., Antelman, S.M. (1980). Electroconvulsive shock: progressive dopamine autoreceptor subsensitivity independent of repeated treatment. *Science.* Nov 14;210(4471):799-801

Chiodo, L.A., Antelman, S.M. (1980). Repeated tricyclics induce a progressive dopamine autoreceptor subsensitivity independent of daily drug treatment. *Nature.* Oct 2;287(5781):451-454.

Chiodo, L.A., Antelman, S.M. (1980). Tricyclic antidepressants induce subsensitivity of presynaptic dopamine autoreceptors. *Eur J Pharmacol.* Jun 13;64(2- 3):203-204.

Cole AJ, Scott J, Ferrier IN, Eccleston D. (1993). Patterns of treatment resistance in bipolar affective disorder. *Acta Psychiatr Scand.* 1993 Aug;88(2):121-3.

Collu, M., Poggiu, A.S., Devoto, P.,Serra, G. (1997a). Behavioural sensitization of D2 mesolimbic dopamine receptors in chronic fluoxetine treated rats. *Eur. J. Pharm.* 322, 123-127.

Collu, M., D'Aquila, P., Gessa, G.L., Serra, G. (1997b). Do antidepressant treatments induce mania by activating dopaminergic trasmission? *Second International Conference on Bipolar Disorders,* Pittsburg, USA, June 19- 21.

Cubeddu A., Serra G. La memantina previene la "depressione" da interruzione del trattamento con imipramina. . Graduate thesis on Pharmacy, University of Sassari AA 2010-2011

Culpepper, L., Ghaemi, N. (2011) Are antipsychotics overprescribed? *Medscape Psychiatric & Mental Illness.* Posted: 02/18/2011

Cuosins, D.A., Butts, K. (2009). Young AH The role of dopamine in bipolar disorders. *Bipolar Disord,* 11: 787-806.

D'Aquila, P..S., Sias, A., Gessa, G.L., Serra, G. (1992). The NMDA receptor antagonist MK-801 prevents imipramine-induced supersensitivity to quinpirole. *Eur. J. Pharmacol.* 224: 199-202.

D'Aquila, P.S., Collu, M., Gessa, G.L., Serra, G. (1997). Dizolcipine prevents the enhanced locomotor response to quinpirole induced by repeated electroconvulsive shock. *Eur. J. Pharm.* 330, 11-14.

D'Aquila PS, Collu M, Gessa GL, Serra G. The role of dopamine in the mechanism of action of antidepressant drugs. *Eur J Pharmacol.* 2000a Sep 29;405(1-3):365-73.

D'Aquila P.S., Collu M., Devoto P., Serra G (2000b): Chronic lithium chloride fails to prevent imipramine-induced sensitization to the dopamine-D2- like receptor agonist quinpirole. *Eur Journ Pharmacol,* 395: 157-160.

D'Aquila, P.S., Peana, A.T., Tanda, O., Serra, G. (2001): Carbamazepine prevents imipramine-induced behavioural sensitization to the D2-like dopamine receptor agonist quinpirole. *Eur Journ Pharmacol* 416: 107-111.

D'Aquila, P.S., Peana, A.T., Panin, F., Grixoni, C., Cossu, M., Serra, G. (2003). Reversal of antidepressant-induced dopaminergic behavioural supersensitivity after long-term chronic imipramine withdrawal. *Eur Journ Pharmacol.* 458, 129- 134.

D'Aquila, P.S., Panin, F., Serra, G. (2004). Long-term imipramine withdrawal induces a depressive-like behaviour in the forced swimming test. *Eur Journ Pharmacol.* 492, 61-63.

D'Aquila, P.S., Panin, F., Serra, G. (2006). Chronic valproate fails to prevent imipramine-induced behavioural sensitization to the dopamine D2- like receptor agonist quinpirole. *Eur Journ Pharmacol.* 535, 208-212.

Demontis F., Serra G. Effetto stabilizzante dell'umoe della Memantina: evidenze sperimentali. Graduate thesis on Pharmacy, University of Sassari AA 2010- 2011.

Diehl, D.J., Gershon, S. (1992). The role of dopamine in mood disorders. *Compr Psychiatry.* Mar-Apr;33(2):115-120.

Diggory, G.L., Buckett, W.R. (1984). Chronic antidepressant administration fails to attenuate apomorphine-induced decreases in rat striatal dopamine metabolites. *Eur J Pharmacol.* Oct 15;105(3-4):257-263

DSM-IV-TR: American Psychiatric Association. 2000

Dumas, R.S. (2004). Role of Neurotrophic factors in the etiology and treatment of mood disorders. *Neuromolecular Medicine.* 5: 11-25.

Dunlop, B.W., Nemeroff, C.B. (2007). The role of dopamine in the pathophysiology of depression. *Arch Gen Psychiatry.* Mar;64(3):327-337.

Dunner, D.L., Fieve, R.R. (1974). Clinical factors in lithium carbonate prophylaxis f ailure. *Arch. Gen. Psychiatry* 30: 229 233.

Dziedzicka-Wasylewska, M. (1997). The effect of imipramine on the amount of mRNA coding for rat dopamine D2 autoreceptors. *Eur J Pharmacol.* Oct 22;337(2- 3):291-.

Emre, M., Mecocci, P., Stender, K. (2008). Pooled analyses on cognitive effects of memantine in patients with moderate to severe Alzheimer's disease. *J Alzheimers Dis.* Jun;14(2):193-199.

EPAR (European Public Assessment Report) (2004) on Ebixa; EMEA 15 Maggio Farlow, M.R., Graham, S.M., Alva, G. (2008). Memantine for the treatment of Alzheimer's disease: tolerability and safety data from clinical trials. *Drug Saf.* 31(7): 577-585.

Fibiger, H.C., Phillips, A.G. (1981). Increased intracranial self stimulation in rats after long-term administration of desipramine. *Science.* Nov 6;214(4521):683-685.

Fibiger, H.C. (1995). Neurobiology of depression: focus on dopamine. In: *Depression and Mania: From neurobiology to treatment. Gessa G.L., Fratta W., Pani L., Serra G. (eds). Adv. Biochem. Psychopharmacol. 49. Racen Press pp: 1-17.*

Freyhan, F.A. (1960). Zur modernen psychiatrischen Behandlung der Depressionen.

Gao Y, Payne RS, Schurr A, Hougland T, Lord J, Herman L, Lei Z, Banerjee P, El- Mallakh RS. Memantine reduces mania-like symptoms in animal models. *Psychiatry Res.* 2011 Jan 25. [Epub ahead of print] PubMed PMID: 21269711.

Gardner, D.M., Baldessarini, R.J., Waraich, P. (Modern antipsychotics drugs: a critical overview. *CMAJ.* June 21, 2005; 172 (13)

Gauthier S, Cummings J, Ballard C, Brodaty H, Grossberg G, Robert P, Lyketsos C. Management of behavioral problems in Alzheimer's disease. *Int Psychogeriatr. 2010 May;22(3):346-72. Epub 2010 Jan 25.*

Gaytan, O., Nason, R., Alagugurusamy, R., Swann, A., Dafny, N. (2000). MK-801 blocks the development of sensitization to the locomotor effects of methylphenidate. *Brain Res Bull.* 51(6):485-492

Geller, B., Williams, M., Zimerman, B., Frazier, J., Beringer, L., Warner, K.L. (1998) Prepubertal and early adolescent bipolarity. *J Affect Disord.* 1998 Nov; 51 (2): 81-91

Gershon, A.A., Vishne, T., Grunhaus, L. (2007). Dopamine D2-like receptors and the antidepressant response. *Biol Psychiatry.* Jan 15;61(2):145-153.

Gessa, G.L., Pani, L., Serra, G., Fratta, W. (1995). Animal models of Mania. In: *Depression and Mania: From neurobiology to treatment.* Gessa G.L., Fratta W., Pani L., Serra G. (eds), Adv. Biochem. Psychopharmacol. 49 Raven Press, pp. 43-66.

Ghaemi, S.N. (2008). Treatment of Rapid-Cycling Bipolar Disorder: Are Antidepressants Mood Destabilizers? *Am/J Psychiatry* 165(3): 300-302.

Gilling, K.E., Jatzke, C., Hechenberger, M., Parsons, C.G. (2009). Potency, voltage-dependency, agonist concentration-dependency, blocking kinetics and partial untrapping of the uncompetitive N-methyl-D-aspartate (NMDA) channel blocker memantine at human NMDA (GluN1/GluN2A) receptors. *Neuropharmacology.* Apr;56(5):866-875.

Goodwin, F.K. (2002). Rationale for long-term treatment of bipolar disorder and evidence for long-term lithium treatment. *J Clin Psychiatry.* 2002;63 Suppl 10:5-12.

Grönig, M., Atalla, A., Kuschinsky, K. (2004). Effects of dizocilpine [(+)-MK-801] on the expression of associative and non-associative sensitization to D- amphetamine. *Naunyn Schmiedebergs Arch Pharmacol.* Feb; 369(2):228- 231.

Heusner, C.L., Palmiter, R.D. (2005). *Expression of mutantNervenarzt* 31:112—118, 1960

NMDA receptors in dopamine D1 receptor-containing cells prevents cocaine sensitization and decreases cocaine preference. *J Neurosci.* Jul 13; 25(28): 6651-6657.

Hoheisel, H.P. (1966) Zur Frage der Verkiirzung von Intervallzeiten psychophar makologisch behandelter phasischer Psychosen. *Nervenarzt* 37:259—263

Heinz, C.R., Grunze, M.D. (2008). Switching, Induction of Rapid Cycling, and Increased Suicidality With Antidepressants in Bipolar Patients: Fact or Overinterpretation? *CNS Spectr.* 13(9):790-795.

Holcomb, H.H., Bannon, M.J., Roth, R.H. (1982). Striatal dopamine autoreceptors uninfluenced by chronic administration of antidepressants. *Eur J Pharmacol.* Aug 27;82(3-4):173-178.

Jeziorski, M., White, F.J., Wolf, M.E. (1994). MK-801 prevents the development of behavioral sensitization during repeated morphine administration. *Synapse.* Feb; 16(2):137-147.

Johnson, J.W., Kotermanski, S.E. (2006). Mechanism of action of memantine. *Curr Opin Pharmacol.* Feb, 6(1):61-67.

Jones, R.W. (2010). A review comparing the safety and tolerability of memantine with the acetylcholinesterase inhibitors. I*nt J Geriatr Psychiatry.* 24: 547– 553.

Lauber, H. (1964) Studie zur Frage der Krankheitsdauer unter Behandlung mit Psychopharmaka. Nervenarzt 35:488 - 491

Li, Y., White, F.J., Wolf, M.E. (2000). Pharmacological reversal of behavioral and cellular indices of cocaine sensitization in the rat. *Psychopharmacology (Berl).* Aug; 151(2-3):175-183.

Kaduszkiewicz,, H., Hoffmann, F. (2008). Review: cholinesterase inhibitors and memantine consistently but marginally improve symptoms of dementia. *Evid Based Ment Health.* Nov; 11(4): 113-120

Kalivas, P.W. (1995). Interactions between dopamine and excitatory amino acids in behavioral sensitization to psychostimulants. *Drug Alcohol Depend.* Feb; 37(2): 95-100.

Kaufman, K.R. (2010) Antiepileptic drugs in the treatment of psychiatric disorders. *Epilepsy Behav.* 2011 May;21(1):1-11.

Keck, P.E., Jr Hsu H.A., Papadakis, K., Russo, J. Jr. (2009). Memantine efficacy and safety in patients with acute mania associated with bipolar I disorder: a pilot evaluation. *Clin Neuropharmacol.* Jul-Aug;32(4):199-204.

Kelsey, J.E., Beer, T., Lee, E., Wagner, A. (2002). Low doses of dizocilpine block the development and subsequent expression of locomotor sensitization to nicotine in rats. *Psychopharmacology (Berl).* Jun; 161(4):370-378.

Kim, H.S., Park, W.K., Jang, C.G., Oh, S. (1996). Inhibition by MK-801 of cocaine- induced sensitization, conditioned place preference, and dopamine receptor supersensitivity in mice. *Brain Res Bull.* 40(3): 201-207.

Kotlinska, J., Bochenski, M., Danysz, W. (2006). N-methyl-D-aspartate and group I metabotropic glutamate receptors involved in the expression of ethanol- induced sensitization in mice. *Behav Pharmacol.* Feb; 17(1):1-8.

Koukopoulos, A., Reginaldi, D. (1973). Does lithium prevent depression by suppressing mania? *Int Pharmacopsychiatry* 8 (3): 152-158.

Kukopulos, A.(variant of Koukopoulos), Reginaldi, D., Laddomada, P., Floris, G., Serra, G., Tondo, L. (1980). Course of the manic-depressive cycle and changes caused by treatments. *Pharmakopsychiatry* 13: 156—167.

Kukopulos, A. (variant of Koukopoulos), Caliari, B., Tundo, A., Floris. G., Reginaldi, D., Tondo, L. (1983) Rapid cyclers temperament and antidepressants. *Compr. Psychiatry* 24, 249– 258.

Koukopoulos A., Reginaldi D., Minnai G., Serra G., Pani L., Johnson F.N. (1995). The long-term prophylaxis of affective disorders. In: Depression and Mania: From neurobiology to treatment; Gessa G.L., Fratta W., Pani L., Serra G. (eds.), Adv. Bichem. Psychopharmacol. 49 Raven Press, pp. 127-147.

Koukopoulos, A., Ghaemi, S.N. (2009). The primacy of mania: a reconsideration of mood disorders. *Eur Psychiatry Mar*;24(2):125-134.

Koukopoulos A, Reginaldi D, Serra G, Koukopoulos AE, Sani G, Serra G. (2010). Antimanic and mood stabilizing effect of memantine as an augmenting agent in treatment resistant bipolar disorder. *Bipolar Disorders.* 12: 348-349

Koukopoulos, A., Serra, G., Koukopoulos, A.E., Reginaldi, D., Serra, G. (2012) The sustained mood stabilizing effect of memantine in the manegement of treatment resistant bipolar disorders: findings from a 12-month naturalistic trial. J. Affective Disorders 136: 163-166.

Kraepelin, E., 1913. Psychiatry, 8th Edition. Barth. Leipzig.

Kupka, R.W., Luckenbaugh, D.A., Post, R.M., Leverich, G.S., Nolen, W.A. (2003) Rapid and non-rapid cycling bipolar disorder: a meta-analysis of clinical studies. *J Clin Psychiatry* Dec; 64 (12): 1483-94.

La Spada, A.R. (2009). Memantine strikes the perfect balance. *Nat Med.* Dec;15(12):1355-1356.

Leverich, G.S., Altshuler, L.L., Frye, M.A., Suppes, T., McElroy, S.L., Keck, .PE. Jr, et al. (2006). Risk of switch in mood polarity to hypomania or mania in patients with bipolar depression during acute and continuation trials of venlafaxine, sertraline, and bupropion as adjuncts to mood stabilizers. *Am J Psychiatry.* Feb; 163(2):232-239.

Macqueen, G.M., Campbell, S., Mcewen, B.S., Macdonald, K., Amano, S., Joffe, R.T., Nahmias, C., Young, L.T. (2003) Course of illness, hippocampal function, and hippocampal volume in major depression. PNAS. 100(3):1387-1392.

Malberg, J.E. (2004) Implication of adult hippocampal neurogenesis in antidepressant action. *J Psychiatry Neurosci*, 29(3):196-205.

Malberg, J.E., Blendy, J.A. (2005). Antidepressant action: to the nucleus and beyond. *Trends Pharmacol Sci.* Dec;26(12):631-638.

Malesa R., Serra G. La memantina previene la sensibilizzazione dei recettori dopaminergici indotta dall'imipramina. Graduate thesis on Pharmacy, University of Sassari AA 2010-2011

Maj, M., Magliano, L., Pirozzi, R., Marasco, C., Guameri, M. (1994). Validity of rapid cycling as a course specifier for bipolar disorder. *Am. J. Psychiatry* 151, 1015 –1019.

Mc Namara J.O.2001 Pharmacotherapy of the Epilepsies. In Goodman & Gilman's, 12 Edition pp: 583-607

Minnai G.P., Salis P.G., Oppo R, Loche A.P., Scano F. Tondo L. (2011) Effectiveness of maintenance electroconvulsive therapy (m-ECT) in rapid cycling bipolar disorder. J ECT Jun; 27(2):123-6.

Muscat, R., Towell, A., Willner, P. (1988). Changes in dopamine autoreceptor sensitivity in an animal model of depression. *Psychopharmacology* (Berl); 94(4):545-550.

Muzina, D.J. (2009). Pharmacologic treatment of rapid cycling and mixed states in bipolar disorder: an argument for the use of lithium. Bipolar Disord. 2009 Jun;11 Suppl 2:84-91. Review.

Ohmori, T., Abekawa, T., Muraki, A., Koyama, T. (1994). Competitive and noncompetitive NMDA antagonists block sensitization to methamphetamine. *Pharmacol Biochem Behav.* Jul; 48(3):587-591.

Ongur, D., Jensen, E., Prescot, A.P., Stork, C., Lundy, M., Cohen, B.M. (2008). Renshaw PF. Abnormal glutamatergic neurotransmission and neuronal-glial interaction in acute mania. *Biol Psychiatry.* 64: 718-726.

Pacchioni AM, Gioino G, Assis A, Cancela LM. A single exposure to restraint stress induces behavioral and neurochemical sensitization to stimulating effects of amphetamine: involvement of NMDA receptors. Ann N Y Acad Sci. 2002 Jun;965:233-46.

Paizanis, E., Hamon, M., Lanfumey, L. (2007). Hippocampal neurogenesis, depressive disorders, and antidepressant therapy. *Neural Plast.* 2007: 1-7..

Peet, M.(1994). Induction of mania with selective serotonin reuptake inhibitors and tricyclic antidepressants. *Br J Psychiatry;* 164:549–550.

Rammes, G., Danysz, W., Parsons, C.G. (2008) Pharmacodynamics of memantine: an update. *Curr Neuropharmacol.* Mar;6(1):55-78.

Randrup, A., Munkvad, J., Fog, R. (1975). Mania, depression and brain dopamine. In: Essman WB, Valzelli Leds. *Current Developments in Psychopharmacology.* New York: Spectrum: 206-248.

Rapoport, S.I., Basselin, M., Kim, H.W., Rao, J.S. (2009) Bipolar disorder and mechanisms of action of mood stabilizers. Brain Res Rev. Oct;61(2):185-209.

Ray, W.A., Chung, C.P., Murray, K.T., Hall, K., Stein, C.M. (2009) Atypical antipsychotic drugs and the risk of sudden cardiac death. N Engl J Med. 2009 Jan 15;360(3):225-35. *N Engl J Med.* 2009 Oct 29;361(18):1814.

Réus, G.Z., Stringari, R.B., Kirsch, T.R., Fries, G.R., Kapczinski, F., Roesler, R., Quevedo, J. (2010). Neurochemical and behavioural effects of acute and chronic memantine administration in rats: Further support for NMDA as a new pharmacological target for the treatment of depression? *Brain Res Bull.* Apr 5;81(6):585-9.

Ritti, A. (1883) Traité Clinique de la Folie à Double Forme. Paris, Octave Doin, Rockhold RW. Glutamatergic involvement in psychomotor stimulant action. Prog Drug Res. 1998;50:155-92.

Rogóz, Z., Skuza, G., Maj, J., Danysz, W. (2002) Synergistic effect of uncompetitive NMDA receptor antagonists and antidepressant drugs in the forced swimming test in rats. Neuropharmacology. Jun;42(8):1024-30.

Rompré, P.P., Bauco, P. (2006). Neurotensin receptor activation sensitizes to the locomotor stimulant effect of cocaine: a role for NMDA receptors. *Brain Res.* Apr 26;1085(1):77-86.

Schildkraut, J.J. (1965) The catecholamine hypothesis of affective disorders: a review of supporting evidence. *Am J Psychiatry.* Nov;122(5):509-522.

Schneck, C. D., Miklowitz, D. J., Miyahara, S., Araga, M., Wisniewski, S., Gyulai, L., Allen, M.H., Thase, M.E., Sachs, G.S. (2008). The Prospective Course of Rapid- Cycling

Bipolar Disorder: Findings From the STEP-BD, *Am J Psychiatry* 165:3, 370-377, March.

Serra, G., Argiolas, A., Klimek, V., Fadda, F., Gessa, G.L. (1979). Chronic treatment with antidepressants prevents the inhibitory effect of small doses of apomorphine on dopamine synthesis and motor activity. *Life Sci.* 25: 415- 424.

Serra, G., Argiolas, A., Fadda, F., Gessa G.L. (1980). Hyposensitivity of dopamine autoreceptore induced by chronic administration of tricyclic antidepressants. *Pharmacol. Res. Comm.* 12: 619-624.

Serra, G., Argiolas, A., Fadda F., Melis, M.R., Gessa, G.L. (1981). Repeated electroconvulsive shock prevents the sedative effect of small doses of apomorphine. *Psychopharmacology* 73: 194-196.

Serra, G., Melis, M.R., Argiolas, A., Fadda, F., Gessa, G.L. (1981). REM sleep deprivation induces subsensitivity of dopamine receptors mediating sedation in rats. *Eur. J. Pharmacol.* 72: 131-135.

Serra, G., Gessa, G.L. (1990) Manuale di Psicofarmacologia Masson Editore.

Serra, G., Collu, M., D'Aquila, P.S., De Montis, G.M., Gessa G.L (1990). Possible role of dopamine D1 receptor in the behavioural supersensitivity to dopamine agonists induced by chronic treatment with antidepressant. *Brain Res.* 527: 234-243.

Serra, G., Collu, M., D'Aquila, P., De Montis, G.M., Gessa, G.L (1991). Chronic imipramine "reverses" B-HT 920- induced hypomotility in rats. *J. Neural Transm.* 84: 237-240.

Serra, G., Collu, M., D'Aquila, P.S., Gessa, G.L. (1992). Role of the mesolimbic dopamine system in the mechanism of action of antidepressants. *Pharmacol Toxicol.* 71 Suppl 1:72-85.

Serra, G., Fratta, W. (2007). A possible role for the endocannabinoid system in the neurobiology of depression. *Clinical Practice and Epidemiology in Mental Healt*; 3: 25.

Serra, G., D`Aquila, P.S. (2008). Do antidepressants induce mania and rapid cycling by increasing dopaminergic transmission? *TDM 2008 International Meeting- Bologna ottobre* 2008. pp. 54-55

Serra, G. (Patent, 2009) Uso della memantina per il trattamento dei disturbi dell'umore. N° MI2009A000174. 11/02/2009

Serra, G. (EP Patent, 2010) Memantine for treating bipolar mood disorders resistant to conventional treatments. EP 2 218 450 A1. Priority: 11.02.2009 IT MI20090174. 18.08.2010 Bulletin 2010/33

Serra, G. (2011) A naturalistic study on antimanic and mood stabilizing effect of memantine in treatment-resistant bipolar disorders. IRBD, Rome 4-6 April, 2011.

Sheline, Y.I., Gado Mokhtar, H., Kroemer, H.C. (2003) Untreated depression and hippocampal volume loss. *Am J Psychiatry.* 160:1516-1518.

Spearing, M.K., Post, R.M., Leverich, G.S., Brandt, D., Nolen, W. (1997) Modification of the Clinical Global Impressions (CGI) Scale for use in bipolar illness (BP): the CGI-BP. Psychiatry Res. Dec 5;73(3):159-71.

Spyraki, C., Fibiger, H.C. (1981). Behavioural evidence for supersensitivity of postsynaptic dopamine receptors in the mesolimbic system after chronic administration of desipramine. *Eur J Pharmacol.* 11;74(2-3):195-206.

Till, E., Vuckovic, S. (1970). Ueber den Einfluss der thymoleptischen Behandlung auf den Verlauf endogener Depressionen. Im Pharmacopsychiatry 4:210—2 19.

Tohen, M., Greil, W., Calabrese, J.R., Sachs, G.S., Yatham, L.N., Oerlinghausen, B.M., Koukopoulos, A., Cassano, G.B., Grunze, H., Licht, R.W., Dell'Osso, L., Evans, A.R., Risser, R., Baker, R.W., Crane, H., Dossenbach, M.R., Bowden, C.L. (2005). Olanzapine versus lithium in the maintenance treatment of Biolar Disorders: a 12-month, randomized, double-blind, controlled clinical trial. *Am J Psychiatry.* Jul; 162(7):1281-90.

Tondo L., Laddomada P., Serra G., Minnai G., Kukopulos A (1981). Rapid cyclers and antidepressants. *Int. Pharmacopsychiat.* 16: 119-123.

Tondo, L., Hennen, J., Baldessarini, R.J. (2003). Rapid-cycling bipolar disorder: effects of long-term treatments. *Acta Psychiatr Scand.* Jul;108(1):4-14.

Tondo, L., Vázquez, G., Baldessarini, R.J. (2010). Mania associated with antidepressant treatment: comprehensive meta-analytic review. *Acta Psychiatr Scand.* 121(6):404-414.

Trujillo, K.A. (2002). The neurobiology of opiate tolerance, dependence and sensitization: mechanisms of NMDA receptor-dependent synaptic plasticity. Neurotox Res. Jun;4(4): 373-391.

Van Dongen, A.M. Editor. (2009). Biology of NMDA receptor. Frontiers in Neuroscience.

Vanelle JM, Loo H, Galinowski A, de Carvalho W, Bourdel MC, Brochier P, Bouvet O, Brochier T, Olie JP. (1994). Maintenance ECT in intractable manicdepressive disorders. Convulsive Therapy. Sep; 10 (3): 195-205.

Vezina P, Queen AL (2000). Induction of locomotor sensitization by amphetamine requires the activation of NMDA receptors in the rat ventral tegmental area. Psychopharmacology. Aug 151(2-3): 184-191.

Videbech, P., Ravnkilde, B. (2004) Hippocampal volume and depression. A meta- analysis of MRI studies. *Am J Psychiatry,* 161(11):1957-1966.

Võikar V, Soosaar A, Volke V, Kõks S, Bourin M, Männistö PT, Vasar E (1999). Apomorphine-induced behavioural sensitization in rats: individual differences, role of dopamine and NMDA receptors. Eur Neuropsychopharmacol. Dec; 9(6): 507-514.

Warnings About Suicidality Risk With Antiepileptic Drugs FDA, Alert 2008, Dec 16.

Wehr, TA, Goodwin, FK. (1979). Rapid cycling in manic-depressives induced by tricyclic antidepressants. Arch Gen Psychiatry. May; 36(5): 555-559.

Westfall TC, Westfall DP. In Goodman & Gilman's The Pharmacological Basis of Therapeutics 11 Edition 237-295.

Willner, P. (1997). The mesolimbic dopamine system as a target for rapid antidepressant action. Int Clin Psychopharmacol. Jul;12 Suppl 3:S7-14.

Wolf ME, White FJ, Hu XT.(1994): MK-801 prevents alterations in the mesoaccumbens dopamine system associated with behavioral sensitization to amphetamine. J Neurosci. Mar; 14(3 Pt 2):1735-1745.

Yap JJ, Covington HE 3rd, Gale MC, Datta R, Miczek KA (2005). Behavioral sensitization due to social defeat stress in mice: antagonism at mGluR5 and NMDA receptors. Psychopharmacology (Berl). Apr; 179(1):230-239.

Yildiz A, Sachs GS., Characteristics of rapid cycling bipolar I patients in a bipolar speciality clinic. J Affect Disord. 2004 Apr; 79 (1-3):247-51

Zdanys, K., Tampi, R.R. (2008). A systematic review of off-label uses of memantine for psychiatric disorders. *Prog Neuropsychopharmacol Biol Psychiatry.* Aug 1;32(6):1362-74.

Zupancic, M.L. (2011). Role of atypical antipsychotics in rapid cycling bipolar disorder: a review of the literature. Ann Clin Psychiatry. 2011 May;23(2): 141- 9.

6

Bipolar Disorder and Suicide

Dagmar Breznoščáková
University of P. J. Šafárik and University Hospital of L. Pasteur,
1st Dept. of Psychiatry, Košice
Slovak Republic

1. Introduction

Bipolar disorder (BD) is a chronic psychiatric disorder characterized by remissions and exacerbations of mood disturbances. BD is associated with increased mortality due to the averse outcomes of medical disorders, accidents and complications of commonly comorbid substance use disorders /Ahrens et al., 1995/. By far, however, the major source of early mortality is a high risk of suicide /Kessler et al., 2005; Tondo et al., 2005/. Suicide behaviour and particularly, committed suicide are among the most tragic events in human life causing a serious changes among relatives and friends as well as imposing a great economic problem for the whole society /Rihmer & Kiss, 2004/. Approximately about one million deaths from suicide are recorded every year and a world-wide number of suicide attempts is estimated to be 10 - 15 millions per annum /Wassermann, 2000/. Reviewing 17 follow-up studies on committed suicide in patients with primary affective disorders, Guze & Robins /1970/ found that about 15% of formerly hospitalized depressed patients would die by suicide. Goodwin & Jamison /1990/ also concluded that 19% of depressed patients (mainly inpatients) died by suicide /Rihmer, 2006/. In their metaanalysis of studies on suicide risk in psychiatric disorders, Harris & Barraclough /1997/ analyzed separately the risk of suicide in unipolar major depression and in BD. They found that the risk of suicide was about 20-fold for patients with index diagnosis of unipolar major depression and the same figure for BD was 15. However, these three studies /Goodwin & Jamison, 1990; Guze & Robins, 1970; Harris & Barraclough, 1997/ cannot provide a precise estimation of separate suicide risk in unipolar and bipolar disorder, that is they overestimate the risk for unipolar depression and underestimate it for BD.

Bipolar disorder is associated with a high frequency of both completed suicides and suicide attempts. Twenty-five percent to 60% of all bipolar patients will have attempted suicide at least once in their lifetime, and 18.9% of deaths among bipolar patients are due to suicide, but with accurate treatment we can prevent suicide risk in 50% of patients /Goodwin & Jamison, 1990/.

2. Epidemiology

2.1 Suicide risk

In the general populations of developed countries, recent rates of completed suicide averaged 0.015±0.007% annually /Tondo et al., 2005/. Risks associated with major affective disorders, including BD, are much greater /Baldessarini et al., 2006/. Suicide rates among patients diagnosed with BD (mostly BD-I), as well as in severe forms of recurrent major depressive disorder, are not only much higher than in the general population but also substantially greater than in association with other psychiatric, substance use, or general medical disorders /Ahrens et al., 1995/. In BD, this risk is annually 0,9% or 30-60 times above the general population rate of 0.015% annually. Suicide accounts for 15% to 20% of deaths among BD patients /Tondo et al., 2005/. In addition, BD is associated with 2-3-fold excess of mortality due to common, stress-sensitive general medical disorders /Hawton et al., 1998/. Suicide risk factors can be classified hierarchically /Rihmer & Kiss, 2004, table 1/.

1. Primary (psychiatric-medical) suicide risk: major psychiatric illness (depression, schizophrenia, substance use disorders)/ comorbid anxiety an/or personality disorders, serious medical illness, feeling of hopelessness and insomnia, concomitant anxiety; previous suicide attempts; communication of wish to die/ suicide intent (direct or indirect; suicide amoung family members (biological and/or social "inheritance"); disregulated serotonergic system, low total serum cholesterol, abnormal dexamethasone suppression test during depression – **competence: health-care**
2. Secondary (psycho-social): childhood negative life-events (separation, parental loss, etc.); isolation, living alone (divorce, separation, widowhood, etc.); lost of job, unemployment; severe acute negative life-events (loss of close relative or friend, recent unemployment, etc.) – **competence: community leaders, teachers, psychologists, religious and civil organizations, health-care**
3. Tertiary (demographic) suicide risk: male gender; adolescent and young males, old age (both genders); vulnerable intervals (spring/early summer, pre-menstrual period, etc.); minority groups (relatives of suicide victims of disasters, bisexuality, same-sex orientation, etc.) – **competence: ?**

Table 1. Hierarchical classification of suicide risk

Suicide attempts are very common, in about a third to a half of these individuals /Scott, 2005/. In the general population, the attempts: suicide ratio ranges between 20:1 and 40:1, with some uncertainty owing to unreliability and probable under-reporting of suicide attempts /Baldessarini et al., 2006; Hawton et al., 1998/. The risk of suicide attempts among BD patients averages 3, 9% annually or nearly 3-time higher than the rate of completed suicides. The suicide ratio may indicate relatively high lethality of suicide attempts in BD, presumably reflecting both level of intent to die and the lethality of methods used.

Recent results showed that about 50% of unipolar depressions were found to be bipolar depressions after careful and skillful probing for past hypomania or mania as well as focusing on not only mood but also overactivity /Benazzi & Akiskal, 2003a; Ghaemi et al., 2000/. Unfortunately, the risk of suicide is particularly high in bipolar patients during the first few years of illness, whereas diagnosis and establishing a sustained program of long-term, mood-stabilizing treatment is typically delayed by 5-10 years from illness-onset /Baldessarini et al., 2006/. Such delays often are longer in women and in bipolar II disorder than in men and in bipolar I disorder /Baethge et al., 2003/. This delay underscores the need for much earlier diagnosis and treatment especially in juveniles, since delay of treatment may increase suicidal risk /Faedda et al., 1995/. Suicide rates in bipolar patients are also characterized by high lethality of suicide attempts: one death out of three attempts, compared to one completed suicide out of 30 attempts in the general population /Aubry et al., 2007/.

Analyzing the clinical characteristics of 230 inpatients with recurrent major depression, Bulik et al. /1990/ found that bipolar II diagnosis was significantly more frequent among the 67 patients who attempted suicide (19%) than in the 163 patients who did not attempt suicide (9%). Dunner et al. /1976/ reported that 3% of the 73 unipolar, 6% of the 68 bipolar I and 18% of the 22 bipolar II patients died by suicide during their 1-9 year follow-up study. Considering all the above, the findings strongly suggest that bipolar II patients might be overrepresented among suicide victims. The two published reports where the prevalence of bipolar II, bipolar I and unipolar depression have been analyzed separately among the suicide victims show that among the 125 consecutive suicide victims with primary major depression at the time of suicide, 44% had bipolar II depression, 2% had bipolar I depression and 54% had first episode or recurrent unipolar depression /Rihmer et al.1995, Rihmer et al. 1990/. Because the lifetime prevalence rates of DSM-III/IV bipolar II illness in the population are relatively low compared with unipolar major depression (2-5% and 15-17% respectively) /Angst, 1998/, these results suggest that bipolar II disorder imparts a particularly high risk of committed suicide among the three different subgroups of major mood disorders. /Rihmer, 2006/.

2.2 Risk factors

In the general population and in unipolar MDD patients, suicide rates are several times higher in men, but in BD patients, the risk is much more similar among men and women /Baldessarini et al., 2006/. Other risk factors for suicide generally, including in BD, in addition to current depression, include white ethnicity, being unmarried, previous severe depression or suicide attempts, current feelings of hopelessness and active abuse of alcohol or illicit drugs, and perhaps limited access to support or clinical services /Tondo et al., 2006/. Stress factors also can contribute to suicidal risk, including deaths, divorce, separations and other major losses, scandals or imprisonment, social isolation.

Interpretation of analyses of risk factors associated with suicide is limited by not knowing of potential suicides that may have been prevented by timely assessment and effective interventions. Although screening of BD patients for the risk factors just summarized has uncertain power to predict specific risk and timing in individuals, their consideration is an important component of clinical assessment of suicidal risk /Jacobs, 2003/. Such clinical unpredictability can be especially challenging with BD patients, given sometimes rapid shifts in mood (lability), strong reactivity to losses, frustrations, or other stressors, impulsivity, disinhibiting effects of commonly abused central depressants, including alcohol, comorbid anxiety disorders, and potential adverse effects of excessive use of antidepressants /Ghaemi et al., 2004/. The most powerful and clinically explorable suicide risk factors in BD are listed in Table 2.

1. Family history of suicide in first degree relatives
2. Past history - previous suicide attempts: high lethality / violent > low lethality / nonviolent
3. Diagnostic subtype: bipolar II > bipolar I (> unipolar); hypomania, mania
4. Clinical features: severe depression, hopelessness, guilt, agitation, thinking of death, suicide intention, etc.; comorbid anxiety disorder, substance use disorder, personality disorder, serious medical illness; mixed states (dysphoric mania, depressive mixed states); aggressive personality features
5. Psychosocial stressors - negative life events: adverse life events (death of close persons, unemployment, marital breakdown, etc.).

Table 2. Suicide risk factors in bipolar disorders

2.2.1 Family history

Family history of suicide is a significant risk factor for suicide behaviour, particularly in persons with BD. If a patient has a family member who suffers from affective disorders or alcoholism, or had one who committed suicide, very careful clinical attention should be given. Inquiries should be made into the patient´s history of affective disorders or alcoholism /Takahashi, 1993/. Bipolar patients with positive family history of suicide in first-degree relatives were found to be significantly more likely to attempt suicide (38%) than those without (14%) / Roy, 1983/ and it has been reported that 0,9% of unipolar depressives, 1,8% of bipolar I and 2,9% of bipolar II patients had a family history of committed suicide /Dunner et al., 1976/. Another study has found a 6,5-fold higher rate of suicide committed among the first-degree relatives of 129 bipolar II (3,9%) than that of the 188 bipolar I (0.6%) patients /Tondo et al., 1998/. Investigating the family history of suicide in first degree relatives among 85 inpatients with DSM-III-R major depression (28 bipolar II, 57 unipolar), Rihmer /2006/ founds that the rate of persons with positive family history of completed suicide was significantly higher in bipolar II than in unipolar patients (21% vs 9%).

2.2.2 Past history – Prior suicide attempt

It is well documented that attempted suicide (particularly in the case of affective disorder) is among the most powerful predictor of committed suicide /Cheng et al., 2000; Wassermann, 2000/. Up to 55% of suicide victims with BD have had at least one previous suicide attempt and every fifth patient with BD dies by suicide /López et al., 2001/. As shown above, with regard to the specific subtypes, it is particularly bipolar II patients that are at the highest suicide risk /Vieta et al., 1997; Moťovský, 2009/. The clinician must not neglect to ask whether the patient has attempted suicide in the past. One out of 10 who have attempted suicide will make another attempt and will die by suicide /Stengel, 1964/. Lay people often insist, "Those who say that they will kill themselves will never do so," which is a commonly believed myth. As stated before, Japan´s suicide rate is 17, 8 per 100 000 among the general population; therefore, among those who have attempted suicide once, the rate is 540 times higher. Some researchers have pointed out that a history of attempted suicide is the most reliable risk factor for suicide. Any form of suicide attempt should be taken seriously.

Detailed information should be collected about any suicide attempt. Was the attempt carried out in a place that would make discovery difficult? Was the detailed plan for suicide premeditated? Were any warning signs conveyed to friends or acquaintances? Did the person write a suicide note? If the person has attempted suicide by violent methods, such as firearms or hanging, the subsequent suicide risk should be considered higher than with nonviolent methods. Persons who have recently experienced another´s suicide have a higher suicide rate. Especially, those who fulfill other risk factors, experience other suicides directly or indirectly through the mass media, and identify themselves with other suicides tend to become suicidal suddenly. The danger of so-called suicide cluster is great, particularly among adolescents / Takahashi, 1993/.

2.2.3 Diagnostic subtype

Analyzing the specific diagnostic subtypes of 69 consecutive (nonviolent) suicide attempters with current DSM-IV major depression in Budapest, Hungary, it has been found that 45 (65%) had unipolar major depression, 19 (28%) had bipolar II depression and 5 (7%) had bipolar I depression /Balász et al., 2003/. Considering the fact that the lifetime prevalence rates of DSM-III-R unipolar major depression, bipolar II and bipolar I disorder in the general population of Hungary are 15, 1%, 2, 0% and 1, 5% respectively /Szádóczky et al., 1998/, this study suggests that bipolar II patients are relatively overrepresented not only among depressed suicide victims but also among depressed suicide attempters. Bipolar patients with comorbid anxiety, personality and substance-use disorders are also at an increased risk of attempted or completed suicide /Chen & Dilsaver, 1996; Vieta et al., 1997/. One of the major sources of the highest suicide risk in bipolar II patients may be the very high rate of comorbid anxiety disorders /Akiskal, 1981; Brieger, 2000/ and depressive mixed states, frequently called "agitated depression" /Benazzi & Akiskal, 2003b/.

2.2.4 Clinical features

Major depressives with history of suicide attempts or committed suicide have a more severe symptomatology in general /López et al., 2001/, are more frequently agitated /Fawcett, 1997/, report more hopelessness, self-blame and guilt /Bulik et al., 1990; Oquendo et al., 2000/ and experience more commonly marital isolation (being single, separated or divorced) and/or loss event than non suicidal depressives /Bulik et al., 1990; Cheng et al., 2000/. It is also well documented that major depressives as well as bipolar patients with comorbid anxiety disorders /Fawcett, 1997; Hawton & van Heeringen, 2000/, substance use disorders /Cheng et al., 2000; Hawton & van Heeringen, 2000/, personality disorders /Cheng et al., 2000; Hawton & van Heeringen, 2000 / or serious medical illness /Cheng et al., 2000; Hawton & van Heeringen, 2000/, are also at an increased risk of attempted or committed suicide. One of the major sources of the highest suicide risk in bipolar II patients may be the very high rate of comorbid anxiety disorders /Akiskal, 1981 / and substance use disorders /Rihmer et al., 2001/. In a study aimed at identifying clinical predictors of suicide attempts in subjects with BD, Dalton et al. /2003/ studied 336 subjects with a diagnosis of bipolar I, bipolar II, or schizoaffective disorder (bipolar type) and examined predictors of suicide in attempters and nonattempters. They found that the lifetime rate of suicide attempts for the entire sample was 26.7%. Bipolar subjects with comorbid substance-use disorder had a 39.5% lifetime rate of attempted suicide, compared with a 23.8% lifetime rate for those without substance-use disorder. The researchers concluded that lifetime comorbid substance-use disorder was associated with a higher rate of suicide attempts in patients with BD.

Leverich et al. /2003/ at the Stanley Foundation Bipolar Network asserted that "to the extent that bipolar illness puts teenagers and young adults at risk for the accumulation of comorbidities, it seems appropriate to pay attention to these and other factors associated with suicide attempts so that some of these elements might be prevented or ameliorated". In an investigation of the association between suicide attempts and the predictive factors previously described in the literature, researchers followed 169 patients identified with bipolar I disorder /Purifacion et al., 2001/. More than one third (56) of the patients had a history of 1 or more suicide attempts. The rate of suicide attempts was much higher in patients with onset of bipolar disorder at or before the age of 25 years than in patients with onset after age 25 (25% vs 10%, respectively). Other factors related to suicide were drug abuse, family history of affective disorders, and severe depressive episodes. The patients who abused drugs had a history of more suicide attempts than those who did not.

2.2.5 Psychosocial stressors – Negative life events

Various life stresses include financial loss, loss of position, physical illness and trauma, death of a relative, and lawsuit, unemployment, major financial problems as well as adverse recent life event (including recent loss of a loved person) have been shown to be a risk factor for attempted and completed suicide/Cheng et al., 2000; Hawton & van Heeringen, 2000/. More than half of completed suicides in both bipolar I and unipolar affective disorders are associated with recent negative life events, but the stressors are commonly dependent on the victim´s own behaviour, particularly in the case of bipolar I disorder /Isometsä et al., 1995/. It might be the consequence of the fact that hypomanic or manic periods easily can result in aggressive behaviour, financial extravagance, episodic promiscuity, generating interpersonal conflicts and marital breakdown /Akiskal, 1981; Goodwin & Jamison, 1990/. Loss of a family member or friend is often followed by grief-reaction, that later can develop into complicated grief /Horowitz et al., 1997/ and into major depression. Lack of social support system is also an important risk factor. People who are single, divorced, separated, or recently left by significant others show a threefold higher suicide rate than those who are married /Takahashi, 1993/. Those who are excluded from a group in society are also reported to have a higher suicide rate. In a study of 32,000 bipolar patients' records /Simon et al., 2007/ the highest risk factor for suicide was being male and having a comorbid anxiety disorder, compared with being young and having a substance-use disorder, which predicted attempts but not necessarily suicide.

3. Specific situations – The high suicide risk

Because specific subtypes of major affective disorder (unipolar, bipolar II and bipolar I) differ from both clinical and research perspective, it is logical to assume that each subgroup might have its own different suicide risk. Current mood states are critical determinants of suicidal risk in BD, particularly depressive and dysphoric-irritable mixed states, which together account for at least three-quarters of suicides in BD, whereas suicide is infrequent in mania and rare in hypomania /Baldessarini et al., 2006; Jacobs, 2003/.

3.1 Mixed states

Mixed episode is a complex syndrome which is difficult to diagnose, has the most prolonged duration of bipolar episodes and more frequent psychotic profile than pure mania with high suicidality and poor response to drugs. Mixed state mania has been well known since Kraepelin in classification systems with criteria that include both a manic and major depressive episode nearly every day for at least a one-week period /Oral, 2005/.

With its relatively high incidence rates, bipolar disorder constitutes a considerable health problem not only for the individual patient and his/her family but also for the economy. Individual patients are in danger of suicidal acts, especially in the depressive phase and with mixed mania which, in its mild form, i. e., with less depressive features, is also known as "dysphoric mania". In studies of this affective mixed state by Strakowski et al. /1996/ 26% or even 55% in another study, of the patients were judged to be in acute danger of suicide. About 25-50% of all bipolar patients make at least one attempt at suicide in the course of their illness. In this respect, women are affected more frequently than men. During the ten-year observation period of a Scottish study, the suicide rate in patients with BD was 23fold higher that of the general population; the most of the suicides occurred relatively early in the illness, namely between the second and fifth year after diagnosis. A particular, additional risk factor was living alone or in separation; also most suicide victims came from the lower social classes. This has not inconsiderable social- political implications against the background that divorces among bipolar patients are three times more frequent than in the general population and that the manifestation of BD is not coupled to any particular social class. On the contrary, on average families of bipolar patients have higher educational and income levels than the families of unipolar depressive patients /Walden & Grunze, 2004/.

The importance of depressive mixed states in predicting suicide in supported in long-term naturalistic follow-up study of moderately to severely ill affective disorder patients (80% of them were inpatients at the index episode) they found that 39% of those who later committed suicide (n=36) as well as 220% of the nonsuicidal comparison subjects (n=373) were in mixed or cycling depressive episode at intake /Maser et al., 2002/.

Among bipolar patients, however, it is not the depressive episode that is the only risk period for suicide. In contrast to classical (i. e. euphoric) mania, where suicidal tendency is extremely rare, suicidal thoughts and attempts are relatively common in dysphoric (mixed) mania or hypomania /Strakowski et al., 1996/.

3.2 Short depressive episode

The so-called recurrent brief depression consists of recurring, short depressive episodes (as rule not longer than two weeks). Frequently 8 or more episodes can occur in 1 year. These short depressions are generally difficult to treat with usual mood stabilizers or antidepressants. The average duration of an episode amounts to 3 days. About 5% of the population is assumed to suffer from these short-lasting depressions. Above all, the irregular rhythm and the unpredictability of the depressions are a serious problem for the afflicted subjects. Accordingly, the danger of suicide is particularly high in this group. It is assumed that this is a separate clinical entity and that the recurrent brief depression does not belong directly to the bipolar spectrum /Walden & Grunze, 2004/.

3.3 Postpartum psychosis

Postpartum psychosis (PPP) is a medical emergency that occurs in 1-4 of every 1000 postpartum women. Though rare, this is the condition that makes for horrifying national headlines: There is a risk of infanticide in 4% of these cases. There is also a significant risk of maternal suicide.

3.3.1 Recognition of PPP

Prompt recognition and treatment of this disorder are paramount. Symptoms of PPP begin 3 to 14 days after delivery 75% of the time. Another 10% of cases will have occurred by one month after delivery. Early symptoms can mimic postpartum blues or postpartum depression, with sleep disturbance, depression, hypomania or fatigue. Rapidly, the symptoms of agitation, hyperactivity, insomnia, confusion, hallucinations, delusions, volatile mood shifts, bizarre behaviour and panic develop. PPP has a strong genetic component. There is controversy over whether it is a distinct entity or a bipolar episode. If the latter, is the bipolar episode genetically prompted by childbirth, simply occurring as a response to stress, or coincidentally occurring after delivery? Whatever the case, if a woman is affected with BD and has a family history of a relative who had PPP, she has a 75% chance of incurring PPP herself. If she is affected with bipolar but has no family history of PPP, her chance of PPP falls to 30%. A large population study in 1987 found that women with BD were hospitalized more often for PPP than were women who had a history of schizophrenia or depression. The identified risk factors included being unmarried, having a first baby, undergoing a caesarean delivery, and infant death.

3.3.2 Treatment of PPP

Treatment of postpartum psychosis begins with hospitalization. Even with the use of medication, hospitalization can be lengthy. Therefore, ECT, which often results in more rapid symptom remission, can be useful, because there´s less time lost for the mother-infant bonding process. In some cases, ECT and medication will be prescribed. ECT is safe, rapidly effective, widely recognized treatment for a medical emergency /Finn, 2007/.

A women who is affected by bipolar disorder at the age of 25 years (which corresponds to the average age): has her life expectancy reduced by the 9 years; loses 12 years of normal, healthy life, an also loses 14 years of normal professional and family life /Walden & Grunze, 2000/.

3.4 Suicide in bipolar children and teens

3.4.1 Symptoms and factors of suicidal behaviour

Suicide is the most severe complication of BD /not only in children and teens/. Death wishes or suicidal thoughts are commonly seen in a young person with BD, particularly when he or she is depressed or is in an episode of mania (especially during a mixed or psychotic state). In fact, high school students with BD have more suicidal thoughts (44%) and poorer functioning ((74,9%) than youth with major depression (22,2% and 83,6%) (without mania or hypomania) or healthy teens (1,2% and 87,5%) /Lewinsohn et al., 1995/. Depressed children may see themselves, their surroundings, and their future negatively.

Some describes depression as "wearing dark glasses" a state in which everything is seen pessimistically. A depressed person may feel hopeless and wish to be dead. A child with a mood disorder may experience a negative event(s) such as abuse; a fight with a friend; the death of the friend, relative or pet; or exposure to violence. This negative event may cause this child´s mood to worsen, and he or she may become very hopeless. If the child does not have the skills to cope with the stress and/or does not have support, he or she may consider suicide as solution for the problem. If the person who is considering suicide becomes desperate and has an available method, like a gun or drugs at home, he or she may use the gun, overdose, or try to commit suicide with other methods. The use of alcohol and drugs; the presence of impulsivity, ADHD, or conduct disorders; or a family history of suicidal behaviours in a depressed person increases the risk of suicide. On the other hand, good coping skills, support, spirituality, consideration of others´ reactions to suicide, and no access to a method of suicide diminishes the risk for suicide.

3.4.2 Recognizing and management of suicidal behaviour

Suicidal ideation includes thoughts about wishing to kill one-self, making plans of when, where and how to carry out the suicide and thoughts about the impact of one´s suicide on others. Suicidal attempts involve any behaviour that is intended to end the person´s life. Suicide attempts are much less common than suicidal ideations. The most common method used to attempt suicide in the United States is by overdosing with over-the-counter or prescribed medications. Other common methods include superficial cutting of the arms or neck, but the method used depends on the availability of the method, opportunity and local customs. Suicidal attempts are more frequent in children and adolescents with psychiatric disorders, especially depression and BD, but it can occur in youth with other psychiatric disorders. Also, it can happen after a stressful situation even if the child does not have psychiatric problems. It is more common in Caucasian males but it has been steadily increasing in African-American males. Approximately 90% of teens who commit suicide have a psychiatric disorder, including BD, major depression, conduct problems and abuse of alcohol and illicit drugs. Sucide is more common in those youth who have tried to attempt suicide, have a family history of mood disorders or suicide and in those who have experienced stressful life situations such as physical and sexual abuse /Shaffer & Pfeffer, 2001/. Even mild suicidal attempts may be indications that the child is at serious risk for committing suicide. It appears that suicidal attempts and completions are more frequent in bipolar disordered youth than in other psychiatric disorders. In a study of high school students /Lewinsohn et al., 1995/, 40% of bipolar teens had at least one suicide attempt in comparison with 20% in teens with unipolar depression and 1% in youth without any psychiatric disorder.

Steve Edwards and colleagues published a simple and straight-forward guide for general medical practitioners to better recognize and respond to suicidal behaviour in youth /Edwards & Pfaff, 1996/. Their core concepts have been developed over many years of clinical practice with young people and have been utilized previously in a wide range of medical and allied health settings. Edwards and colleagues recommend the 4R´s principles – a "ticket of entry" for parents and others wanting to help an emotionally distressed youth by recognizing and responding to suicidal behaviours. The 4R´s include: A/ Recognizing of signs, B/ Raising the issue, C/ Realizing the risks, D/ Responding.

a. The following signs or symptoms should alert parents (teachers and other relatives) that their children may be at risk to develop suicidal thoughts or attempts: persistent symptoms of depression, BD, conduct problems, anxiety disorders, abuse of illicit drugs or alcohol, psychosis (hallucinations or delusions), borderline personality disorder, past or present attempts of suicide, recurrent suicidal ideation, male sex and older than 14 years old. The risk of suicide further increases if in addition to one or more of the above noted signs your child has or more of the following problems: psychological, social or academic problems: poor school performance, employment problems, few or no friends, legal problems, frequent conflict with others (friends, family, teachers), poor coping and problem-solving skills, family or environmental stressors: physical or sexual abuse, neglect, interpersonal loss (girl-or boyfriend, death in the family), rejection by others, family history of mood disorders or suicide, exposure to stress and conflicts in his/her neighborhood/school, exposure to media (e. g. television, radio or newspapers) glorifying suicide, physical health: poor physical health (chronic illness, AIDS, cancer), availability of method (e. g. guns at home, pills).

b. Some "myths" regarding suicide that you should know: Talking about suicide give the child ideas about suicide.

   Not talking about suicide or minimizing the talk about the ideas of suicide will distract the child and he or she forgets about it.

   Children or adolescents who talk about suicide are manipulating or seeking attention.

   It is important show children that our main goal is to help and protect them.

c. The lack of hope or "no light at the end of the tunnel" indicates that the child is thinking that there is no way out of his troubles and he or she may be seriously considering committing suicide. Youth that cannot cope with stressful situations and cannot find solutions for their problems may become more hopeless when confronted with a stressful situation and try to commit suicide. In contrast, certain protective factors such as religious beliefs, caring about the effect of suicide on his/her parents, siblings and friends, and having a strong support system may delay or stop the suicidal behaviour. Finally, having an available method (e. g., guns, pills) increases the risk that a person who is thinking about suicide will use the method. Additionally, the suicide risk is greatest for bipolar patients: during a depressive or mixed episode, while transitioning from the manic/hypomanic state to the depressive state, immediately following psychiatric hospitalization, particularly if the admission to the hospital was due to suicidal ideation or attempt /Birmaher, 2004/.

d. It is important to carefully monitor child. If child was hospitalized due to suicidal behaviours, after his or her discharge from the hospital it is important to have prompt follow-ups because most suicide attempts occur during the first months after discharge from the hospital. The treatment of suicide includes psychosocial therapy and medications depending on the underlying psychiatric disorder and the circumstances around the suicidal attempt. Specific studies to have shown that cognitive-behavioural therapy (CBT), interpersonal psychotherapy (IPT) and psychodynamic psychotherapy are ball options for the management of a suicidal youth but further studies are necessary /Brent et al., 1997/. Studies on the efficacy of lithium for suicide in children and adolescents are needed. Importantly, lithium (and other medications) should not be self-administered by suicidal youth because it is very dangerous in case of an overdose.

The use of medications that may increase disinhibition or impulsivity, such as the benzodiazepines, should be prescribed with caution because they may increase the risk for suicide and are dangerous in the case an overdose. It is important to mention that in cases where a child has committed suicide, his or her relatives, friends and teachers may benefit from psychotherapy to facilitate grieving, reduce guilt and depression, and decrease the effects of guilt. Having the traumatic experience of a significant other committing suicide is often helped through psychotherapy. Moreover, the psychosocial interventions may minimize the risk of imitative or copycat suicides /Shaffer & Pfeffer, 2001/.

4. Treatment

Since suicide is a multicausal human behaviour with many biological, psychological and cultural components, its prevention should also be complex, even in the case of bipolar disorder which is the ″most biological″ illness in the field of psychiatry and which requires long-term pharmacotherapy in the majority of the cases. Since bipolar disorders usually show a peak onset between 15 and 25 years of age, but there is 8-10 years of delay in correct diagnosis /Akiskal, 2002; Ghaemi et al., 2000; Goodwin & Jamison, 1990/, early detection of bipolarity, the nature of the disorder, including the soft manifestations as well, is the first step in suicide prevention. Misdiagnosis of bipolar depression as unipolar depression results in treatment with antidepressants alone, and this can have negative effects on the course of the illness, because of inducing mixed depressive episodes, hypomanic or manic switches, rapid cycling and therefore increasing the chance of suicidal behaviour /Rihmer, 2006; Benazzi & Akiskal, 2003/.

Effective acute and long-term treatment has a strong protection against suicide and probably against other complications (secondary substance-abuse disorders, marital instability, loss of job, cardiovascular mortality, etc.) /Rihmer, 2006/. Intensive treatment including pharmacotherapy, psychotherapy, especially cognitive-behavioural therapy and electroconvulsive therapy should be conducted.

Schwartz & Thase /2011/ examined all randomized trials evaluating the use of pharmacotherapy in the treatment of acute bipolar II depression. Studies with mixed samples of bipolar I and II or bipolar II and unipolar depression were examined as well. Twenty-one randomized trials were identified and reviewed. Therapeutic agents were rated according to the quality of evidence supporting their efficacy as treatments for bipolar II depression. Ninety percent of relevant trials were published after 2005. Quetiapine was judged as having compelling evidence supporting its efficacy. Lithium, antidepressants, and pramipexole were judged as having preliminary support for efficacy. Lamotrigine was considered to have mixed support. Although progress has been made, further research on bipolar II depression is warranted.

4.1 Psychopharmacological therapy

While successful acute pharmacotherapy of depressive or mixed episodes can only prevent the risk of suicide connected with a given episode, it is adequate prophylactic therapy that can provide long-term results in patients with BD.

4.1.1 Lithium

Long-term treatment with lithium provides more consistent and compelling evidence of reduced suicidal risk than any other treatment. The efficacy of lithium in the treatment of manic states and in prevention of recurrences of bipolar patients is well documented /Goodwin & Jamison, 1990; Maj et al., 2000; Bowden, 2002/, and recent data indicate that combination of lithium (and other mood-stabilizers) with antidepressants reduces the chance of hypomanic or manic switching when bipolar depression is treated with antidepressants only (Henry et al., 2001). However, about 50% of bipolar patients do not show satisfactory prophylactic response to lithium. The identification of certain factors predicting a good response versus a partial or non-response to lithium treatment has been made possible by the data that has emerged from research protocols as well as from clinical practice /Bowden, 1998; Swann et al., 1999/. These predictors include a good response to lithium a previous treatment for a manic state, a 'pure' (that is to say, euphoric) manic episode, only a small number of episodes previous to the current episode, absence of rapid cycling, absence of psychotic characteristics, no substance abuse, a manic episode that is for the most part moderate, positive family history of bipolar illness, early onset, mania-depression interval type of course predicts a good prophylactic response, while higher frequency of episodes, depression-mania interval type of course, dysphoric mania, rapid cycling, comorbid substance-use disorders, late onset of manic episodes and mania secondary to a somatic problem; indicate partial or non-response /Aubry et al., 2007; Rihmer, 2006; Bowden, 2002; Maj et al., 2000/. Tondo et al /1997/ concluded that the risk of suicide is seven times higher in patients who are not taking lithium than in those who are. Even emphasizing that the patients were not assigned to the two groups in a random manner in these studies and differences can be observed in the severity of the disorder and/or the compliance of the patients in the two groups, the results are still impressive. In another study, Tondo et al. /1998/ evaluated the risk of suicide before, during and after lithium treatment in 310 bipolar patients. Again the authors demonstrated that the risk of suicide was about 6,5 times lower when a lithium maintenance treatment was administered /review in Baldessarini et al., 2002/. In a meta-analysis of 32 randomized trials comparing lithium, placebo and other active treatments, Cipriani et al. /2005/ found that patients who received lithium were less likely to die by suicide. However, this conclusion has been challenged because one study giving a strong argument for a protective effect of lithium was misunderstood /Connemann, 2006/. Kessing et al. /2005/ reported the results of a nationwide survey in Denmark including about 13 000 patients who purchased at least one the rate of suicide decreased with the number of prescriptions of lithium. Although this study adds to prior evidence that continued lithium treatment is associated with reduced suicide risk, this risk is still considerably higher (about 10 fold) than the rate for the general population. To date, it remains difficult to determine when lithium's preventative effect on suicide takes effect, but it has been reported that this effect was already noticeable in the months following the start of lithium treatment /Dunner, 2004/. Some studies /Kessing et al., 2005; Baldessarini et al., 2006/ include cases of recurrent major depression or schizoaffective disorders as well as a majority of bipolar disorder patients and so are represent manic depressive disorders broadly. There is a nearly four-fold lower relative risk of suicide attempts, and over nine-fold lower risk of suicides, during lithium treatment, and the attempt rate with lithium is only twice that estimated for the international general population, whereas the suicide rates remains 10-times above the general population risk.

Of some importance, the ratio of attempts to suicides, a proposed index of lethality (of intent or means) of suicidal behaviour, is low in manic depressive disorders patients without lithium treatment, and 2,5-times higher with lithium treatment, suggesting decreased lethality with lithium treatment. Moreover, risk increases soon after stopping lithium, especially abruptly /Baldessarini et al., 2006/.

We found the presence of suicidal thoughts and attempts in inpatients with BD (DSM-IV) and assessed changes of treatment with lithium over the period of time /Breznoščáková et al., 2009/. It was retrospective survey of 125 in-patient's file hospitalized at the 1st Dept. of Psychiatry, University of P. J. Šafárik, Košice (1997 - 2007) with typical limitations for restrospective case survey. The average age was 31 years at all, but during first hospitalization 24 years /Tab. 3/. The first episode was depressive in 62% of patiens and average number of episodes was 7, 6 /Tab. 4/. It was trend of decrease in use of lithium over the time (68% vs 84% in men, 29% vs 60% in women).

N=125	men	women	together
Average age during last hospitalization in years	35,4	26,2	30,8
First hospitalization - average age	25,2	23,3	24,1

Table 3. Average age and the 1st hospitalization

	men	women	together
Average number of episodes	8,8	6,7	7,6
The number of depressive episodes	5,0	3,3	4,1
The number of manic episodes	3,3	1,9	2,6
The number of mixed episodes	0,5	1,5	1,0

Table 4. The number of episodes

The average dose of lithium in men was 1, 7g and serum lithium concentration: 0, 65 mmol/l; in women average dose of lithium: 1, 3g and serum lithium concentration: 0, 7 mmol/l /Fig. 1/. During depressive episode given average dose of lithium was 1, 29 g and average serum lithium concentration was 0, 98 mmol/l in men. During depressive episode was given average dose of lithium 0, 57g and average serum lithium concentration was 0, 64 mmol/l in women/Fig. 2/. Average dose of lithium was 1, 37g and average serum lithium concentration was 0, 48 mmol/l during manic episode in men. Any woman had been treatment of lithium during manic episode, what can be possible to explain only hypothetically that there was preferred treatment with atypical antipsychotics for better and faster onset of effect or for higher risk of side effects.

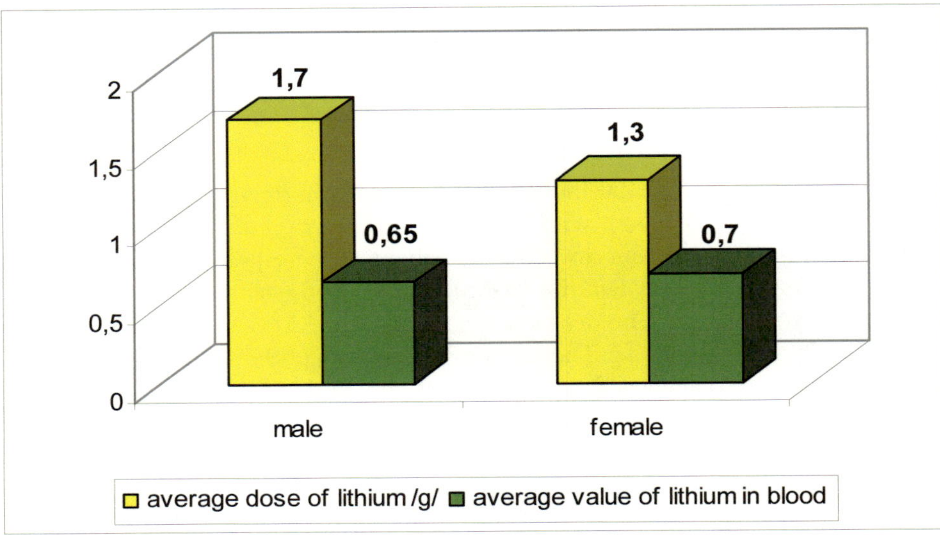

Fig. 1. Dose of lithium/day(g) and serum lithium concentrations(mmol/l)

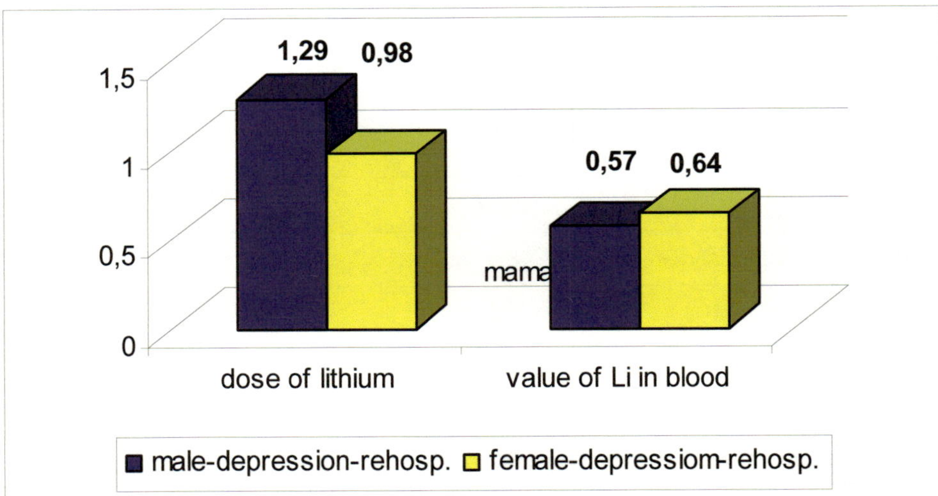

Fig. 2. Dose of Li/Li concentration - depressive episode

The occur of suicidal thoughts was frequent during depressive episode (in 50% of women and 18% of men).

During this time suicidal thoughts occurred in 22 men at all and suicidal attempts in 5 men (the same patient could have suicidal thoughts and attempts, too). We found suicidal thoughts in 62 women and suicidal attempts in 14 women /Fig. 3/. Suicidal thoughts occurred in 24 women without treatment of lithium /from 34/. On the other hand we found higher occur of suicidal thoughts in men with treatment of lithium /19 from 34/. In 20 men /60%/ and in 40 women /79%/ without treatment of lithium were occurred suicidal attempts /Fig. 4/.

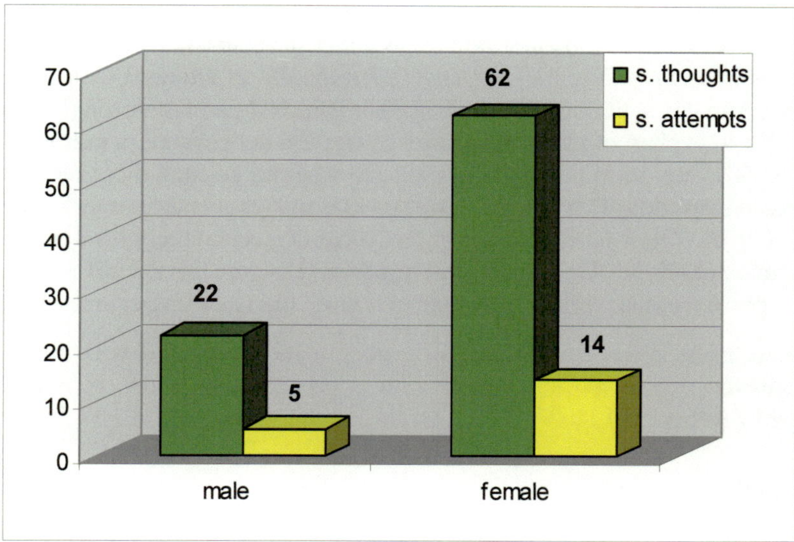

Fig. 3. Frequency of suicidal thoughts and attempts (N=125)

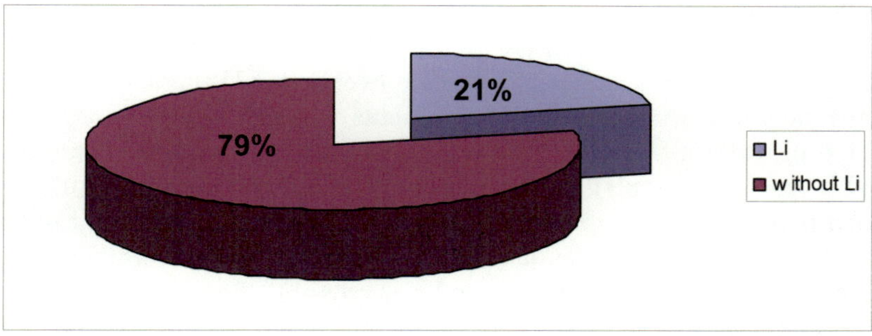

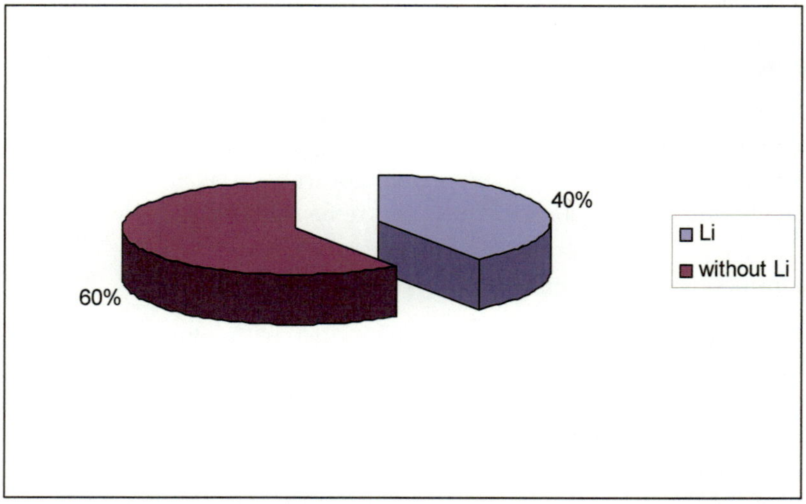

Fig. 4. Suicidal attempts and treatment /men and women/

We can conclude that higher frequency of episodes BD was in men than in women (9 versus 7). The men were taking higher doses of lithium at all /1, 7 versus 1, 3g/ and had lower serum lithium concentrations /0, 65 mmol/l against 0, 7 mmol/l/ as women. We attended trend of decreasing of treatment of lithium in women (29% versus 60%) and men, too (68% versus 84%), what is possible to explain of increasing of use of atypical antipsychotics in role of new mood stabilizers. In 74% of women and 54 % of men are occurred suicidal thoughts, more frequent during depressive episode. Three times more suicidal thoughts and attempts were in women than in men with BD. We found 1, 5 times higher occur of suicidal thoughts in women with BD without treatment of lithium. One of the most important assignments is fourfold higher occur of suicidal attempts in women without treatment of lithium than in with treatment with lithium.

To summarize, even if the suicide rate in bipolar patients treated with lithium remains higher than in the general population, lithium' s preventative effect on suicide is widely acknowledged /Aubry et al., 2007/.

4.1.2 Other anticonvulsants

Unfortunately, vey few data are available on the antisuicide effects of other anticonvulsants in BD. A few years ago, Goodwin et al. /2003/ published a retrospective study based on the follow-up of more than 20 000 patients treated for BD. They showed that the risk of fatal suicide was 2, 7 times higher in patients treated with valproate than in patients treated with lithium. In a randomized, open-label, perspective, 2, 5-year follow-up study Thies-Flechtner et al. /1996/ investigated the number of suicide events in 175 bipolar, 110 schizoaffective and 93 recurrent depressive patients. The patients were randomly assigned to lithium, carbamazepine or amitriptyline. There were 14 very serious suicide events (9 completed suicides and 5 serious attempts) during the study and 7 out of the 14 suicide acts were among bipolar patients. Most of the 14 suicide acts happened in the carbamazepine group (4 suicides and 5 attempts), and none of the 14 suicidal patients were taking lithium. Most anticonvulsants remain largely unexamined for possible beneficial effects on suicidal behaviour. Nevertheless, there is some evidence that other anticonvulsants may be less effective than lithium against suicidal behaviour /Baldessarini et al., 2006/.

4.1.3 Antipsychotics

Clozapine is the first treatment to receive FDA approval for reducing risk of suicidal behaviours, though only for schizophrenia patients, and without clear evidence of reduced risk of completed suicides /Hennen & Baldessarini, 2005/. This time, ability of antipsychotics to limit risks of suicidal behaviour remains untested.

4.1.4 Antidepressants and electroconvulsive therapy

Suicidal behaviour is particularly strongly associated with acute depressive illness and antidepressants have been proved effective, at least in the short-term treatment of acute, non-psychotic, nonbipolar major depressive disorder of moderate severity /Baldessarini & Tarazzi, 2005/. Adverse behavioural responses to antidepressant treatment in mood disorder patients, including insomnia, restlessness, irritability, and agitation, may well increase risk of aggressive-impulsive acts, perhaps including suicidal behaviour, in some adults and children, and particularly in depressive or mixed phases of BD /Fawcett J, 1997/.

The indication for electroconvulsive therapy (ECT) may exist, besides for refractoriness to drug treatment, also for pregnancy which places restrictions on drug treatment and for patients at high risk for suicide in whom a rapid improvement is required, such as, e. g. in mixed mania. In these cases the early use of ECT can be considered, as well as for severe and, especially, psychotic manias and depressions /Walden & Grunze, 2004/. Nevertheless, effectiveness of ECT for sustained suicide prevention has not been proved, and requires further study, including in BD patients, specifically /Jacobs, 2003/.

4.2 Psychosocial interventions

Rucci et al. /2002/ investigated the lifetime rates of suicide attempts among 175 bipolar I patients during a 2-year period of intensive pharmacotherapy (lithium, valproate, carbamazepine), and one of two adjunctive psychosocial interventions (psychotherapy specific to bipolar disorder or nonspecific intensive clinical management). They found that the patients experienced threefold reduction during maintenance. There was no significant difference regarding suicide attempts between the two subgroups with different psychosocial interventions. It has been reported that successful episode-preventive medication with mood stabilizers in bipolar patients, counteracted dysfunctional cognitions (including lowered self-esteem), and adjunctive cognitive therapy could help to optimize the long-term course of bipolar illness /Wolf & Müller-Oerlinghausen, 2002/. Psychotherapeutic techniques based on problem-solving, rehearsal, and cognitive or behavioural methods may limit suicidal risk in relatively mildly ill depressed patients. It is high encourage to make a so-called “emergency plan” for patients, especially with BD /Breznoščáková et al., 2008/.

5. Conclusion

Leverich et al /2003/referred to bipolar disorder as "the most complex, highly comorbid, and potentially lethal illness among the major psychiatric disorders and far higher than in general population". They concluded that without the healthcare support required to deal with the illness, its associated morbidity and suicidality will continue to have devastating effects on large numbers of individuals and their families. Suicidal acts are likely to occur early in BD illness, and in association with severe depressive or mixed states /Baldessarini et al., 2006/. Acute, serious negative life events can precipitate suicide behaviour in depressed patients especially, who are particularly vulnerable for suicide behaviour. Evidence of long-term effectiveness of treatments against mortality risks in BD is limited, with the notable exception of lithium prophylaxis, which clearly reduces risk of death from suicide and may also limit risk of death from natural causes /Ahrens et al., 1995; Cipriani et al., 2005/. As for suicide prevention, the main target should not only be persons who are acutely suicidal, but all depressive (and other psychiatric) patients. Their treatment should start as early as possible in order to prevent the progression of the illness and the development of suicidal thoughts and behaviour /Rihmer & Kiss, 2004/.

We are, of course, unable to prevent all suicides. However, our present pharmacological and psychosocial intentions are effective enough to minimize the chance of suicide in patients with BD that represents the highest risk of self-inflicted death /Rihmer, 2006/.

6. References

Ahrens B, Miller-Oerlinghausen B, Schou M et al. (1995). Excess cardiovascular and suicide mortality of affective disorders maz be reduced by lithium profylaxis. J Affect Disord; 33: 67-75

Akiskal HS (1981). Subaffective disorders: Dysthymic, cyclothymic and bipolar II disorders in the „borderline" realm. Psychiatr Clin North Amer, 4: 25-46

Akiskal HS (2002). Classification, diagnosis and boundaries of ipolar disorders: A review. In Bipolar disorder, Maj M, Akiskal HS, Lopez-Ibor JJ, Sartorius N, John Wiley & Sons, Chichester

Angst F., Stassen HH., Clayton PJ., Angst J. (2002). Mortality of patiens with mood disorders: follow-up over 34-38 years. J Affect Disord; 68: 167-181

Angst J (1998). The emerging epidemiology of hypomania and bipolar II disorder. J Affect Disord; 50: 143-151

Aubry JM, Ferrero F, Schaad N, bauer MS. (2007). Pharmacotherapy of Bipolar Disorders, Hohn Wiley & Sons, 978-0-470-05823-7, 0 Chichester, England

Baethge C, Tondo L, Bratti IM et al. (2003). Prophylaxis latency and outcome in bipolar disorders. Can J Psychiatry; 48: 449-457

Balázs J, Lecrubier Y, Csiszér N, Kozsták J, Bitter I (2003). Prevalence and comorbidity of affective disorders in persons making suicide attempts in Hungary: Importance of the first depressive episodes and of bipolar II diagnoses. J Affect Disord; 76: 113-119

Baldessarini RJ, Pompili M, Tondo L (2006). Suicide in Bipolar Disorders: Risks and Management, CNS Spect 11:6, 465-471

Baldessarini RJ, Tarazzi FI (2005). Drugs and the treatment of psychiatric disorders. In: Brunton LL, Lazo JS, parker KL, Goodman and Gillman´s. The Pharmacological Basis of Therapeutics 11th ed., McGrav-Hill Professional, New York

Baldessarini RJ, Tondo L, Hennen J, Viguera AC (2002). Is lithium still worth using ? An update of selected recent research. Harv Rev Psychiatry, 10: 59-75

Benazzi F, Akiskal HS (2003a). refining the evaluation of bipolar II: Beyond the SCID-IV guidelines for hypomania. J Affect Disord; 73: 33-38

Benazzi F, Akiskal HS (2003b). Clinical and factor-analytic validation of depressive mixed states: A report from the Ravenna-San Diego collaboration. Curr Opi Psychiatry; 16 (Suppl 2), 70-78

Birmaher B (2004). New Hope for children and teens with bipolar disorder, Three Rivers Press, 0-761-52718-4, New York

Bowden CL (1998). Key treatment studies of lithium in manic-depressive illness: efficacy and side effects. J Clin psychiatry, 59 Suppl 6: 13-19; discussion 20

Bowden CL (2002). Pharmacological treatment of bipolar disorder: A review. In Bipolar disorder, Maj M, Akiskal HS, Lopez-Ibor JJ, Sartorius N, John Wiley & Sons, Chichester

Breznoščáková D, Novák T, Stopková P (2008). How to under control mania and depression (Ako sa vyrovnať s mániou a depresiou). Maxdorf, 978-80-7345-132-5, Praha

Breznoščáková D, Pálová E, Bodnár B et al. (2009). Suicide in inpatients with bipolar disorder (Suicidalita u hospitalizovaných pacientov s bipolárnou afektívnou poruchou), Psychiatr. prax, 10 (1); 29-31

Brieger P (2000). Comorbidity in bipolar affective disorder. In Bipolar disorders: 100 years after manic depressive insanity, Marneros A, Angst J, Kluwer Academic Publishers, Dordrecht

Bulik CM, Carpenter LL, Kupfer DJ, Frank E. (1990). Features associated with suicide attempts in recurrent major depression. J Affect Disord, 18: 29-37

Chen YW, Dilsalver SC (1996). Lifetime rates of suicide attempts among subjects with bipolar and unipolar disorders relative to subjects with other axis I disorders. Biol Psychiatry; 3: 896-899

Cheng ATA, Chen THH, Chen CC, Jenkins R (2000). Psychological and psychiatric risk factors for suicide: Case-control psychological autopsy study, Brit J Psychiat, 177: 360-365

Cipriani A, Pretty H, Hawton K, geddes JR (2005). Lithium in the prevention of suicidal behavior and all-cause mortality in patients with mood disorders: a systematic review of randomized trials. Am J Psychiatry, 162: 1805-1819

Connemann BJ (2006). Lithium and suicidality revisited. Am J Psychiatry, 163: 550; author reply 550-551

Dalton EJ, Cate-Carter TD, Mundo E, Parikh SV, Kennedy JL. (2003). Suicide risk in bipolar patients: the role of co-morbid substance use disorders. Bipolar Disord.; 5(1):58-61. (1398-5647)

Dunner DL (2004). Correlates os suicidal behavior and lithium treatment in bipolar disorder. J Clin Psychiatry, 65 Suppl 10: 5-10

Dunner DL, gershon ES, Goodwin FK (1976). Heritable factors in the severity of affective illness. Biol Psychiatry; 11: 31-42

Faedda GL, Baldessarini RJ, Suppes T, Tondo L, becker I, Lipschitz D (1995). Pediatric-onset bipolar disorder: a neglected clinical and public health problem. Harv Rev Psychiatry; 3: 171-195

Fawcett J (1997). The detection and consequences of anxiety in clinical depression, J Clin Psychiatry, 58 (Supll. 8): 35-40

Finn KK. (2007). Bipolar and Pregnant: how to manage and succeed in planning and parenting while living with manic depression, Health Communications, Inc., 13-978-0-7573-0683-9, FL, USA

Ghaemi SN, Boima EE, Goodwin FK (2000). Diagnosing bipolar disorder and the effect of antidepressants: A naturalistic study. J Clin Psychiatry; 61: 804-808

Ghaemi SN, Rosenquist KJ, Ko JY, Baldessano CF, Kontos NJ, Baldessarini RJ (2004). Antidepressant treatment in bipolar versus unipolar depression. Am J P01sychiatry; 161: 163-165

Goodwin FK, Jamison KR. (1990). Manic-Depressive Illness: Oxford University Press, New York, NY

Goodwin FK, Fireman B, Simon GE, Hunkeler EM, Lee J, Revicki D (2003). Suicide risk in bipolar disorder during treatment with lithium and divalproex. JAMA, 290: 1467-1473

Guze SB, Robins E (1970). Suicide and primary affective disorders. Brit J Psychiatry; 117: 437-438

Harris EC, Barraclough B (1997). Suicide as an outcome for mental disorders. Brit J Psychiatry; 170: 205-228

Hawton K, Arensman E, Wassermann D et al. (1998). Relations between attempted suicide and suicide rates among young people in Europe. J Epidemiol Community Health; 52: 191-194

Hawton K, van Heeringen C (2000). International handbook of suicide and attempted suicide. John Wiley and Sons, Chichester

Hennen J, Baldessarini RJ (2005). Suicidal risk during treatment with clozapine: a metaanalysis. Schizophr Res; 73: 139-145

Henry C, Sorbara F, Lacoste J. Gindre C, Leboyer M (2001). Antidepressant-induced mania in bipolar patients: Identification of risk factors. J Clin Psychiatry; 62: 249-255

Horowitz MJ, Siegel B, Holen A, Bonanno GA, Milbrath C, Stinson CH (1997). Diagnostic criteria for complicated grief disorder. Amer J Psychiatry, 154: 7, 904.910

Isometsä E, Heikkinen M, Henriksson M, Aro H, Lönnqvist JK (1994). Recent life events and completed suicide in bipolar affective disorder. A comparison with major depressive suicides. J Affect Disord, 33: 99-106

Jacobs DG (2003). Practice guideline for the assessment and treatment of patients with suicidal behaviors. Am J Psychiatry; 160(11 Suppl): 1-60

Kessing LV, Sondergard L, Kvist K, Andersen PK (2005). Suicide risk in patients treated with lithium. Arch Gen Psychiatry, 62: 860-866

Kessler RC, Chiu WT, Demler O (2005). Prevalence, severity, and comorbidity of 12-month DSM-IV disorders in the National Comorbidity Survey Replication. Arch Gen Psychiatry; 62: 617-627

Leverich GS, Altshuler L, Frye M, et al. (2003) Factors associated with suicide attempts in 648 patients with bipolar disorder in the Stanley Foundation bipolar disorder in the Stanley Foundation bipolar network. J Clin Psychiatry; 64:506-515.

Lewinsohn PM, Klein DN, Seely JR (1995), Bipolar disorders in a community sample of older adolescents: prevalence, fenomenology, comorbidity, and course. J Am Child Adolesc Psychiatry 34 (4): 454-63

López P, Mosquera F, de León J, Gutiérrez M, Ezcurra J, Ramírez F (2001). Suicide attempts in bipolar patiens. J Clin Psychiatry, 62: 963-966

Maj M, Tortorella A, Bartoli L (2000). Mood stabilizers in bipolar disorder. In Bipolar disorders. 100 years after manic-depressive insanity, Marneros A, Angst J, Kluwer Academic Publishers, Dordrecht

Maser JD, Akiskal HS, Schettler P, Scheftner W, Mueller T, Andicott J, Solomon D, Clayton P (2002). Can temperament identify affectively ill patients who engage in lethal or nonlethal suicidal behaviour? A 14-year prospective study. Suicide and Life-Threatening Behavior; 32: 10-32

Moťovský B (2009). Bipolar affective disorder II (Bipolárna afektívna porucha II), Psychiatr. prax, 10 (5): 212-215

Oquendo MA, Waternaux C, Brodsky B, Parsons B, Haas GL, Malone KM (2000). Suicidal behavior in bipolar mood disorders: clinical characteristics of attempters and nonattempters. J Affect Disord, 59: 107-117

Oral ET. (2005). Treatment of acute mania, In: Treatment of Bipolar Disorders with Second Generation Antipsychotic Medications, Jarema M., 9-10, Society of Integrated Sciences, 0172-780X

Purifacion L, Mosquera F, deLeon J, et al. (2001). Suicide attempts in bipolar patients. J Clin Psychiatry; 62(12):963-966. (0160-6689)

Rihmer Z (2006). Suicide Prevention, In: Bipolar Psychopharmacotherapy: Caring for the Patient, Akiskal HS, Tohen M, John Wiley & Sons, 13 978-0-470-85607-9, Chichester, England

Rihmer Z, Barsi J, Arató M, Demeter E (1990). Suicide in subtypes of primary major depression. J Affect Disord; 18: 221-225

Rihmer Z, Kiss K (2004). Risk factors for suicide in bipolar disorders: In Bipolárna porucha (Bipolar Disorder). Vavrusova L, Osveta, 80-8063-136-0, Martin

Rihmer Z, Rutz W, Pihlgren H (1995). Depression and suicide on Gotland. An intensive study of all suicides before and after a depression-training programme for general practitioners. J Affect Disord; 35: 147-152

Rihmer Z, Szádoczky E, Fűredi J, Kiss K, Papp Z (2001). Anxiety disorders comorbidity in bipolar I, bipolar II and unipolar major depression: results from a population-based study in Hungary. J Affect Disord, 67: 175-179

Roy A (1983). Family history of suicide. Arch Gen Psychiatry; 40: 971-974

Rucci P, Frank E, Kostelnik B, Fagiolini A, Malinger AG, Schwartz HA, Thase ME, Siegel L, Wilson D, Kupfer DJ (2002). Suicide attempts in patients with bipolar I disorder during acute and maintenance phases of intensive treatment with pharmacotherapy and adjunctive psychotherapy. Am J Psychiatry; 159: 1160-1164

Scott J. (2005). Psychological treatments: does the evidence stack up? In Bipolar Disorder: The upswing in research and treatment. McDonlad C et at., 165-167, Taylor&Francis, 1-84184-501-9, Abingdon, Oxon, UK

Schwarz HA, Thase ME (2011). Pharmacotherapy for the treatment of acute bipolar II depression: current evidence. J Clin Psychiatry; 72(3): 356-66 (1555-2101)

Simon GE, Hunkeler E, Fireman B, Lee JY, Savarino J. (2007). Risk of suicide attempt and suicide death in patients treated for bipolar disorder. Bipolar Disord. 9:526-530

Stengel E (1964). Suicide and attempted suicide. Penguin, London

Strakowski S., McElroy S., Keck Jr P., West S. (1996). Suicidality among patiens with mixed and manic bipolar disorder. Amer. J. Psychiat.; 153: 674-676

Swann AC, Bowden CL, Calabrese JR, Dilsaver SC, Morfia DD (1999). Differential effect of numer of previous episodek of affective disorder on response to lithium or divalproex in acute mania. Am J Psychiatry, 156: 1264-1266

Szádóczky E, Papp Z, Vitrai J, Rihmer Z, Fűredi J (1998). The prevalence of major depressive and bipolar disorders in Hungary. J Affect Disord; 50: 153-162

Takahashi Y (1993). Depression and Suicide. In Affective Disorders: perspectives on Basic Research and Clinical Practice, Kariya T, Nakagawara M, Seiwa Shoten Publishers, 0-87630-674-1, Tokyo, Japan

Thies-Flechtner K, Müller-Oerlinghausen B, Seibert W, Walther A, Greil W (1996). Effect of prophylactic treatment on suicide risk in patients with major affective disorders. Pharmacopsychiatry; 29: 103-107

Tondo L, Albert M, Baldessarini RJ (2006). Suicide rates in relation to health-care access in the United States. J Clin Psychiatry; 67: 517-523

Tondo L, Baldessarini RJ (2005). Suicidal risk in bipolar disorder. Clin Neuropsychiatry; 2: 55-65

Tondo L, Baldessarini RJ, Hennen J., Floris G, Silvetti F, Tohen M (1998). Lithium treatment and risk of suicidal behavior in bipolar disorder patiens. J Clin Psychiatry, 59: 405-414

Tondo L, Jamison KR, Baldessarin RJ (1997). Effect of lithium maintenance on suicidal behavior in major mood disorders. Ann N Y Acad Sci, 836: 339-351

Vieta E, Benabarre A, Colom F, Gastó C, Nieto E, Otero A (1997). Suicidal behavior in bipolar I and bipolar II disorder. J Nerv Ment Disord, 185: 407-409

Walden J., Grunze H. (2000). Bipolar Affective Disorder: Etiology and Treatment, Georg Thieme Verlag, 3-13-105611-8, Stuttgart

Walden J., Grunze H. (2004). Bipolar Affective Disorders, Georg Thieme Verlag, 3-13-105612-6, Stuttgart

Wassermann D (2000). Suicide. An unecessary death. Martin Dunitz, London

Wolf T & Müller-Oerlinghausen B (2002). The influence of successful prophylactic drug treatment on cognitive dysfunction in bipolar disorders. Bipolar Disorders; 4: 263-270

Part 3

Neurodevelopmental Aspects

7

Correlations Between the Monoaminergic Status and the Psychoneuroendocrine Typology in a Murine Model – Possible Biomolecular Predictions for an Individualized Pharmacotherapy

Andreea Letitia Arsene, Niculina Mitrea and Dumitru Lupuliasa
Carol Davila University of Medicine and Pharmacy,
Faculty of Pharmacy Bucharest,
Romania

1. Introduction

The progress in the molecular pharmacology area continues to maintain a very alert rhythm at the beginning of the 3rd millennium. The volume of information accumulated in this area, with boundaries constantly expanding, as modern technology applied in current research which allows linking the imbalances manifested in the central nervous system metabolism with its various pathologies.

The neurons represent unique anatomical fractions, with the ability to forward the information in a network system, which justifies the experience-dependent mechanisms such as memorization, learning or consciousness. In order to fulfill these functions, the neurons are, structurally and functionally polarized. This aspect is obvious in their tripartite composition: cell body, axon and dendrites. While the body comprises the biosynthesis structure (nucleus, ribosomes, endoplasmic reticulum, Golgi apparatus and mitochondria for energy storage), the axon is equipped with molecular and under-cellular components for the propagation of the action potential from the cell body to distant targets, the dendrites represent a set of branched endings which prolong from the cell body and have as result an increase of it, in order to receive the signal. The structural connections for the signal transmission are the basis of the neural architecture. Inside the mature nervous system, normally, the neural circuit (neuro-architecture) supports adaptive changes. Inside an old synapses neural cells can involute (until their disappearance), with simultaneous formation of new branches (new neuritis) which will establish new synapses. These changes in the neurotic and synaptic architecture prove the great plasticity (capacity) of the nervous system to adapt to the environmental conditions. Today we consider that there are two major intracellular classes of factors (signals) which adjust the neuronal development and adaptation: growth factors (the growth factor - fibroblasts, the ciliary neutrophil factor, etc.) and the neurotransmitters factors (dopamine, serotonin, acetylcholine, glutamate, etc.). The loss of balance between the two systems of factors leads to the development of the pathological states.

The investigation of the central nervous system remains a fascinating field; it constitutes an inexhaustible source of new discoveries. The diversity of human nature and the uniqueness of the individual cerebral architecture lead to the exploration in this area on very subtle and difficult ways to analyze.

The individuality cell status, immune, genetic, etc., raised many questions, over time, and opened gates for new researches on human *polytypism*. Thus, under the action of identical factors, physical and emotional, the human beings develop different physiological, behavioral and emotional responses. This phenomenon was called **polytypism.** The phenomenon was inferred in 1980 by Rosenman RH who described, for the first time, the adrenergic behavioral type (type A).Lately, in the 1990s there was identified and described the opioid type (type O), the "non-A" type, opposite to the adrenergic type from the behavior point of view, and in 2000 the behavioral oncology has introduced the concept of C type, with genetic predisposition to the development of the biopsychosocial cancer.

In this context, the individualization of pharmacotherapy based on the psychoneuroendocrine typology, could be an important step for a specific medication, targeted and with fewer side effects. For this purpose, the molecular pharmacology researches in the world should bring evidences regarding the cellular, molecular processes which are the base of the pathophysiology for different psychoneuroendocrine types.

2. The psychoneuroendocrine behavior typology, factor of the biological and pharmacological variability

The differentiation of the adrenergic typology was first realized in 1978 by RH Rosenman, by describing some specific behavioral characteristics that predispose it to the emergence and the development of psychiatric and cardiovascular diseases: competitiveness, sharp ambition, continuous involvement in multiple and diverse activities, with a sense of haste and time urgency, irritability, impulsivity, reduced ability to disconnect and relaxation. Later, in the 1990s there was identified and described the "non-A" type, opposite to the adrenergics from the behavior point of view. Presently it is defined as the opioid type (O type), with the psychoneuroendocrine predominance of the endogenous opioid system. It has the following characteristics: defensive, calm, relaxed, non-aggressive, introverted, resistant to pain, but with predisposition to the hiperalgia post-stress syndrome.

The specialty literature describes several methods for identifying the A type of behavior in humans:

- **The structured interview (SI),** developed by Rosenman and Friedman (in 1978), in studies on mid-range employees. This interview contained a series of questions and followed: the volume of voice, the speed of response, the rhythm of the words, the latency of the responses, the gestures, the mimics, and the signs of hostility.
- **Self-assessment methods,** where the specialty literature includes:
    - **Jenkins Questionnaire (JAS)** contained 52 questions, similar with the structured interview and was developed by Jenkins in collaboration with Rosenman and Friedman;

- **Framingham Scale** contained 10 questions about the time urgency, the competitiveness and the motivation to work, the strong need to excel, the measures in which it feels dominated, the speed of eating;
- **Bortner grading scale** consisted of 14 questions and hass been extensively used in the European epidemiological studies;
- **Special Scale** useful for the identification of some particular features of A type (e.g. Interpersonal Communication).

3. Clinical and murine studies for investigating the adrenergic behavioral psychoneuroendocrine typology

The specialty literature describes numerous clinical and murine studies that have attempted to correlate the behavioral characteristics of A type with different physiological and biochemical parameters responsible for the onset of the psychiatric and cardiovascular diseases.

3.1 The cardiovascular reactivity

Numerous clinical studies (Appleton K.M) have been performed to correlate the characteristic features of A type with the cardiovascular responses, especially in stress. Thus, after the differentiation of individuals taken into study in groups of type A and type B (based on the Structured Interview validated by Rosenman, they were submitted to stressful situations (e.g. mental arithmetic exercises or unpleasant images displayed on a screen). They determined the following parameters: electrocardiogram, blood pressure, heart rate, peripheral vasoconstriction (the fingers). All studies have shown for the A type subjects exaggerated cardiovascular responses, in stress conditions.

Jones et al. investigated the variability of blood pressure in type A, related to the pleasure of victory. Thus, it was found that both victory and defeat, determined significantly higher cardiovascular responses for type A versus type B. The type A winners were distinguished from all other subjects by maintaining high levels of blood pressure and the type A losers have lost interest in competition.

It was also pursued in a study of 81 volunteers (published by Meesters C.M.G), the associatian between hostility and the risk of death by acute myocardial infarction. For comparison there were used a lot of subjects which had suffered a first myocardial infarction and a group which had no previous event. Following this study there was shown that the hostility was significantly associated with the risk of myocardial infarction, in people aged over 50 years.

In a clinical study conducted in the U.S.A., there was studied the generally individual sympathetic tonus of the type A individuals versus the ones of type B, by determining the pupil's diameter and also the platelet catecholamine concentrations (Powell L.H).

The study was conducted on 112 volunteers, differentiated on the basis of the Structured Interview. The individuals involved in the behavioral A typology were evidenced by a significantly higher adrenergic tone than on type B: higher pupillary diameter and high concentrations of platelet adrenaline.

In a study conducted in Canada there were followed, at individuals of type A, variations of the cardiovascular parameters after the administration, in bolus, a 50µg CCK-4 (cholecystokinin-tetrapeptid). This compound, administered in the shown manner, has the ability to produce symptoms of panic (panicogen agent). The variations of the heart rate, after the administration of CCK-4, were significantly higher in type A versus type B (Le Melledo J.M).

3.2 Pain sensitivity (endogenous analgesia)

Clinical studies (Cristea A) were conducted for evaluating analgesia and also the post-stress syndrome, in acute and chronic stress in type O compared with type A. Data reported in literature evoked the following:

- the basal pain sensitivity was significantly different in the two psychoneuroendocrine types (**hypersensitivity to pain for type A, compared with a painful hypo-sensibility in type O**);
- in chronic stress, the opioid becomes hypersensitive to pain and the adrenergic manifests depressive symptomatology.

3.3 Nervous behavior of types A and O

In a psychiatric hospital from Norway *(Haukeland University Hospital)* a clinical trial was conducted on 99 patients admitted for behavioral disturbances associated with depression (Oedegaard K.J). The patients were distinguished in lots A and B, based on the Jenkins questionnaire and there was followed the frequency of unipolar depression, bipolar depression and behavior depending on the type of migraine. The results of the study showed the following:

- the patients classified in the behavioral adrenergic typology were diagnosed with bipolar depression (in proportion of 65%);
- type A is characterized by a ciclotimic temper;
- the frequency of the migraine crisis was not significantly correlated with either type A or type B.

Comparative experimental studies have been conducted on animal behavior of type A and O, in pharmacological tests: the actometry test (for investigation the spontaneous motor activity), the platform test, the inclined plane test and the plate with holes test (to research the evasion-investigation behavior), the cross-maze test (for investigating the anxiety). These tests were performed on animal communities distributed on both psychoneuroendocrine typology and on gender (male of type A and male of type O, respectively on female of type A and O). The motor activity and the behavior of investigation were significantly higher in type A, the best results being recorded for the male animal communities (Cristea A). The cross-maze test revealed a significant predisposition to anxiety of the adrenergic type, regardless the gender.

Some authors speculated that the pathophysiological hallmark of type A individuals is the hyperactivity of the sympathetic nervous system, although the molecular basis of these findings have not been established, yet. It is therefore necessary to discern the intimate, molecular mechanisms determining both the susceptibility to certain diseases and a pharmacotherapy of choice and with few side effects. From this point of view the pathogenic picture caused by the imbalance of the monoaminergic system becomes very important. Precise information regarding the different concentrations of monoamines in the cerebral tissue can bring considerable benefits to highlight the dynamics of these neuromediators. In addition, various therapies can change depending on the monoaminergic status for each individual.

3.4 Aim of the study

Based on these considerations, the researches conducted and presented in this paper were aimed to investigate the cerebral monoaminergic status of the adrenergic psychoneuroendocrine typology. Results were compared with the opposite psychoneuroendocrine type (type O) and with the intermediate, balanced, normal type (type N).

The paper presents an experimental model, on mice, to investigate the possible correlations between the behavioral psychoneuroendocrine typology and the brain levels of noradrenaline (NA), dopamine (DA), serotonin (5-HT) and γ -amino butyric acid (GABA). We used two different experimental contexts:

- correlations between the cerebral levels of the monoamines with the behavioral psychoneuroendocrine typology in **basal state**;
- correlations between the cerebral levels of the monoamines with the behavioral psychoneuroendocrine typology after exposure to **acute stress**.

4. Matherials and methods

4.1 Animals

Studies were conducted on 120 male, Albino Swiss mice, weighing 20-22 g. They were housed in a room and maintained at 25 ± 2°C and 45-55% relative humidity, with an alternating 12h light-dark cycle. They had free access to food and water untill the morning of the experiment. All animals used in this study were maintained in facilities fully accreditated and the experiments described here were performed in compliance with the European Communities Council Directive of 24 November 1986 (86/609/EEC) and Ordinance No. 37 of the Romanian Government from 2nd February 2002.

4.2 Identification of the murine behavior type

Mice were divided in three behavioral groups, according to their reactivity to painful stimulus (endogenous analgesia): type A (associated with hypersensitivity to pain) and type B ("non-A", the opposite type of behavior) which exhibits hypo-reactivity to pain.

For the identification of the murine behavior type we used the hot-plate test (Ugo Basile apparatus). The plate was heated at 60 °C and the animals were divided, based on their reactivity to painful stimulus (endogenous analgesia, expressed as *jumping time off the heated plate*), into three working groups: the adrenergic "A" type, the equilibrated, intermediate,

"N" type and the "O" type, according to Gauss normal distribution curve. The average value (M) of the jumping time off the plate was established. Mice that possessed a value of the jumping time of M ± 1SD were selected as intermediate, "N" type. Mice that possessed the value of the jumping time less than M - 1SD (<M - 1SD) were selected as adrenergic "A" type, while the jumping time of more than M + 1SD (>M + 1SD) were selected as the "non-A" type ("O" type).

Acute stress was induced to animals using the classical forced swimming test. (Petit-Demouliere B.)

4.3 The assessment of cerebral monoaminergic status

Mice were sacrificed by decapitation and the whole brains were rapidly removed, weighed and kept at -80°C until analyzed. Cerebral monoaminergic status (neuronal monoamines levels) within the behavioral model was evaluated by measuring the murine brain levels of noradrenaline (NA), dopamine (DA), serotonin (5-HT) using HPLC with UV detection, and γ-aminobutiric acid (GABA), using a fluorimetric assay. Results were expressed as µg neurotransmitter/mg wet tissue.

4.4 Simultaneous, HPLC evaluation of endogenous NA, DA and 5-HT

Samples were homogenized and deproteinized in 0.2M perchloric acid containing 100µM EDTA-Na_2. The homogenate was left for 30min. to deproteinize. Then, the homogenate was centrifuged at 10,000×g for 15min at 0°C (Janetzki K24 cooling centrifuge). After centrifugation, the supernatant was adjusted to pH=3.35 by adding 1M acetic acid. 20µl were injected into an HPLC reversed-phase system (VARIAN - PROSTAR) with a Chromsep-Inertsil 5 OSD2, 250x 4.6 mm column. The mobile phase consisted of 0.8 mM EDTA-Na_2, 0.12M $NaH_2PO_4 \times H_2O$, 0.646g sodium heptane sulphonate and 17% methanol. Monoamines were detected simultaneously, using an UV detector (210nm).

4.5 Determination of endogenous GABA concentrations

Brain tissue samples were separately homogenized in 10 volumes of 0.01 M HCl using a glass homogenizer. The determination of endogenous GABA concentration is based on a fluorimetric assay that depends on the formation of a fluorescent product from the reaction between GABA and ninhydrin at alkaline pH and in the presence of glutamate. The reagents used in the assay were 0.05 M glutamic acid in 0.2 M sodium phosphate buffer, pH 6.4, 14 mM ninhydrin in 0.5M sodium carbonate buffer, pH 9.9-10 and copper tartrate reagent consisting of 1.6g Na_2CO_3, 329 mg tartric acid and 300 mg $CuSO_4 \times 5H_2O$, all made up in 1 liter of distilled water. 0.25 ml of homogenate was diluted with 0.25 ml of 0.01 M HCl and 0.5 ml of 10% trichloroacetic acid. This last reagent was used to precipitate the proteins. After the samples were centrifuged, 100 µl aliquots of the supernatant were added to 15 µl of glutamate solution and 200 µl of the ninhydrin solution. This mixture was incubated at 60°C for 30 minutes and allowed to cool before the addition of 5 ml copper tartrate reagent. We also prepared two internal standards by adding to the samples of homogenate known amounts of GABA (50 µg and 100 µg GABA per sample) with the trichloroacetic acid.

5. Results and discussion

5.1 Assessment of the basal neuronal concentrations of noradrenaline, dopamine, serotonin and gamma-amino butyric acid on the adrenergic and opioid types of behavior

5.1.1 Distribution on psychoneuroendocrine murine groups

The researches were conducted on a community consisting of 100 mice. In the first stage was established the individual reactivity to pain of the entire group (n = 100) using the hot plate test. There have been registered the results of the individual reaction to pain expressed by the times of jump. Based on these experimental data there could be noticed that for the collectivity of animals tested for reaction to pain, the individual values of the times of jump ranged between 4 and 60 seconds, the average time of jump (Jt medium) was of 36.02 sec, and the standard deviation (SD) value was of 13.9 sec.

Depending on the individual reactivity to pain, the types of behavior were defined as follows:

- the adrenergic type (A), hypersensitive: Jt = 4-22 sec;
- normal, balanced type (N), normo-sensitive: Jt = 29-42 sec;
- the opioid type (A), hipo-sensitive: Jt = 50-60 sec.

The distribution of the animals from the researched collectivity, depending on the pain sensitivity is shown in Figure 1.

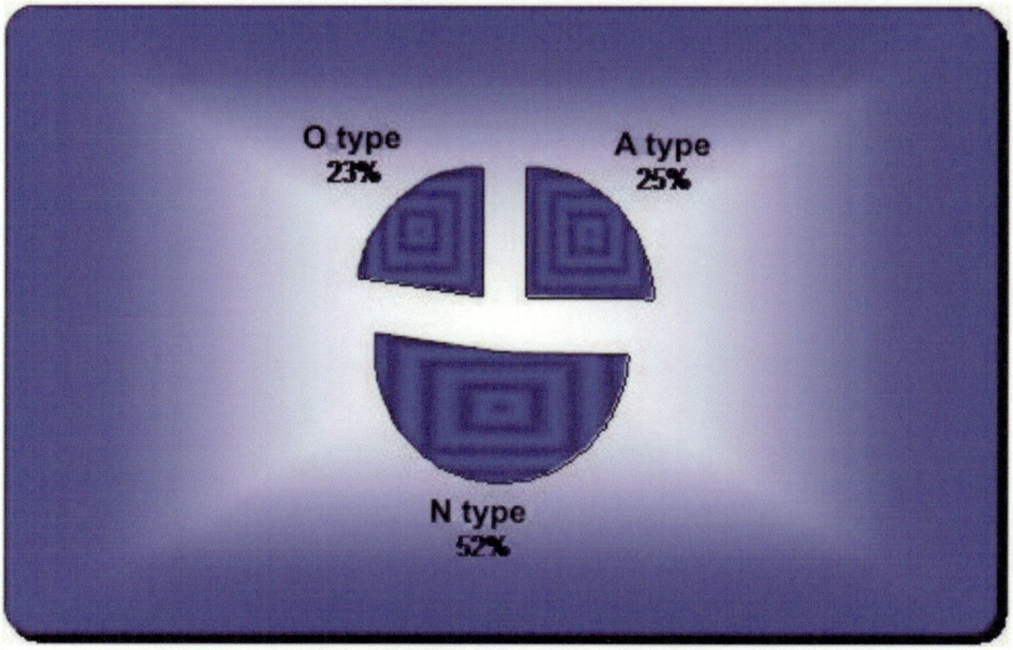

Fig. 1. The distribution of animals (n = 100) according to their pain sensitivity

The groups average values of the neuronal endogenous concentrations of noradrenaline (NA), dopamine (DA), serotonin (5HT) and GABA are shown in Table 1.

Behavior type	Cerebral concentration of the studied neuromediators in the basal state (µg/ mg wet tissue)			
	NA	DA	5HT	GABA
A type	1.01±0.189	0.67±0.2	0.12±0.005	1.14 ± 0.5
N type	1.04±0.11	1.02 ± 0.13	0.36 ± 0.09	0.70 ± 0.33
O type	1.52±0.26	1.1±0.01	0.53±0.01	0.54 ± 0.18

Table 1. The average values of the studied neuromediators, in basal state

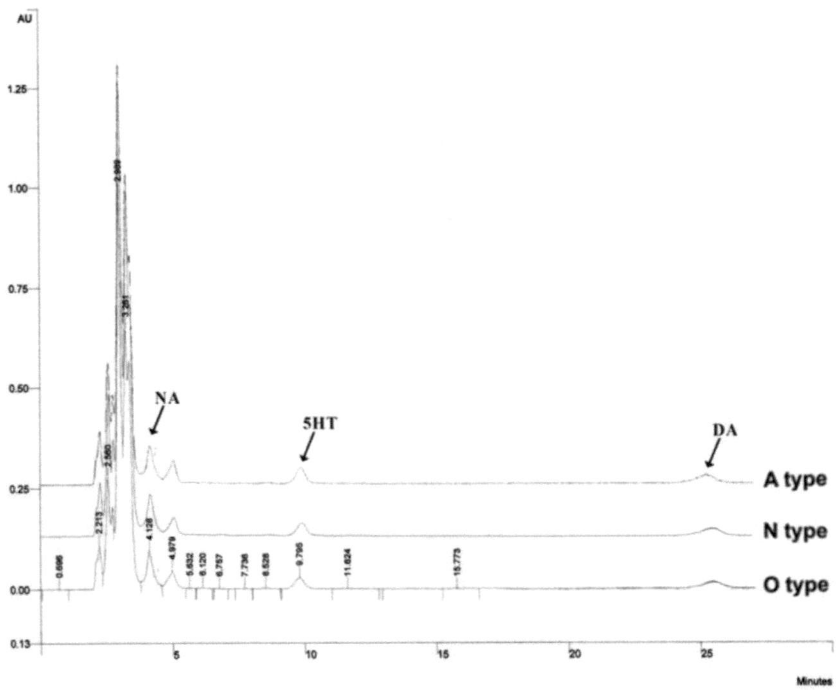

Fig. 2. Examples of chromatograms corresponding to types A, N and O in basal state

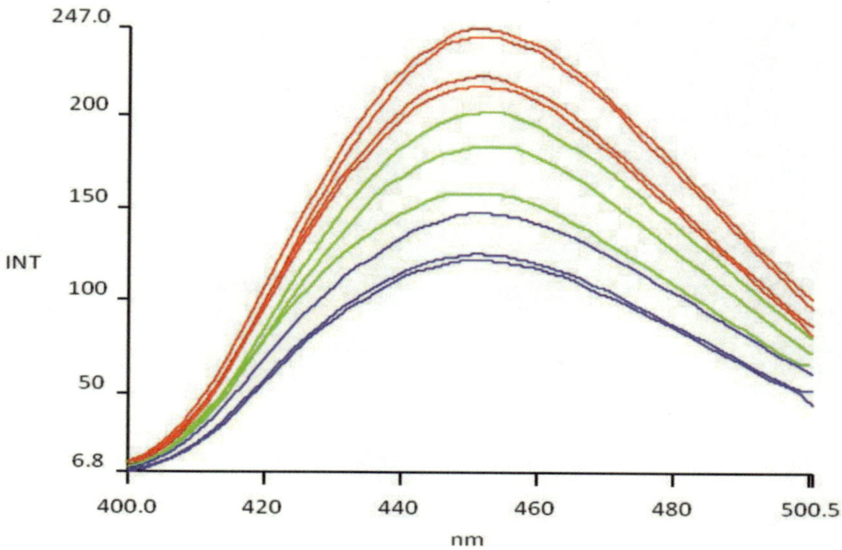

Fig. 3. Examples of fluorescence spectra corresponding to the neuronal endogenous concentrations of GABA in type A (___), type N (___) and type O (___)

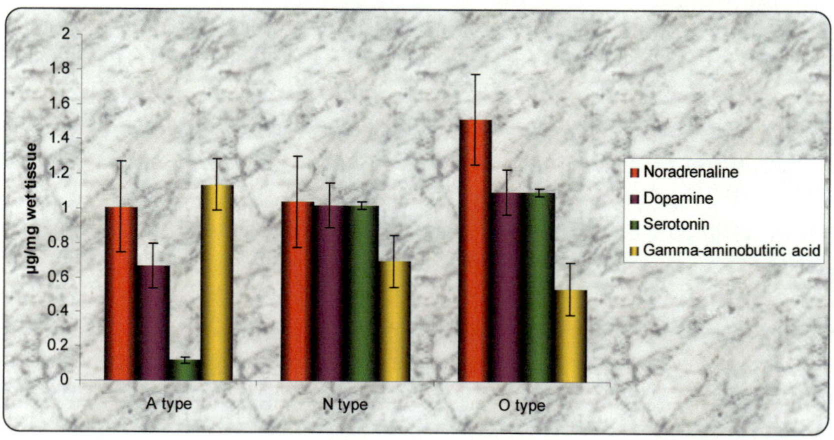

Fig. 4. The brain levels of the studied neuromediators, for the three murine behavioral types, in basal state

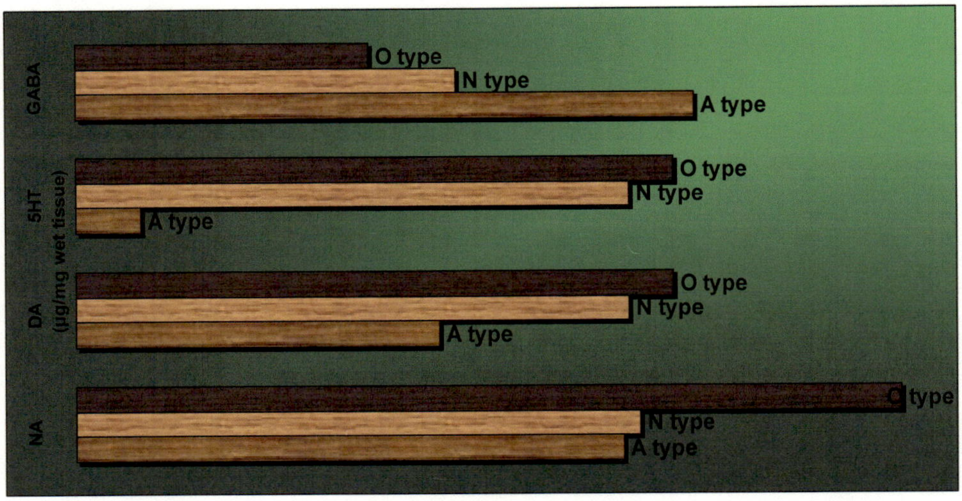

Fig. 5. Comparative assessement of the neuromediators neuronal concentrations studied, in brain, depending on the psychoneuroendocrine typology. Basal state

The status of the three types of basal monoaminergic psychoneuroendocrine studied concentrations was assessed by quantifying the neuronal noradrenaline, dopamine, serotonin and γ-amino butyric acid. Analyzing the experimental results highlighted the following issues:

- the neural dynamics of noradrenaline in basal state, has a maximum concentration in the opioid type (1.52 ± 0.26µg/mg wet tissue);
- there is a positive correlation (Pearson correlation coefficient r = 0.957) between the jumping time (a physiological parameter that describes the behavior typology) and a cerebral biochemical marker, namely the concentration of noradrenaline;
- both the adrenergic type and the balanced type releases much smaller amounts of noradrenaline, compared with type O (statistically significant): 1.01 ± 0.189µg/mg wet tissue, $p < 0.05$(type A) and the 1.04 ± 0.11µg/mg wet tissue, $p < 0.05$ (type N) (fig 5);
- the neural dynamics of dopamine in basal state, has a maximum concentration in the balanced type (1.02 ± 0.13µg/mg wet tissue), the recorded values being significantly higher than those recorded for the type A (0.67 ± 0.2µg/mg wet tissue , $p < 0.01$) and type O (1.1 ± 0.01µg/mg wet tissue, $p < 0.01$);
- the neural dynamics of serotonin in basal state, has a maximum concentration in the balanced type (0.36 ± 0.09µg/mg wet tissue), the recorded values being significantly higher compared with type O (0.53 ± 0.01µg/mg wet tissue, $p < 0.001$), especially compared with type A (0.12 ± 0.005µg/mg wet tissue, $p < 0.05$);
- the adrenergic psychoneuroendocrine type has a much higher concentration of GABA (fig 5), statistically significant (1.14 ± 0.50 µg GABA/mg wet tissue) compared with both the opioid type (0.54 ± 0.18 µµmols GABA / mg wet tissue, $p < 0.01$) and the intermediate type, balanced (0.70 ± 0.33 µµmols GABA / mg wet tissue, $p < 0.05$).

The GABA-ergic status varies inversely with the sensitivity to pain of the individuals from a population (Fig. 7), recording the maximum values for hypersensitive typology (Type A), Pearson correlation coeffient having a value of r = -0.9758.

It can be said therefore that the GABA-ergic transmission modulates, balances, in the basal state, the behavioral type "of warning", the adrenergic type.

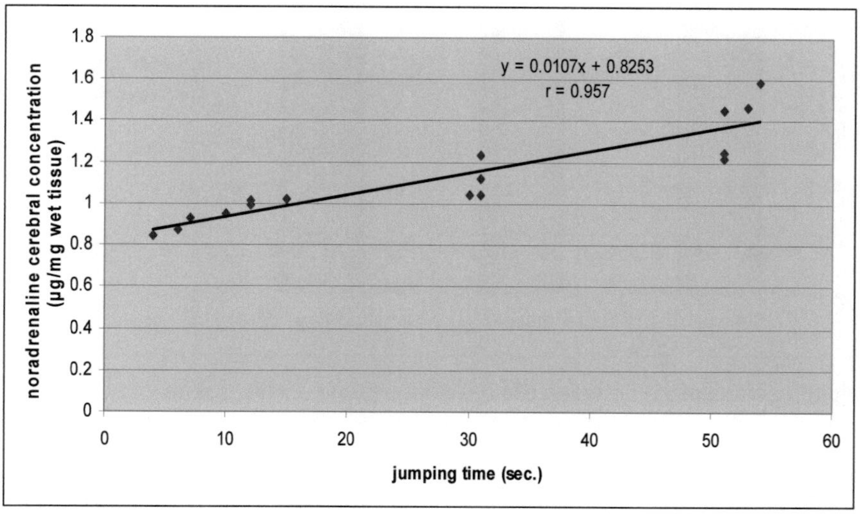

Fig. 6. Correlation between the jumping time and NA concentration for the three behavioral types

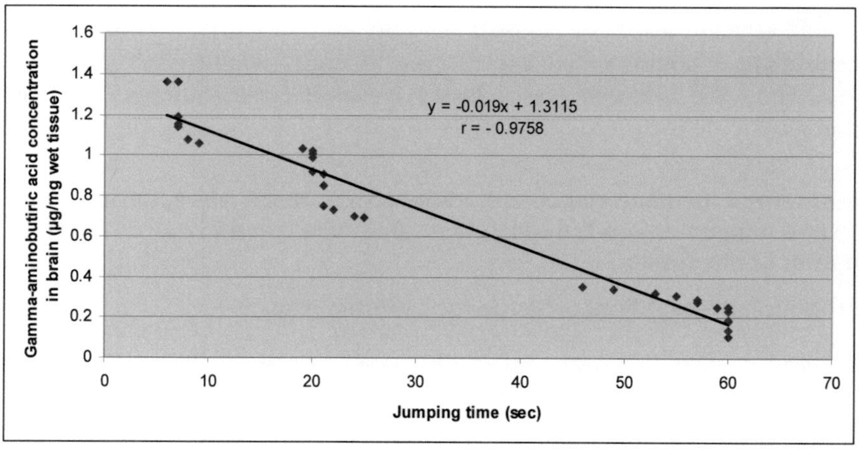

Fig. 7. Correlation between the jumping time and GABA concentration for the three behavioral types

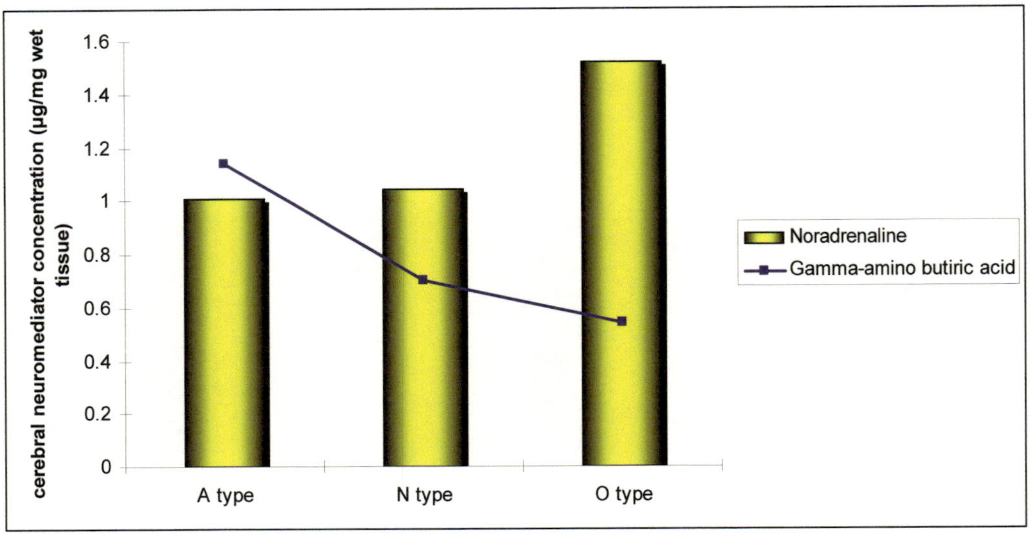

Fig. 8. Negative correlation between the basal neuronal concentrations of noradrenaline and γ-amino butyric acid , in basal state, for the the three murine behavioral types described (r = -0.9577)

The comparative analysis of the followed parameters (painful sensitivity - physiological parameter, expression of the psychoneuroendocrine typology and the neuronal concentrations of some biologically active monoamines - molecular biochemical markers) was able to reveal the following:

- in basal state, the adrenergic type (hypersensitive to painful stimuli) is not highlighted by a extremely high monoamine-ergic status, but is revealed as a very well balanced type, in this regard (high concentrations of both noradrenaline and γ-amino butyric acid);
- the opioid type, hipo-sensitive to pain, has also, unexpectedly, in basal state, the highest concentrations of noradrenaline in neurons, and large amounts of serotonin;
- in basal state, the balanced type releases the most large amounts of dopamine and serotonin.

5.2 Assessment of the neuronal concentrations of noradrenaline, dopamine, serotonin and gamma-amino butyric acid on A adrenergic and opioid types of behavior after acute stress

5.2.1 Distribution on psychoneuroendocrine murine groups

The researches were conducted on a community consisting of 100 mice. In the first stage was followed the individual reactivity to pain of the entire group (n = 100) using the hot plate test. There have been registered the results of the individual reaction to pain expressed by the times of jump. Based on these experimental data there could be noticed that for the collectivity of animals tested for reaction to pain, the individual values of the times of jump ranged between 4 and 60 seconds, the medium time of jump (average medium Jt) was of 36.04 sec and the standard deviation (SD) value was of 15.85 sec.

Depending on the individual reactivity to pain, the types of behavior were defined as follows:

- the adrenergic type (A), hypersensitive: Jt = 4-20 sec;
- the normal, balanced type (N), normo-sensible: Jt = 29-42 sec;
- the opioid type (A), hipo-sensible: Jt = 52-60 sec.

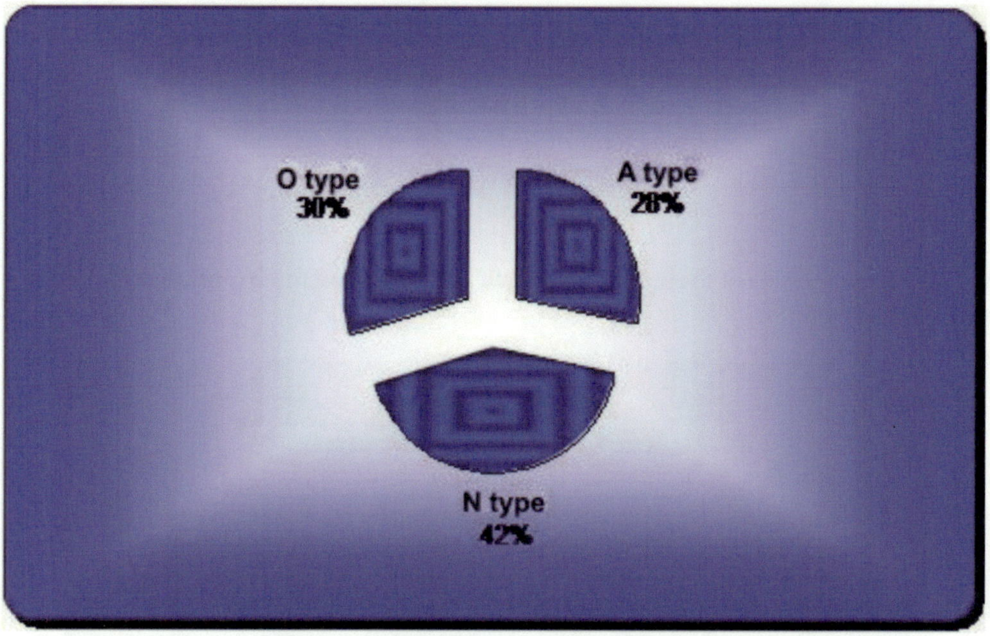

Fig. 9. Distribution of animals (n = 100) according to pain sensitivity

The acute stress was induced through the forced swimming test, the test of "desperation". After the 6 minutes of forced swim, each animal was sacrificed, the cerebral tissue was isolated and the biological material was processed to determine the concentration of the neuromediators NA, DA and 5HT by HPLC with UV detection.

Behavior type	**Cerebral concentration of the studied neuromediators after inducing acute stress (µg/ mg wet tissue)**			
	NA	DA	5HT	GABA
A type	3.41 ± 1.14	0.13 ± 0.03	0.12 ± 0.05	1.55 ± 0.77
N type	1.59 ± 0.24	0.32 ± 0.12	0.45 ± 0.13	1.52 ± 0.55
O type	1.55 ± 0.13	0.25 ± 0.05	0.47 ± 0.1	3.15 ± 0 .68

Table 2. The average values of the studied neuromediators, after acute stress

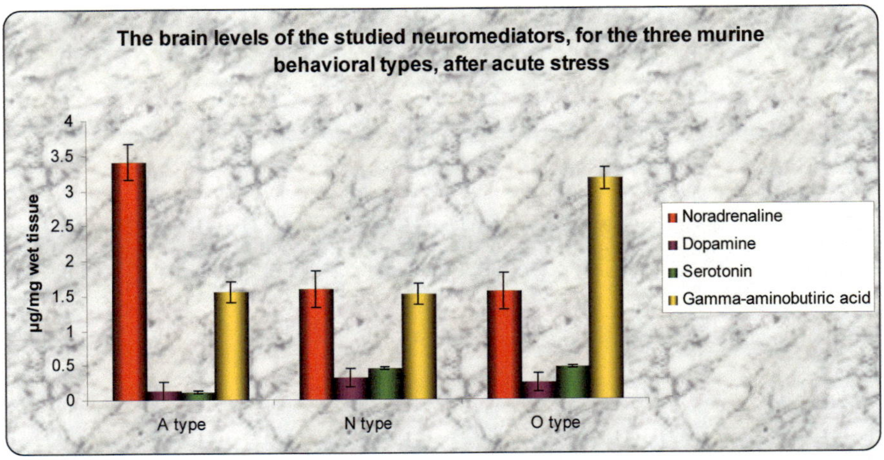

Fig. 10. The brain levels of the studied neuromediators, for the three murine behavioral types, after acute stress

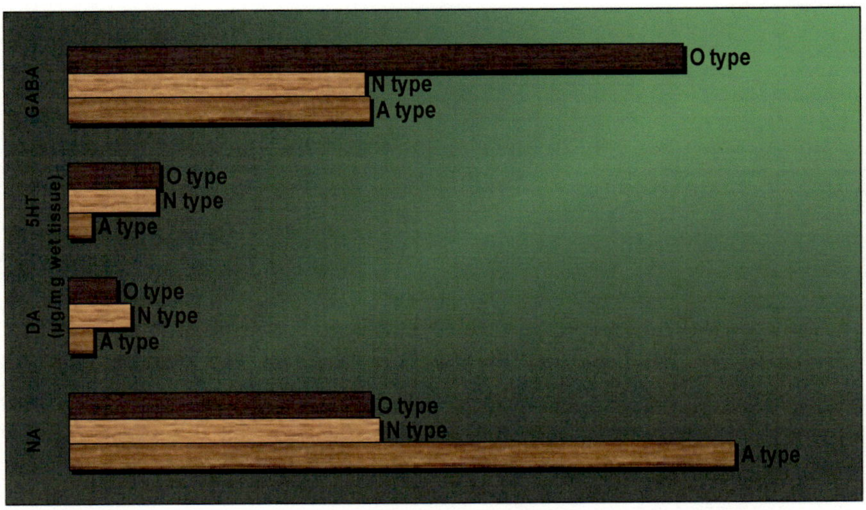

Fig. 11. Comparative assessement of the neuromediators neuronal concentrations studied, in brain, depending on the psychoneuroendocrine typology. Acute stress

The assessment of the monoaminergic status of the three psychoneuroendocrine types after acute stress highlights the following:

- in acute stress, the adrenergic type releases, in neurons, the largest amounts of noradrenaline (statistically significant) (3.41 ± 1.14µg/mg wet tissue) compared with type N (1.59 ± 0.24µg / mg wet tissue), but especially compared with type O (1.55 ± 0.13µg/mg wet tissue, p <0.01);
- type A it also developes the lowest amounts of γ-amino butyric acid , its concentrations being negatively correlated with those of noradrenaline (r=-0.9345, fig.y);
- the balanced type releases, in acute stress, the highest quantities of dopamine, the dynamic of this neuromediators varying in the following order: type A (0.13 ±

0.03µg/mg wet tissue, p <0.000001) <type O (0.25 ± 0.05µg/mg wet tissue , p <0.001) <type N (0.32 ± 0.12µg/mg wet tissue);

- the neuronal dynamics of serotonin, in stress state, has a maximum concentration in the opioid type (0.47 ± 0.1µg/mg wet tissue), the recorded values were significantly higher compared with type A (0.12 ± 0.05µg/mg wet tissue; p <0.01); the balanced type, in turn, revealed higher values of serotonin concentration (0.45 ± 0.13µg/mg wet tissue) compared with type A.
- the opioid psychoneuroendocrine type has a much higher concentrations of GABA, statistically significant (3.15 ± 0,68 µg GABA/mg wet tissue) compared with both adrenergic type (1.55 ± 0.77 µg GABA/mg wet tissue, p <0.001) and the intermediate balanced type (1.52 ± 0.55 µ/mg gwet tissue, p <0.001).
- our data showed a positive correlation (r = 0.9584) between the neuronal levels of serotonin and GABA after inducing acute stress in the O type of behavior. (fig. 13).

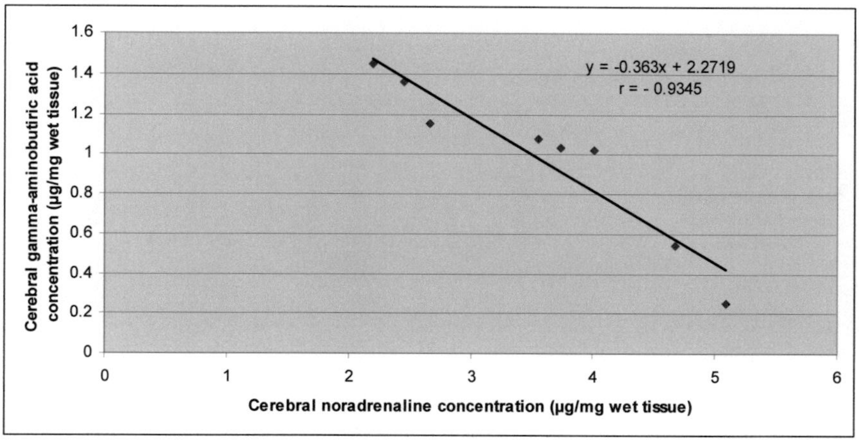

Fig. 12. Negative correlation between neuronal concentrations of of noradrenaline and γ-amino butyric acid , for the A type, after acute stress

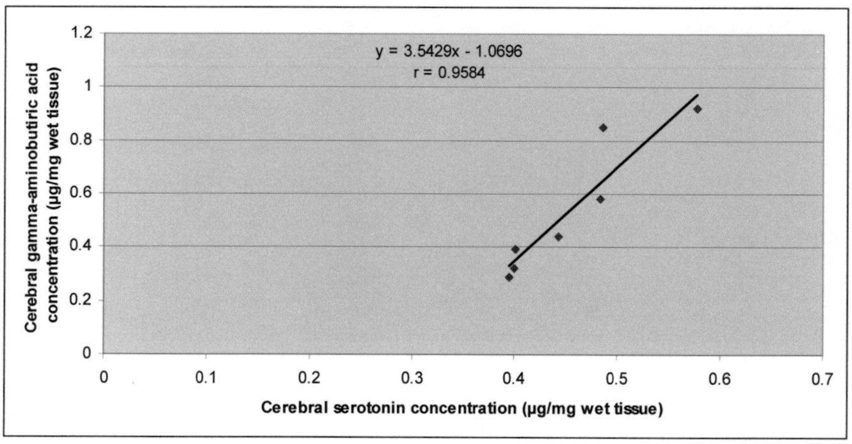

Fig. 13. Positive correlation between neuronal concentrations of of serotonin and γ-amino butyric acid , for the O type, after acute stress

After the induction of acute stress, the two followed parameters (painful sensitivity - physiological parameter, expression of the psychoneuroendocrine typology, respectively, the status of the monoaminergic-molecular biochemical marker) revealed some notable observations:

- in acute stress, the adrenergic type biosynthesis and releases the largest quantities of noradrenaline;
- the opioid type releases, in stress, gamma-amino butyric acid and serotonin.

The comparative assessment of the dynamics of the neuronal concentrations of noradrenaline (NA), dopamine (DA), serotonin (5-HT) and γ-amino butyric acid (GABA) in basal state and after acute stress, is shown in Figures 14-17.

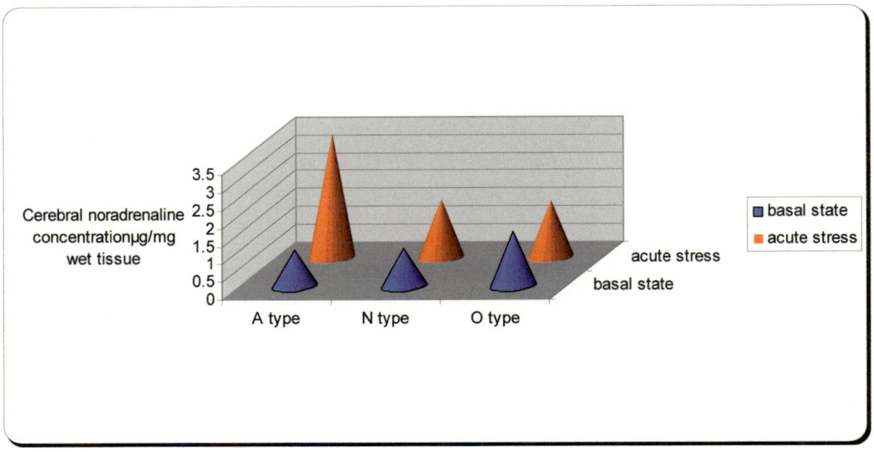

Fig. 14. Dinamics of brain concentrations of noradrenaline in basal state and after acute stress, for the studied behavioral typologies

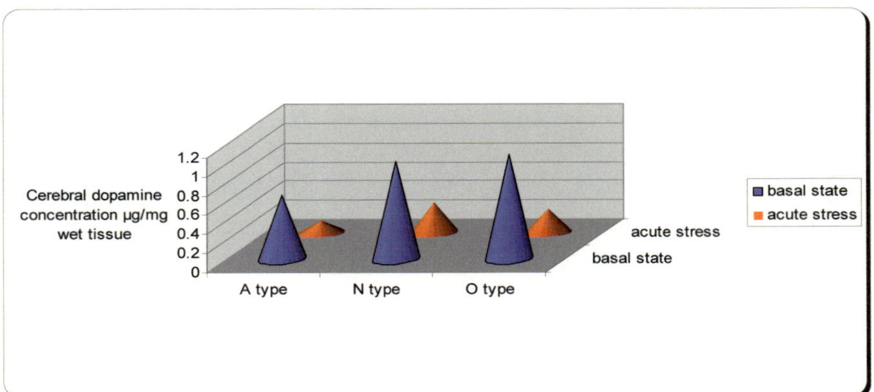

Fig. 15. Dinamics of brain concentrations of dopamine in basal state and after acute stress, for the studied behavioral typologies

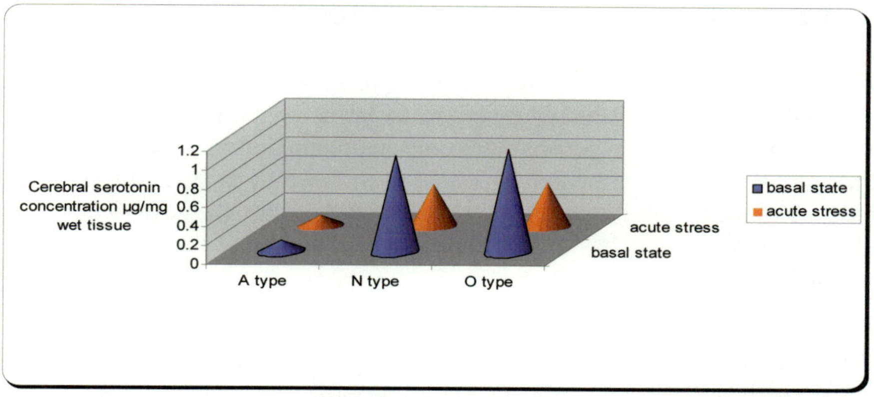

Fig. 16. Dinamics of brain concentrations of serotonin in basal state and after acute stress, for the studied behavioral typologies

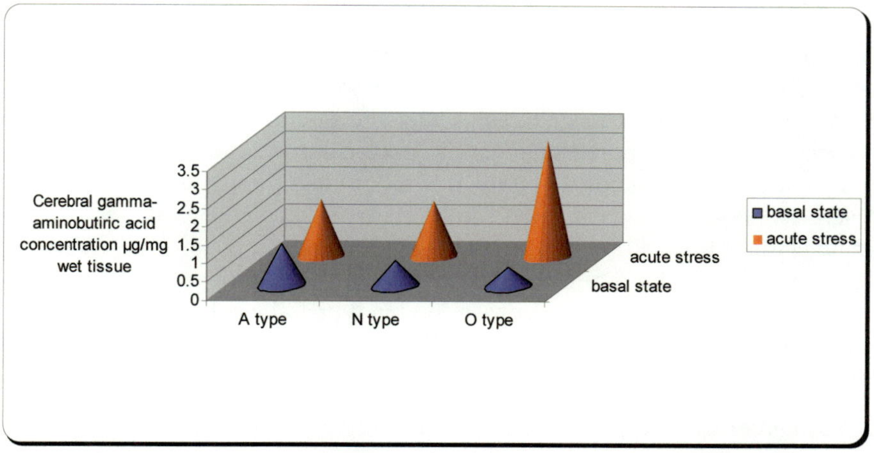

Fig. 17. Dinamics of brain concentrations of γ-amino butyric acid in basal state and after acute stress, for the studied behavioral typologies

6. Conclusion

The experimental study presented proves, once again, the great biological variability and brings new information on the molecular mechanisms that maintain the homeostasis of the central nervous system. From this point of view, the psychoneuroendocrine typology reveals itself as a major factor of the pharmacological responses to drugs with action on the monoaminergic transmission.

The different therapies might suffer changes (might be improved) to optimize the pharmacological response by the criterion of the psychoneuroendocrine typology. The clinical observations reported in the literature, pathologies associated to typologies and the pharmacologic response to various drugs possess a high degree of subjectivity, requiring intimate, molecular discerning mechanisms which cause both susceptibility to various diseases, and appropriate therapy.

From this point of view, the researches presented reveal molecular information, which can maintain and complete the clinical and experimental data reported in the literature. Thus, the type A, characterized by a behavioral point of view by alert, rush, speed, hostility, aggressiveness, is proving extremely balanced in terms of basal neural mechanisms. The basal monoaminergic tone of type A is stable and maintained in moderate limits (high levels of noradrenaline, together with high levels of γ-amino butyric acid–the main inhibitory neuromediator of the CNS). We can say that, in the basal state, the adrenergic type, though alert and active in terms of behavior, is balanced and protected at the molecular level by the inhibitory GABA-ergic mechanisms.

The intervention of the stress factors distorts, however, in the adrenergic type, these balance mechanisms (activating/inhibiting), against the inhibition mechanisms (protection), by releasing extremely large quantities of noradrenaline, together with low amounts of γ-amino butyric acid.

This could be one of the possible observations of molecular pharmacology which prones the A type vulnerability to stress (with the pathological consequences reported in literature).

In contrast to the adrenergic type, the O type of behavior offers surprising observations. It developed high neuronal concentrations of NA in basal state (much higher than for type A), but with no psychosomatic effect. It seems that the opioid type does not use important cellular concentrations of noradrenaline, its molecular mechanisms being modulated, probably by endogenous opioids. In stress, it is modulated (beside the endogenous opiates) through serotoninergic pathways (high serotonin levels) together with an extremely high GABA-ergic status. In addition, the acute stress does not lead, for type O, to substantial fluctuations in the amount of cerebral noradrenaline, compared with basal condition. It is quite possible for the opioid psychoneuroendocrine type to be behavioral modulated, by an interrelation between the endogenous opioid system with the triptaminergic (serotoninergic) system.

These experimental data lead, naturally, to the following question: *why some individuals cellulary use more noradrenaline, while others use endogenous opioids? Is there, perhaps, a "preference" of the neurons for certain neuromediator?* A possible answer might be found in modern theories of the substrate receptors resistance (e.g. the assumption of the resistance of the insulin receptors, which explains the etiopathology of type 2 diabetes). Thus, it is quite possible that the adrenergic receptors of type O are resistant to noradrenaline, while the opioid receptors of type A are resistant to endogenous opioids.

Our hypothesis is sustained by the researchers of the serotonin type of behavior. by using the modern technology of the positron emission tomography (TEP), they have shown that these individuals framed by questionnaires of personality in the serotoninergic typology, have a large number of $5HT_{1A}$ receptors. Extrapolating this observation to types A and O, it might appear the hypothesis of the existence of a significant variation of the adrenergic receptor density, respectively, opioids among the individuals of a population.

The neural dynamics of the monoaminergic mediators for extreme typologies (A and O) is confirmed by the results of the intermediate, balanced type, which biosynthesizes and releases, in stress, all the researched neuromediators, in large quantities. There is also a special preference of the intermediate N type for using dopamine as a biomolecule of stress.

The paper provides molecular evidences that support and argue the theory of the pharmacological variability, due to the psychoneuroendocrine typology (as a part of the biological variability), and bipolar disorder is included within. A review seeking to identify the more consistent findings suggested that there are several genes involved in the development of bipolar disorder (Serretti A., 2008), such as those for serotonin (SLC6A4 and TPH2), dopamine (DRD4 and SLC6A3) and excitatory neuromediators (DAOA and DTNBP1). There is also a the theory asserting that people developping bipolar disorder experience a series of stressful events, each of which lowers the threshold at which mood changes occur (Brian Koehler et al). There are also reported evidences of neuronal abnormalities in bipolar disorder due to stress. Published clinical studies, reported the prevalence of bipolar disorder within the adrenergic behavioral type.

In this sense, the imbalances of the monoaminergic systems, which may lead and/or sustain bipolar disorder require, certainly, an individualized pharmacotherapy on the criterion of the psychoneuroendocrine typology.

7. References

Appleton K.M., et al. (2007). Type A behaviour and consumption of atherogenic diet: No association in the PRIME study. *Appetite*. pp. 46:378-386

Borsa C. (2002). Gerovital new formula and Aslavital effect on monoamine oxidase activity in mitochondrial fraction from rat liver and brain. *Rom. J. Geront. Ger.* pp. 24:7-23

Brian Koehler, (2005). Bipolar Disorder, Stress, and the HPA Axis. Available from http://www.isps-us.org/koehler/bipolar_stress.htm

Cristea A. (1998). Sindromul hiperalgic al tipului comportamental opioid, în stresul cronic. *Durerea acută şi cronică*, Vol 6, No. 2. 42-45

Cristea A. et al. (1997). Comportamentul nervos la tipurile adrenergic şi opioid în testele de farmacologie experimentală.*Farmacia*. Vol. 45 No. 2. pp. 71-79

Cristea A., et al. (1994). The evolution of the painful sensitivity in acute and chronic stress.*Rom. J. Phisyol.* Vol. 31 No.1-4. pp. 75-79

Cristea A., et al. (1996). Catecholamine levels in opioid and adrenergic psychoneuroendocrine types. *Rom. J.Endocrinol.* Vol. 34 No.1-4. pp. 95-101

Cristea A.N. & Ştefănescu A-M. (2002). Depressive syndrome of the psychoneuroendocrine adrenergic „A" type, in chronic stress. 7th ECNP (European College of Neuropsychopharmacology) Regional Meeting, Bucharest, Romania, Abstr. Book, pp.01-11

Haynes S.G., et al. (1978). The relationship of psychosocial factors to coronary heart disease in the Framingham Study: Methods and risk factors.*Am.J.Epidemiol.* pp. 107:362-383

Herman S. et al. (1981). Self ratings of type A (coronary prone) adults. *Psychosom. Med.* Vol. 43 No.5, pp. 405-413

Jenkins C.D. et al. (1971). *Manual for the Jenkins Activity Survey*. Psychosological Corp., New York, pp. 374-421

Jones K.V. (1985). The thriel of victory: blood pressure, variability and type A behavior pattern.*J.Behav.Med.* Vol. 8 No. 3. pp. 277-285

Le Melledo J.M., et al. (1998). The influence of Type A behavior pattern on the response to the panicogenic agent CCK-4.*J. Psychosom. Res.* pp. 51:513-520

Meesters C.M.G. & Smulders J. (1994). Hostility and myocardial infarction in men.*Journal of Psychosomatic Research*. Vol. 38 No. 7. pp. 727-734

Oedegaard K.J. et al. (2006). Type A behaviour differentiates bipolar II from unipolar depressed patients. *J.Behav.Pharmacol*. pp. 90:7-13

Petit-Demouliere B, Chenu F, Bourin M. (2005). "Forced swimming test in mice: a review of antidepressant activity.". *Psychopharmacology (Berl)* Vol. 177 no.3, pp 245–55.

Powell L.H. et al. (1984). Can type A behavior be altered after myocardial infarction? A second year report from the Recurreent Coronary Prevention Project. *Psychosom. Med*. pp. 46:293-331

Rhodewalt F. (1984). Self involvment, self attribution and type A coronary-prone behavior pattern.*J.Pers.Soc.Psychol*. Vol. 47 No. 3. pp. 662-670

Rosenman R.H. (1978). *The Interwiew method of assessment of the coronary prone behavior pattern*. in Dembronski, T.M., *et al*. eds., *Coronary prone behavior*, New York, pp. 55-70

Serretti A. & Mandelli L. (2008). The genetics of bipolar disorder: genome 'hot regions,' genes, new potential candidates and future directions. *Molecular Psychiatry*, Vol. 13 No. 8, pp. 742–71

8

Paediatric Bipolar Disorder – Are Attachment and Trauma Factors Considered?

Peter I. Parry
Flinders University
Australia

1. Introduction

1.1 Debate over the boundaries of bipolar disorder

Significant debate and controversy surrounds the boundaries of Bipolar Disorder (BD). Proponents of a broader category for BD within psychiatric nosology (e.g. Akiskal, 2007) argue that more limited episodes of mood instability in both time and severity belong on a broader bipolar spectrum. Others (e.g. Paris, 2009) contend that hypomanic symptoms that fail to meet full DSM-IV or ICD-10 criteria for time or severity for BD-I and some BD-II disorders are more likely to represent reactive affective states related to environmental and relational stressors and/or personality traits or disorders. A widening of what constitutes BD beyond traditional concepts of manic-depressive illness has been related to historical and social factors impacting on psychiatric nosology (Healy, 2010).

In this context probably the most intense controversy has been over the way the borders of BD have been extended into childhood. Paediatric Bipolar Disorder (PBD), synonymous with "Juvenile Bipolar Disorder", has been described in an editorial (Ghaemi & Martin, 2007) in the *American Journal of Psychiatry* as "notoriously controversial, with the epicentre of the debate being whether the condition can be diagnosed in pre-pubertal children at all."

1.2 Historical perspective on PBD

1.2.1 Pre-1995 perspectives

In antiquity the term "mania" historically was applied to any state of frenzied madness or marked behavioural dyscontrol and, as Healy (Healy, 2008 p.7) illustrates, the manic states described by Hippocrates were essentially states of delirium accompanied by fever. According to Healy (2008, p.56) mania was not described in the context of manic-depressive illness until the mid 19th century by Baillarger in France and it was not until Kraepelin at the dawn of the 20th century that the term gained its widespread modern psychiatric usage.

Kraepelin noted amongst his 900 cases of manic-depressive psychosis that the disorder could have onset in adolescence but cases with onset prior to age 10 were sporadic with a rate of 0.4% (Silva et al., 1999). Traditionally BD has been viewed as having its onset in late adolescence to young adulthood. Rare sporadic pre-pubertal cases were described, but it wasn't until the 1980s that articles appeared raising the question that childhood onset cases of BD may present atypically and could be being missed (Carlson, 1984). However clinical practice did not alter until after the appearance of a series of articles in the mid 1990s.

1.2.2 Post-1995: The "narrow" and "broad" PBD phenotypes

Two articles published in 1995 sought to redefine mania and BD as presenting in atypical but reliably measurable ways in children and adolescents. Researchers at Washington University in St Louis (WUSL) characterised mania in children as presenting with prolonged episodes of "ultradian" (several times per day) cycling of mood episodes (Geller et al., 1995), meanwhile a group from the Massachusetts General Hospital affiliated with Harvard (MGH/Harvard) (Wozniak et al., 1995) characterised mania in children as presenting with chronic irritability generally without distinct time limited mood episodes.

The *Journal of the American Academy of Child and Adolescent Psychiatry* has given PBD prominence in major reviews (Geller & Luby, 1997; Pavuluri et al., 2005; Kowatch et al., 2005; Liu et al., 2011) and a report on the National Institute of Mental Health (NIMH) "research roundtable on pre-pubertal bipolar disorder" (Nottelman, 2001). The NIMH research roundtable defined the two subtypes as "narrow phenotype" (WUSL) and "broad phenotype" (MGH/Harvard).

1.2.3 Rise in diagnostic rates of PBD

Following this academic lead the number of children and adolescents diagnosed with PBD in the USA skyrocketed. Community rates of BD diagnosis in the paediatric range increased 4,000% from 1994-5 to 2002-3 (Moreno et al., 2007) and PBD became the most common diagnosis in US preadolescent psychiatric inpatient units by 2004 (Blader & Carlson, 2006).

Blader and Carlson cited "diagnostic upcoding" as a major driving force for the increased rate of PBD diagnosis. "Diagnostic upcoding" occurs when factors extraneous to the patient's condition provide benefit for a particular diagnosis. These factors mainly involve the way health insurers fund health care based on diagnosis rather than clinical need. Thus a child with ADHD and Oppositional Defiant Disorder (ODD) or Conduct Disorder (CD) may be having serious problems relating to his family and school and need an inpatient evaluation, but the inpatient evaluation might only be funded if there is a diagnosis of BD. There has been less diagnosing of PBD outside the USA (Parry et al., 2009), perhaps because most other developed countries do not link mental health care so directly to DSM diagnoses.

The epidemiology of PBD is worthy of further study in itself as vastly differing rates of diagnosis have been found for mainly cross-sectional and retrospective recall studies. The diagnostic rate depends greatly on the criteria used by the researchers no matter where the studies are done (Van Meter et al., 2011) and thus reflects differing viewpoints and does little to assist resolution of the controversy. However a retrospective recall study of adults with BD reflected the international divergence: 2% of Dutch and German subjects reported pre-teen onset, whilst 22% of the USA cohort reported pre-teen onset (Post et al., 2008).

1.2.4 "Severe mood dysregulation" (SMD)

Follow-up studies of youth diagnosed with "broad phenotype" PBD have shown they are no more likely to progress to adult BD than the general population. This has led to a renaming of this group as exhibiting "Severe Mood Dysregulation" (SMD) (Brotman et al., 2006; Dickstein et al., 2006; Stringaris, 2009 & 2011; Leibenluft, 2011).

1.3 A controversial diagnosis

The validity of PBD has been subject to vigorous debate in the literature and media (Parry & Allison, 2008) and described as a "fad diagnosis" in "epidemic" proportions by the head of the former DSM-IV task force (Frances, 2010). Psychiatrists have published books for parents both for (e.g. Papolos & Papolos, 2000; McDonnell & Wozniak, 2008) and against (Kaplan, 2011) the diagnosis. Kaplan argues diagnoses such as ADHD and ODD/CD often suffice without recourse to a "comorbid" PBD diagnosis.

The relationship of the pharmaceutical industry and psychiatry has been a focus of concern in recent years (Freedman et al., 2009). The PBD diagnosis has been a particular focus of this debate (Frances, 2010; Parry & Levin, 2011; Levin & Parry, 2011; Robbins et al., 2011).

The controversy surrounding PBD intensified following a much publicised and tragic medication related death of a 4 year old girl, Rebecca Riley, in Boston in 2006. In the wake of the tragedy, the Boston Globe reported that both Rebecca's 6 year old sister and 11 year old brother and both her parents were also diagnosed with PBD and BD. Also there was a litany of child protection notifications, including the battering of her brother by their father and that her 13 year old half-sister had been removed by child protection services due to alleged sexual abuse also by Rebecca's father (Cramer, 2007). In the wake of the tragedy vigorous debate about PBD amongst researchers and clinicians spilled into the public media. Van der Kolk, a Harvard professor prominent in traumatology research, was quoted saying: "the (PBD) diagnosis is made with no understanding of the context of their life" (Carey, 2007).

1.4 Alternative perspective: Attachment insecurity and developmental trauma

Thus one of the main critiques of the construct of PBD is that it has arisen from and compounded a neglect in psychiatric nosology of attachment insecurity and developmental trauma in the lives of children and adolescents (McClellan, 2005; Harris, 2005; Carlson & Meyer, 2006; Parens & Johnston, 2010; Parry & Levin, 2011).

To date there has not been any systematic literature review to test whether in fact this is the case. This chapter therefore explores whether developmental contextual factors have been neglected, through a systematic literature review of the presence of attachment theory and developmental trauma and maltreatment concepts in the PBD literature.

2. Methods

A systematic review of the literature was conducted using the Scopus academic search engine. Scopus allows for searches for specific words within large numbers of selected articles, which aids this type of literature review. Searches can be in various fields such as title, abstract and/or keywords. In particular an "All Fields" search with Scopus should detect a word when it is in the article's title, keywords, abstract and list of citations/references titles. The search covered the period from January 1995 to June 2010.

2.1 Defining a body of PBD literature

A body of PBD literature was defined by a Scopus search in "Title-Abstract-Keyword" fields for [*pediatric* or *paediatric* or *juvenile* or *early-onset* or *adolescen** or *teenage** or *child** or *youth* or *kids*] and [*bipolar* or *mania* or *manic* or *hypomania* or *hypomanic* or *manic-depression* or *manic-depressive*] for publications since 1995 to 15 June 2010. This gave rise to 7,257 articles, though with low specificity for PBD articles. In Scopus an "All Fields" search detects a word in the article's list of citations as well as in title, keywords, and abstract. From the 7,257 articles an "All Fields" search for the word "attachment" found 165 articles of which 15 were PBD oriented. Full texts of these 15 articles were examined for context of the word "attachment".

To obtain a more specific body of PBD literature a Scopus search was conducted in "Title-Abstract-Keyword" fields for permutations of: [*pediatric* or *paediatric* or *juvenile* or *youth* or *child** or *early* or *adolescen** or *teenage**] (with and without *"onset"* or *-onset*) and [*bipolar* or *mania* or *hypomania* or *"manic depression"*] also [*bipolar* or *manic* or *hypomanic*] and [*child** or *teen** or *"adolescen**" or *youth* or *kids*] also [*bipolar* or *mania* or *hypomania* or *"manic depression"*] and [*"in a"* - *child* or *boy* or *girl* or *adolescent*] also [*child* or *boy* or *girl* or *adolescent* - *"with"*] and [*bipolar* or *mania* or *hypomania* or *"manic depression"*].

As of 15 June 2010 the search found 1,113 publications. Perusal indicated high specificity to articles relating to PBD. This subset of PBD literature was then subjected to a Scopus "All Fields" search. To ascertain whether attachment theory and trauma aspects were considered, a search for the terms *attachment*, *trauma* (also detects posttraumatic/traumatized etc) or *PTSD* or *maltreatment* or *abuse* was conducted.

2.1.1 PBD literature from "narrow phenotype" and "broad phenotype" researchers

From the PBD literature of 1,113 articles, two subsets of literature were defined by affiliation with the two academic child psychiatry departments that first promoted PBD: WUSL and MGH/Harvard. Given the question of how much the PBD literature considered attachment theory and trauma factors, literature from institutions that had historically most influenced the PBD literature should give some important indication as to the question of incorporation or otherwise of attachment theory and trauma concepts. There were 64 articles affiliated with WUSL, and 137 articles affiliated with MGH/Harvard. No articles were affiliated with authors from both institutions. Full texts of 198 of these 201 publications were downloaded and manually searched for the terms - *attachment, trauma, PTSD, maltreatment, abuse,* and *neglect*. Only 3 articles were accessible by just abstract and citation list.

2.1.2 Attachment theory literature

A body of attachment theory related literature was defined by Scopus search in "Title-Abstract-Keywords" for [*"attachment theory"* or *"attachment security"* or *"attachment insecurity"* or *"avoidant attachment"* or *"secure attachment"* or *"insecure attachment"* or *"ambivalent attachment"* or *"disorganised attachment"* or *"reactive attachment"* or *"resistant attachment"* or *"attachment disorganisation"* or *"developmental psychology"* or *"developmental trauma disorder"* or *"developmental neurobiology"* or *"developmental psychopathology"* or *Bowlby*] resulting in 4,583 publications from 1995 to 13 June 2010. To aid specificity the above terms were searched in "Title" field only, to give a sample of 746 publications.

This "attachment theory related literature" was searched for the presence of PBD terms by searching within "All Fields" for [*"pediatric bipolar"* or *"pediatric onset bipolar"* or *"pediatric onset bipolar"* or *"paediatric bipolar"* or *"paediatric onset bipolar"* or *"juvenile bipolar"* or *"juvenile onset bipolar"* or *"early-onset bipolar"* or *"child* onset bipolar"* or *"child* bipolar"* or *"adolescen* bipolar"* or *"adolescen* onset bipolar"* or *teenage* bipolar"* or *"teenage* onset bipolar"* or *"pediatric mania"* or *"pediatric hypomania"* or *"paediatric mania"* or *"paediatric hypomania"* or *"juvenile mania"* or *"juvenile hypomania"* or *"early-onset mania"* or *"early-onset hypomania"* or *"child* mania"* or *"child hypomania"* or *"adolescen* mania"* or *"adolescen* hypomania"* or *"teenage* mania"* or *"teenage* hypomania"* or *"youth mania"* or *"youth hypomania"*]. Specific terms such as these were used to define publications that specifically referred to PBD rather than publications relating to offspring of adults with bipolar disorder. Only 8 articles were found.

3. Results

3.1 "Attachment", "PTSD/trauma" and "maltreatment/child abuse" in PBD literature

In 1,113 articles on PBD there were just 14 publications with the word "attachment"; 29 publications with "trauma/PTSD"; and 64 publications containing at least one of "maltreatment/child abuse/sexual abuse/physical abuse/emotional abuse" in an "All Fields" search. With overlap this amounted to 84 publications in total (Figure 1).

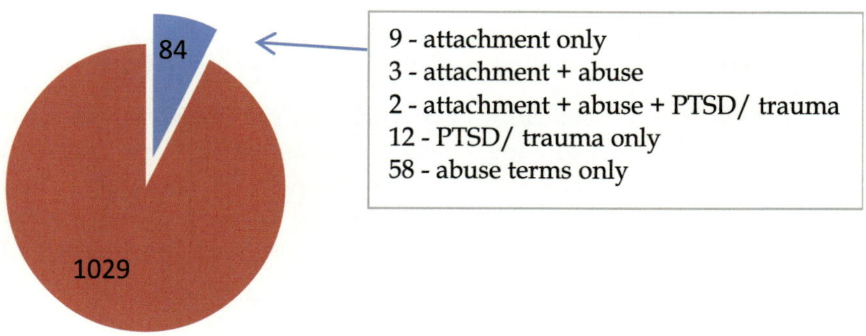

Fig. 1. Attachment and maltreatment/trauma terms in the PBD literature.

3.2 Fifteen PBD articles mentioning "attachment"

10 of the 14 articles that found "attachment" in the "all fields" search were PBD oriented, 4 related to (non-PBD) offspring of bipolar parents' studies. A further 5 articles were found from the less specific list of 7,257 articles. Thus 15 full-text articles were examined and the word "attachment" was used in the following contexts:

3.2.1 Attachment related concepts as a significant theme (3 articles)

A case study (Bar-Haim et al., 2002) of a 7 year old boy with multiple neurodevelopmental delays and diagnoses of PBD, ADHD and ODD included an attachment perspective. An article on family therapy for PBD children (Miklowitz et al., 2006) accepted the validity of PBD phenotypes but promoted family therapy approaches. A review of PBD (Carlson & Meyer, 2006) was critical of over-diagnosis of PBD, noting that PBD research "would benefit from a developmental psychopathology perspective".

3.2.2 Attachment in text as minor theme (5 articles)

An American Academy of Child and Adolescent Psychiatry (AACAP) research forum on early-onset bipolar disorder (Carlson et al., 2009) contained a passage on contextual issues, maltreatment and family dysfunction. The article mentioned "insecure attachment" as a "risk factor for emotional dysregulation and externalizing disorders" among offspring of parents with bipolar disorder. This was one of the very few documents to use the term "maltreatment" and "insecure-attachment", though there was no specific mention of neglect or PTSD. The research forum also noted: "low socioeconomic status, stressful life events, cognitive style, negative hostile parenting as reflected in low maternal warmth, poor social supports, parent divorce and conflict and physical and sexual abuse have all been identified as risk factors for development of EOBP (early onset bipolar disorder)."

Dickstein and Leibenluft (2006) reviewed differences between "narrow phenotype" PBD and "severe mood dysregulation", including neuroimaging differences and referred to attachment theory based neurobiology research. The article mentions concepts from the attachment theory based literature e.g. the importance of facial gaze in mother-infant dyads.

A personal perspective on a career in child psychiatry (Cytryn, 2003) noted "insecure attachment" was found in a small prospective study of offspring of mothers with bipolar disorder. The offspring developed psychiatric disorders but not PBD.

McClure et al. (2002) expressed caution about the validity of PBD diagnoses and advocated for attachment perspectives in history taking and observations of child-family interactions.

A summary (Parens & Johnston, 2010) of a workshop on "controversies surrounding bipolar disorder in children" had "attachment" in a citation title and once in the text: "...workshop participant and child psychiatrist Mary Burke speculated that, in the underprivileged community where she practices, one of the most effective ways to help children now receiving the BP diagnosis would be to promote attachment and reduce stress on families."

3.2.3 "Attachment" only in a citation title (5 articles)

A review (Post & Leverich, 2006) of psychosocial stress as a risk factor for earlier onset and worsened course of bipolar disorder, discussed the ameliorating influences of psychotherapy and psychoeducation. "Attachment" was mentioned in the title of a reference (Insel) which was used in a text description of animal attachment oriented studies, noting that these studies: "should make one extremely cautious in ascribing what appear to be genetic predispositions to genes, as opposed to familial/environmental influences that can themselves determine lasting neurobiological and behavioral traits."

A study (Meyer et al., 2006) of the Wisconsin Card Sorting Test in adolescent offspring of mothers with bipolar disorder had "attachment" in a citation title (Cicchetti) which was referenced in the passage: "Our results suggest that early exposure to extreme levels of maternal negativity appears to increase the risk for apparent frontal lobe dysfunction, which in turn, heightens vulnerability for the development of bipolar illness. This suggests that prevention efforts with high-risk families should go beyond children's symptomatology to focus on ways of improving the environments in which they are developing."

An article (Costello et al., 2002) that discussed abuse and parenting as minor themes had "attachment" in a citation title (Nachmias) which was used as a reference for: "evidence suggests that responsive caretakers may buffer the risk for depression and other forms of psychopathology". Another (Hirshfeld-Becker et al., 2003) had "attachment" in a citation title (Mannassis), which was referenced with others to say "some studies find an association between behavioral inhibition and anxiety disorders", and a fifth (Petti et al., 2004) had "attachment" in a citation title which was referenced in relation to a life events checklist that did not address attachment concepts, though social relationships were discussed.

3.2.4 "Parent-child relationship" as a keyword synonym for "attachment" (1 article)

The keyword "parent-child relationship" as a synonym for "attachment", appears to have led Scopus to choose an article (Schenkel et al., 2008) that stated: "Compared to controls, parent–child relationships in the PBD group were characterized by significantly less warmth, affection, and intimacy, and more quarreling and forceful punishment."

3.2.5 "Reactive Attachment Disorder" (1 article)

One article (Marchand et al., 2005) did not refer to attachment theory, but to "Reactive Attachment Disorder" in the DSM-IV sense. However the article focused on trauma and complex PTSD as differential diagnoses to PBD, noting: "children with symptoms suggestive of bipolar disorder must be carefully screened for exposure to adverse events."

3.3 Full text searches of two academic centres prominent in PBD research

The above search for attachment, trauma and maltreatment terms was in "All Fields" so would not detect terms if in articles' text but not in title, abstract, keywords or citations. There were 201 articles from authors affiliated with WUSL (research centre to first propose "narrow phenotype" PBD) and MGH/Harvard (research centre to first propose "broad phenotype" PBD). These were full text searched.

3.3.1 PBD literature affiliated with WUSL

Eleven of 64 articles contained at least one of the searched terms except for "maltreatment".

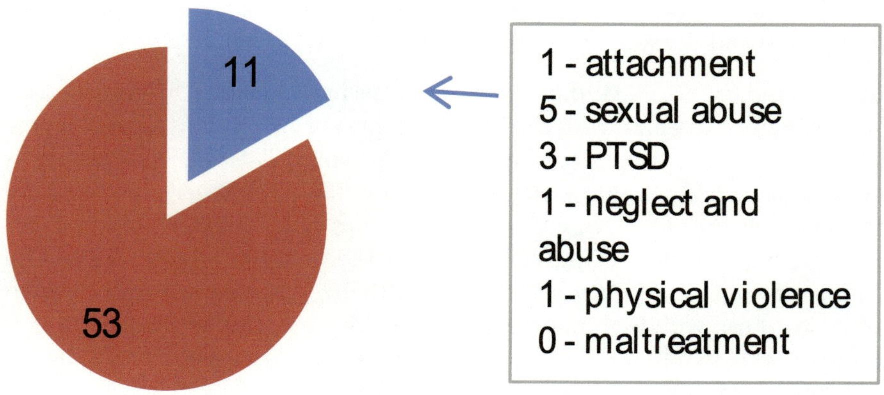

Fig. 2. Attachment and maltreatment/trauma terms in WUSL PBD literature.

As previously mentioned (Petti et al., 2004) contained "attachment" in a citation title. However though discussing the child subjects' social and family relationships, the article didn't address attachment per se in the text.

Five articles (Geller & Luby, 1997; Geller et al., 2000; Geller et al., 2002; Craney & Geller, 2003; Geller, Tillman, Badner, & Cook, 2005) contained the term "sexual abuse". These referred to "sexual abuse" as a differential diagnosis for "manic hypersexuality". They concluded that as only 1.1% in the cohort of 93 children with PEA-BP (prepubertal and early adolescent onset bipolar disorder) had "sexual abuse or overstimulation", whereas 43% (particularly the children who had hit puberty) had "manic hypersexuality" this "strongly supports hypersexuality as a symptom of mania" (Geller et al. 2002).

PTSD was mentioned (Geller et al., 2004) in a list of potential differential or comorbid diagnoses for the cohort of 93 (86 at follow-up), noting no cases of PEA-BP had PTSD. Another article (Geller et al., 2009) also mentions zero cases of PTSD in a diagnostic list for forty-seven 14 year old PBD subjects in a neuroimaging study. A further article (Luby & Navsaria, 2010) had PTSD in a citation title but PTSD/trauma was not mentioned in the text.

The terms "physical violence" and "sexual abuse" were listed in a "Life Events Checklist" and noted that with the cohort of 93 PEA-BP children there were significantly more adverse life events than for both ADHD and normal control groups (Tillman et al., 2003). The authors concluded: "Because there was no a priori reason to expect significantly more independent life events in the PEA-BP compared to the ADHD and NC groups, these results warrant further research into the role of life events in the onset of PEA-BP."

A study (Luby & Beldon, 2003), of 21 "Bipolar I" depressed preschoolers compared with 54 unipolar depressed preschoolers diagnosed by the PAPA (Preschool Age Psychiatric Assessment that is based on DSM-IV), mentioned "neglect" and "abuse" in the following context: "adverse environmental outcomes include neglect and/or abuse as well as psychosocial stressors and trauma". They concluded: "the finding that preschoolers with this bipolar syndrome did not experience greater trauma or adverse life events than other groups is also of importance. While this does not confirm the syndrome is a bipolar disorder, it does suggest that it cannot be explained by developmental deviation secondary to trauma, as has been widely speculated. However, longitudinal follow-up data will be needed to more definitively clarify this nosologic issue". The authors did note a limitation of the study was: "Findings are also limited by sole reliance on parent report of symptom states, frequencies and duration".

One article (Craney et al., 2003) didn't mention attachment theory by name, but did note that 2 year follow-up research with the PEA-BP cohort of 93 children found "low maternal warmth" the only predictive factor for relapse of mania. The risk was strong: "subjects with low maternal-child warmth were 4.1 (95% CI ¼ 1.7–10.1) times more likely to relapse after recovery (19). No other baseline characteristics (e.g. MDD, CGAS, mixed mania, continuous cycling, psychosis, ODD/CD) predicted recovery or relapse." In fact there was a 100% relapse over 2 year follow-up for those with low maternal warmth compared with 40% relapse for those with high maternal warmth. They concluded that this was a similar effect to high expressed emotion (EE) in schizophrenia, and stated: "These data from the PEA-BP sample strongly point toward the need for research on non-pharmacological modalities".

3.3.2 PBD literature affiliated with MGH/Harvard

The Massachusetts General Hospital in Boston is affiliated with Harvard University and has been the main research centre proposing "broad phenotype" PBD. Of 137 articles, 23 contained one of the searched for terms somewhere in the full text.

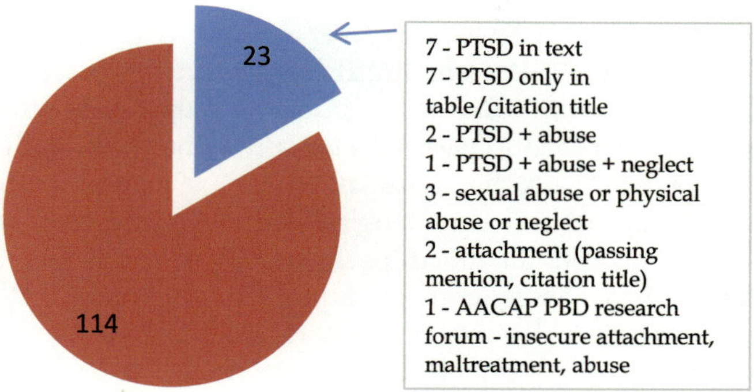

Fig. 3. Attachment and maltreatment/trauma terms in MGH/Harvard PBD literature.

The word "attachment" appears in 2 articles. One article (Henin et al., 2005) mentioned "attachments in infancy" in the passage: "the few studies that have examined the psychosocial functioning of children at risk for mood disorders have suggested that these children display poorer social skills and attachments in infancy (Zahn-Waxler et al 1984), as well as deviant school behaviors (Weintraub et al 1975, 1978), impaired academic performance (McDonough-Ryan et al 2000, 2002), suicidality (Klimes-Dougan et al 1999), and poorer peer social networks (Pellegrini et al 1986) in childhood. Taken together, these findings suggest that bipolar disorder may be characterized by extensive premorbid social and academic maladjustment."

The other (Biederman et al., 1998) mentioned "reactive attachment disorder" in a passage: "…a key limitation of our work: neither the structured interview diagnoses nor the clinical chart ratings can be accepted as unequivocal evidence for the diagnosis of bipolar disorder. For example, some of our patients met criteria for PTSD, and we did not assess for other disorders such as reactive attachment disorders that might present with manic symptoms. Thus, although our results demonstrate a link between mood stabilizer treatment and maniclike symptoms, they are not definitive as regards the treatment of bipolarity."

Neither paper elaborates upon attachment theory beyond those statements. Also the statement from Biederman et al. (1998) is somewhat at odds with the reported findings in the 16 other articles that mention PTSD. Nine of these articles only mentioned PTSD in a diagnostic list or table: in a diagnostic list of anxiety disorders (Spencer et al., 1999; Hirschfeld-Becker et al., 2006); as a comorbid diagnosis with 14% of preschool and 10% of under age 10 PBD diagnosed children (Wilens et al., 2003); as 1 comorbid PTSD case in a cohort of 18 PBD diagnosed children (Moore et al., 2007a) and 2 of 32 PBD diagnosed children (Moore et al., 2007b) and another article on the same cohort listed 2 of 28 PBD diagnosed children comorbid for PTSD (Frazier et al., 2007). Another study (Harpold et al., 2005) found high rates of all anxiety disorders within a PBD cohort and that PTSD had the highest odds ratio of correlating with PBD and concluded "our results indicate that BPD (bipolar disorder) significantly and robustly increased the risk of a broad range of anxiety disorders in youth." A recent study (Joshi & Wilens, 2009) also found high comorbidity rates with PBD. Wozniak (2003) did refer to PTSD in the text, noting that PBD research has been criticized amongst other things for "difficulty in distinguishing bipolar disorder (BD) from other conditions marked by irritability such as attention deficit/hyperactivity disorder (ADHD) and posttraumatic stress disorder (PTSD)."

Three papers (Biederman et al., 2000; Biederman et al., 2003; Wozniak et al., 1999) that dealt with the issue of trauma and PTSD more directly in the text concluded that PBD precedes trauma and PTSD. A child with PBD is so disruptive that they create traumatic situations and family relationships that then impact traumatically upon them. Both Biederman et al. (2000, 2003) articles refer to the earlier Wozniak et al. (1999) study and contain the same passage that states: "Using data from a longitudinal sample of boys with and without ADHD, Wozniak et al. (1999) identified paediatric bipolar disorder as an important antecedent for, rather than consequence of, traumatic life events... When traumatized children present with severe irritability and mood lability, there may be a tendency by clinicians to attribute these symptoms to having experienced a trauma. To the contrary, longitudinal research suggests the opposite: mania may be an antecedent risk factor for later trauma and not represent a reaction to the trauma (Wozniak et al 1999)."

Wozniak et al. (1999) had reported: "Our results showed that the diagnosis of bipolar disorder at baseline assessment in children with ADHD was the most significant predictor of the development of later trauma during the 4 year follow-up period. Although not entirely surprising, this finding, to our knowledge, has not been previously reported. Considering that mania is a very severe disorder with high rates of explosiveness, aggression, impulsivity, and poor judgement (Wozniak et al 1995a), it could predispose an affected child to trauma exposure...If confirmed, these results could help dispel the commonly held notion that mania like symptoms in youths represent a reaction to trauma."

The Wozniak et al. (1999) study was a 4 year follow-up study of 128 boys with ADHD (of whom 14 were diagnosed on structured interview with comorbid PBD at baseline and a further 7 diagnosed with PBD at the 4 year follow-up) plus 109 normal controls of whom 2 were diagnosed with comorbid PBD at follow-up. Fifteen of the 128 experienced a traumatic event and 4 (27%) of these 15 had comorbid PBD compared to 10 (9%) rate of comorbid PBD in the 113 ADHD boys without traumatic events during the follow-up period. The authors noted limitations - "our number of trauma-exposed subjects (including controls) was relatively small (n=23), and a very small number of traumatized subjects (n =2) went on to develop PTSD. ...our results should be viewed as preliminary until confirmed with larger samples." They also noted they did not assess for PTSD at baseline: "the findings reported in this study must be seen in light of methodological limitations. Since we assessed trauma only for the 4 year follow up period and did not make a lifetime assessment of trauma, we cannot rule out the possibility that trauma could have predated or contributed to the development of bipolar disorder in some children. However, if trauma were to lead to mania rather than the other way around, we should have found that children without mania traumatized during the follow-up period would be more likely to go on to develop mania. This was not the case in our study." Additionally whilst the study reported 1 child (out of 237) had experienced "physical abuse" and 3 children experienced "sexual abuse", the study does not report on any verbal or emotional abuse in the "types of trauma" examined.

The ages of the boys at the 4 year follow-up was peripubertal on average (ADHD 10.3 SD 2.9, ADHD + Trauma 12.3 SD 3.1, Control 11.5 SD 3.6, Control + Trauma 12.0 SD 4.1) and it is not reported as to what extent early life attachment factors were assessed. Wozniak et al. (1999) stated: "The literature suggests that protective factors operating at various stages of development may buffer children from posttraumatic suffering. For example, in a study of children and adults surviving Scud missile attacks in Israel, symptoms in children correlated with symptoms in their mothers. These authors concluded that maternal stress-buffering capacity plays a crucial role in minimizing suffering in traumatized preschool children (Lahor et al 1997)." Despite this passage Wozniak et al. (1999) do not appear to elaborate on parent-child relationships as mediating stress in their study. Also only parents and not children were interviewed if the child was under age 12, therefore presumably nearly all children were not interviewed at baseline.

A more recent article (Steinbuchel et al., 2009) affiliated with MGH/Harvard found an increased rate of PTSD in adolescents with PBD, though also tended to view PBD as a risk factor for PTSD. Subjects with both PTSD and PBD developed significantly more substance use disorders (SUD) and the authors concluded that "follow-up studies need to be conducted to elucidate the course and causal relationship of BPD, PTSD and SUD." Another article (Althoff et al., 2005) cautious in tone, stated: "In 2005 the idea is clearly not 'nature v nurture' but 'nature and nurture and how they interact'. Recent discoveries have shown the interaction between the serotonin transporter gene and trauma affecting likelihood of MDD and reduced by presence of positive social support. Thus far there have not been studies of specific G X E interactions with JBD." Further caution was expressed in a study (Faraone et al., 2001) of girls with ADHD and bipolarity, noting: "We did not assess for post-traumatic stress disorder (PTSD), which often is expressed with symptoms of ADHD and bipolarity. Thus, we cannot determine if cases of PTSD may have obscured our results."

A recent article (Doyle et al., 2010) reported lack of specificity in the Child Behaviour Checklist for diagnosing JBD (Juvenile Bipolar Disorder - equivalent PBD): "The items on the three scales that contribute to the CBCL-JBD profile reflect emotional and behavioral lability and distractability, i.e., items that index the capacity for self-regulation across a wide range of domains (i.e., cognitive, behavioral and affective). Further evidence for this conceptualization comes from Ayer et al. who found that the CBCL–JBD phenotype can be modeled as sharing a single latent trait with a different secondary CBCL scale purported to measure post-traumatic stress problems (PTSP). Like the CBCL-JBD phenotype, the PTSP scale is associated with suicidality and poor outcome and features a number of items overlapping with the CBCL-JBD that relate to self-regulation. Based on this analysis, the authors suggest both scales index a single dysregulatory syndrome. The fact that the CBCL-JBD phenotype taps into a trait relevant to a range of psychiatric disorders may help to explain the profile's lack of diagnostic specificity to juvenile-onset BPD in clinical studies."

Six articles mentioned the term "abuse": physical and sexual abuse were listed in a trauma list (Wozniak et al., 1999); brief mention of sexual abuse as a differential to manic hypersexuality (Soutullo et al., 2009); physical and sexual abuse briefly mentioned in relation to PTSD (Steinbuchel et al., 2009); abuse in a citation title which is referenced in the text: "findings in the pediatric (Ackerman et al., 1998) and adult (Kessler et al., 1995) literature document high rates of comorbid PTSD in bipolar subjects" (Harpold et al., 2005); a study (Baldessarini et al., 2004) reported "no history of physical or sexual abuse was found in any case" in a cohort of 82 PBD children (73% prepubertal with 74% having "onset of first symptoms" under age 3).

Another article (Bostic et al., 1997) mentioned infants being depressed in "abusive and neglectful situations". Otherwise "neglect" is not mentioned by MGH/Harvard authors except in the context of "neglect of PBD" as a diagnosis. The term "maltreatment" is not mentioned. The AACAP 2006 Research Forum (Carlson et al., 2009) had co-authors from the MGH/Harvard group and as above did mention maltreatment and abuse by name.

3.4 PBD terms in the attachment theory oriented literature

Just 8 papers were found by Scopus search for "attachment" in "All Fields" from a body of 746 articles. However on close examination not all these articles were strong on attachment theory based themes. The main focus for 7 of these was on anxiety and depression arising out of parent-child relationships. PBD was only a major theme in an editorial (Miklowitz & Cichetti, 2006) that was more in the context of the PBD literature (the journal, *Developmental Psychopathology*, issue was devoted to PBD) rather than attachment theory. It was possibly selected by Scopus as attachment oriented because of the phrase "developmental psychopathology" in the title and text. "Sexual abuse" is in the title of a reference. The editorial doesn't contain the word "attachment", nor "PTSD/trauma" or "maltreatment".

4. Comparison of neuroimaging reviews

Given that research in both developmental traumatology and amongst PBD investigators has focussed on neuroimaging in recent years, a comparison (but in this case not a systematic review) of neuroimaging reviews from both the PBD literature and attachment/trauma literature is of interest.

Schore is a prominent author in the attachment and developmental trauma literature who has reviewed neuroimaging research data in two books (Schore, 2003a; Schore, 2003b) and a review article (Schore, 2002). The indexes of each book do not contain the word "bipolar", and "mania" is mentioned only once in each book - in reference to right orbitoprefrontal cortex (ROPFC) dysfunction. However both books focus on ROPFC dysfunction as primarily relating to impaired modulation of subcortical limbic structures and manifesting as affect dysregulation and behavioural impulsivity relating to disorders of attachment and trauma, disruptive behaviour disorders and personality disorders. The terms "bipolar", "mania/c" or "hypomania/c" do not occur in the review article.

A recent review (McCrory et al., 2011) of the neurobiological, genetic and epigenetic factors associated with childhood maltreatment also reports amygdala hyper-reactivity and reduced frontal cortical control of subcortical limbic structures. In particular fMRI studies of emotional processing of human faces in both adults and children revealed: "hyperactivity of the amygdale in response to negative facial affect." The review covers epigenetic changes that appear to underpin such neurobiological findings and the importance of secure-attachment to promote resiliency against such effects of maltreatment. Specifically it appears that "an early hostile environment contributes to stress-induced changes in the child's neurobiological systems that may be adaptive in the short term but which reap long term costs." Additionally cognitive deficits, particularly deficits of working memory are correlated with maltreatment and institutionalization.

Interestingly the more recent PBD literature increasingly includes neuroimaging studies comparing PBD diagnosed cohorts with, for example, normal controls (e.g. Pavuluri et al., 2009a). None of the terms for attachment, PTSD or maltreatment/abuse appear in this article. Yet it describes similar findings concerning the right pre-frontal cortex and limbic system, including right amygdala reactivity and impaired right prefrontal cortical functioning, that the above reviews from an attachment and developmental trauma/maltreatment perspective describe.

5. Discussion

5.1 Attachment and trauma/maltreatment terms generally overlooked in PBD literature

A systematic review of the PBD literature via searching for the term "attachment" lends credence to critics' claims that the PBD literature in general does not address or consider attachment theory concepts. The almost complete absence of attachment theory concepts makes interpretation of trauma and maltreatment/abuse events in childhood problematic as there is evidence that attachment security/insecurity mediates the effects of trauma and abuse upon children (Cook et al., 2005). Furthermore developmental trauma, maltreatment/abuse and PTSD related concepts receive infrequent coverage in the PBD literature. The two research institutions that first promoted PBD illustrate this: researchers from WUSL report a virtual absence of PTSD in their cohort; researchers from MGH/Harvard suggest PTSD mainly arises secondary to PBD, though more recent publications from the group are more cautious.

The very low rate of sexual abuse and no cases of PTSD in the WUSL research is remarkable in any clinical cohort. It is also at odds with research (Rucklidge, 2006) on a cohort of adolescents in New Zealand that found 29.2% reported sexual abuse on the same diagnostic instrument used in diagnosing the WUSL cohort, and over 50% of the New Zealand PBD sample had a trauma history compared with 10% of controls.

The MGH/Harvard group propose that PTSD where it does occur comorbidly with PBD arises secondary to PBD itself. However the main reference for this, a study (Wozniak et al., 1999) of 128 peripubertal boys with ADHD, of whom 14 had comorbid PBD, noted several limitations of their study including that it was of low power and that trauma and PTSD were not assessed at baseline. Nor from the article does it appear that early attachment histories had been taken in depth. Nonetheless if there was increased risk for experiencing trauma in the 4 year follow-up period for the boys with ADHD and a comorbid PBD diagnosis, an alternative hypothesis, not explored in the article, would be that the boys with ADHD and comorbid PBD at baseline were in fact exhibiting symptoms of earlier developmental trauma. Such earlier developmental trauma, mediated by psychodynamic, family dynamic, behavioural learning and other environmental contextual factors, could mean the 14 boys were more vulnerable to traumatic events over the 4 year follow-up period than those with ADHD but without PBD as defined in the study's methodology. More recent articles (Althoff et al., 2005; Steinbuchel et al., 2009; Doyle et al., 2010) from authors affiliated with MGH/Harvard are more open to the possibility of trauma factors causing or exacerbating symptoms, yet still conceptualise these symptoms in terms of PBD.

5.2 SMD articles also limit mention of attachment and trauma factors

As noted above, "broad phenotype" PBD has effectively been renamed "Severe Mood Dysregulation" (SMD) (Brotman et al., 2006; Dickstein et al., 2006; Stringaris, 2009 & 2011; Leibenluft, 2011). However a reading of these 5 papers suggests attachment and trauma related factors appear to be only a limited focus thus far, of research into SMD. Furthermore SMD is likely to feature in DSM-5 under the title: "Disruptive Mood Dysregulation Disorder" (DMDD). This proposed diagnosis has drawn some intense criticism, particularly from Frances, head of the former DSM-IV task force, who has described DMDD as one of the "worst ideas" for a new DSM diagnosis (Frances, 2011). Frances also notes that DMDD has likely been accepted as "a lesser evil replacement for childhood bipolar disorder—less stigmatizing and less likely to result in reflex long term antipsychotic use." But he suggests:

> "DMDD will capture a wildly heterogeneous and diagnostically meaningless grab bag of difficult to-handle kids. Some will be temperamental and irritable, but essentially normal and just going through a developmental stage they will eventually outgrow without a stigmatizing diagnosis and a harmful treatment. Others will have conduct or oppositional problems that gain nothing by being mislabelled as mood disorder. Yet others will have serious, but not yet clearly defined psychiatric disorders that require careful and patient monitoring before an accurate diagnosis can be made."

However attachment, developmental trauma and maltreatment are still not mentioned.

5.3 Attachment theory based literature fails to mention PBD

Some PBD authors (e.g. Biederman, 2003) have strongly argued that mania and bipolar disorder is not considered by researchers who come from a more traditional child psychopathology perspective. A search of the attachment theory based literature, as outlined above, does in fact suggest unawareness or dismissal of the concept of PBD. However in defence of attachment oriented studies, it could be argued that most work to date has been in infancy and early child development prior to the onset of typical DSM clinical syndromes, at least as classically defined. Much of the attachment and developmental traumatology literature is in psychology, general science and neuroscience journals, whereas the PBD literature is primarily in US based psychiatry journals. To some extent this supports the hypothesis that there are differing paradigms governing the way children with severe emotional and behavioural problems are assessed and diagnosed.

5.4 Neuroimaging: PBD or developmental trauma/maltreatment

The specific case of neuroimaging in PBD research and attachment-trauma oriented research is an example where similar findings in the attachment-trauma oriented literature appear to be interpreted differently by authors from the PBD literature, and without cross-referencing.

Neuroimaging of children with disorganized attachment and trauma histories has, amongst other findings, revealed impaired right prefrontal cortex control over a hyperactive right amygdala. This can be explained in terms of the function of these structures in attachment relationships and for survival in the face of threat (Schore, 2002a). Neuroimaging of children diagnosed with PBD (Pavuluri et al., 2009) found essentially the same findings but made no reference to attachment and trauma factors. When this very interesting data from a technically sophisticated study was presented at the AACAP 2009 conference (Pavuluri, 2009), I and others asked during the presentation why the children could not simply have been labelled as "affect dysregulated" rather than as having bipolar disorder? The presenter agreed they could well have, but stated that if they were not described as suffering bipolar disorder then research funding would be unlikely. At the same conference similar neuroimaging findings delineated an ADHD cohort from a PBD cohort (Delbello, 2009). The research again appeared to have high technical quality, but once again it is possible that a-priori assumptions may have governed the scope of possible conclusions. When I asked during the presentation if PTSD or disorganized attachment had been considered in addition to ADHD and PBD, the presenter replied that they had not been investigated.

The rise of new and exciting technological developments in neuroimaging and epigenetics hopefully will help develop understanding of childhood developmental psychopathology. But accurate understanding is likely to only grow if a wide range of hypotheses are maintained and all contextual factors, both historical and current, in a child's life are considered. McCrory et al. (2011) in their recent review, whilst acknowledging the high likelihood of trauma preceding brain changes, advocate for this and state that longitudinal studies are needed that "allow changes in the child's environment and behavior to be measured alongside changes in brain structure and function…if we are to make even tentative inferences regarding causality."

5.5 Perspectives from different paradigms

It has been argued that science proceeds not just in terms of applying the scientific method, but within a historical and sociocultural context with implicit assumptions and belief systems that set the parameters of the research, in other words according to a prevailing paradigm (Kuhn, 1962). The prevailing paradigm governs what is considered for study and treatment and what is not. Thus even research of high intellect, internal consistency and technical quality can lead to false conclusions if the paradigm is too restrictive. Furthermore differing paradigms can co-occur and be operative in the same era.

Based on this systematic literature review plus a selective review of neuroimaging research, there does indeed appear to be a communication gulf between two different paradigmatic approaches in child & adolescent psychiatry and developmental psychopathology.

Developmental Trauma Disorder (DTD) (Van der Kolk & Courtois, 2005) is another proposed diagnosis for DSM-5. DTD has been proposed as a more accurate descriptor for many children diagnosed with PBD (Levin 2009). However DTD is not officially within the DSM-IV, whereas ADHD is and PBD has been given semi-official status under the rubric of BD-NOS (BD Not Otherwise Specified). The DSM-IV diagnoses are used to guide and constrain much of the funding for therapy and research, particularly in the USA. Thus in the neuroimaging research presented at the AACAP 2009 conference, ADHD and PBD receive consideration, but DTD and attachment and contextual factors seemingly did not.

It has been argued that one root cause of this problem lies with the atheoretical symptom focused approach incorporated within DSM-IV (Denton, 2007) and consequently mainstream psychiatry has become too detached from attachment theory, psychoanalysis and traumatology and the progress made in these fields (Dignam et al., 2010).

These factors, in conjunction with "diagnostic upcoding" pressures, the influence of the pharmaceutical industry and a societal tendency to repress recognition of trauma have been argued as fuelling the rise in PBD diagnosis rates (Parry & Levin, 2011). Rather than existing in parallel, researchers in PBD and other DSM diagnoses may likely benefit from increased dialogue with researchers from attachment theory and developmental traumatology perspectives. Furthermore attachment theory oriented research would be advanced by exploring attachment and trauma influences in DSM-IV and ICD-10 syndromes.

5.6 Signs of increasing attention to attachment and trauma factors

In 2005 the first "treatment guidelines" for PBD (Kowatch et al., 2005) did not mention attachment or trauma factors and focussed almost exclusively on pharmacotherapy algorithms in treatment for PBD, although there was an accompanying critical commentary (McClellan, 2005). In 2006, although still labelling the phenomenology as BD, the AACAP 2006 research forum (Carlson et al., 2009) did list a range of contextual environmental adversity factors as implicated in the aetiology of PBD. In 2007 an official AACAP "practice parameter" publication (AACAP, 2007) included authors who have published articles sceptical about PBD. This AACAP practice parameter combined both paradigmatic perspectives with quite differing views within the one document. It contained a section on the "diagnostic controversy" which, referencing work from both WUSL ("narrow-phenotype" PBD) and MGH/Harvard ("broad phenotype" PBD), noted that "although symptoms of early-onset bipolar disorder appear stable over time (Biederman et al., 2004b; Geller et al., 2004) [citations in original], juvenile mania has not yet been shown to progress into the classic adult disorder." The practice parameter also listed "psychotherapeutic interventions" as important in treatment and noted "dialectical-behavioural therapy may be helpful for youths with mood and behavioural dysregulation."

At the 2010 AACAP conference there were two symposia (AACAP, 2010a & 2010b) each with several papers highlighting contextual factors and stressing a more non-aetiological descriptor of "affect dysregulation" rather than using the bipolar or mania label for children with mood swings. Also in 2010, a report (Parens & Johnson, 2010) of a 2 day workshop, involving researchers in the field of PBD, records vigorous debate over the validity of the PBD diagnosis. Attachment and trauma are mentioned and paradigmatic aspects of the issue are also canvassed. Also as illustrated in the literature review above, more recent articles from researchers affiliated with MGH/Harvard have drawn attention to the need for more research into PTSD related factors.

In contrast a recent review and meta-analysis of pharmacotherapy in PBD (Liu et al., 2011) made no mention of psychotherapy, nor of psychosocial factors in diagnosis. The review noted limited efficacy of traditional mood stabilizers (Lithium and anticonvulsants) in PBD, whereas second generation antipsychotics (SGAs) had more efficacy and speculated "such results are consistent with the hypothesis that pediatric-onset bipolar disorder may represent a different subtype of bipolar disorder that could respond to different treatments than those observed in adult-onset cases." It has been often argued however that SGAs simply exert their effect via sedation (e.g. Ghaemi & Martin, 2007; Frances, 2010; Kaplan, 2011) and do not confirm a particular diagnosis as for example juvenile mania.

5.7 Implications for therapy

The debate about whether a child with severe emotional and behavioural problems has PBD, versus DTD or ADHD plus/or ODD or CD is far from academic. The choice of treatment, the risk of suffering side-effects, the child's perception of self, the family's perception of their child and the perception and behaviour of relevant others such as teachers are strongly influenced by the diagnostic label. The controversy over PBD has become impassioned because of such consequences.

Treatment guidelines (Kowatch, 2005; Liu et al., 2011) for PBD strongly promote use of psychotropic agents. PBD has been blamed for leading to an explosion, particularly in the USA, in the use of atypical antipsychotic agents and polypharmacy approaches for children (e.g. Frances, 2010; Parry & Levin, 2011; Robbins et al., 2011; Kaplan 2011).

5.8 Limitations

This systematic literature review relied on one academic search engine, Scopus, albeit one that aids this form of literature search. Defining a body of literature in a sensitive yet specific enough manner proved somewhat challenging. A full reading of all 7,257 publications would be needed to make the searches more thorough. Nonetheless the hypothesis being tested pertains to a broad trend over the past decade and a half within child and adolescent psychiatry, rather than specific researchers or scientific articles. In that sense the use of Scopus in this manner to examine broad trends can be justified. Furthermore it can be argued that full text searching the literature from two US child and adolescent academic centres most strongly associated with developing the PBD phenotypes should give a strong indication of how attachment, maltreatment and trauma factors are considered in the wider PBD literature. The systematic literature review covered the 15 ½ years to June 2010 and it is quite possible that more may have been written on attachment and trauma/maltreatment factors in the very recent PBD literature. However the PBD phenotypes have become entrenched in research and clinical practice, at least in the USA, during the time frame since the germinal articles in 1995.

6. Conclusion

Intense controversy over the validity of PBD remains despite a decade and a half of research into the postulated PBD phenotypes. A main criticism of the PBD constructs is that they fail to consider attachment theory and maltreatment and developmental trauma factors.

A systematic search of the PBD literature presented here found this to generally be the case. There was a virtual absence of consideration of attachment theory. Trauma and PTSD was described as likely secondary to pre-existing childhood mania by researchers associated with the "broad phenotype" PBD construct. Maltreatment factors were relatively absent in findings from cohorts in both "broad phenotype" and "narrow phenotype" PBD research. Furthermore attachment, maltreatment and trauma factors do not appear to be a focus of research that reconceptualises "broad phenotype" PBD as SMD.

A comparison of neuroimaging studies from attachment/developmental traumatology and PBD research shows remarkably similar findings interpreted quite differently. Two different paradigms appear operative within the field. Increased dialogue across these paradigmatic perspectives is likely to help resolve the controversial nature of PBD. To quote Carlson & Meyer (2006), PBD research "would benefit from a developmental psychopathology perspective". This involves greater consideration of attachment insecurity and a child's psychodynamic defences against traumatic contextual factors.

7. Acknowledgment

The author is grateful to Gareth Furber for the figures, and to Stephen Allison, Jon Jureidini, David Healy, Anja Kriegeskotten and Joy Elford for comments on the manuscript.

8. References

AACAP (2007). AACAP Official Action: Practice parameter for the assessment and treatment of children and adolescents with bipolar disorder. *Journal of the American Academy of Child and Adolescent Psychiatry,* Vol.46, No.1, (January 2007), pp.107-125, ISSN 0890-8567

AACAP (2010a). American Academy of Child and Adolescent Psychiatry annual meeting 2010. Symposium 30: Dysregulated but NOT bipolar: New Insights into childhood problems with self-regulation. 13.01.2011, Available from http://aacap.confex.com/aacap/2010/webprogram/Session5573.html

AACAP (2010b). American Academy of Child and Adolescent Psychiatry annual meeting 2010. Clinical Perspectives 29: When the diagnosis is bipolar: Are there other explanations?". 13.01.2011, Available from http://aacap.confex.com/aacap/2010/webprogram/Session4853.html

Akiskal, H. (2007). The Emergence of the Bipolar Spectrum: Validation Along Clinical-Epidemiologic and Familial-Genetic Lines. *Psychopharmacology Bulletin,* Vol.40, No.4, (November 2007), pp. 99-115, ISSN 0048-5764

Althoff, R.R., Faraone, S.V., Rettew, D.C., Morley, C.P., & Hudziak, J,J. (2005). Family, twin, adoption, and molecular genetic studies of juvenile bipolar disorder. *Bipolar Disorders,* Vol.7, No.4, (August 2005), pp. 598-609, ISSN 1398-5647

Bar-Haim, Y., Prez-Edgar, K., Fox, N.A., Beck, J.M., West, G.M., Bhangoo, R.K., Myers, F.S., & Leibenluft, E. (2002). The emergence of childhood bipolar disorder: a prospective study from 4 months to 7 years of age. *Journal of Applied Developmental Psychology,* Vol.23, No.4, (November 2002), pp. 431-455, ISSN 0193-3973

Biederman, J., Mick, E., Bostic, J.Q., Prince, J., Daly, J., Wilens, T.E., Spencer, T., Garcia-Jetton, J., Russell, R., Wozniak, J., & Faraone, S.V. (1998). The naturalistic course of pharmacologic treatment of children with maniclike symptoms: a systematic chart review. *Journal of Clinical Psychiatry,* Vol.59, No.11, (November 1998) pp. 628-637, ISSN 0160-6689

Biederman, J., Mick, E., Faraone, S.V., Spencer, T., Wilens, T.E., & Wozniak, J. (2000). Pediatric mania: a developmental subtype of bipolar disorder? *Biological Psychiatry,* Vol.48, No.6, (15 September 2000), pp. 458-466, ISSN 0006-3223

Biederman, J., Mick, E., Faraone, S.V., Spencer, T., Wilens, T.E., & Wozniak, J. (2003) Current concepts in the validity, diagnosis and treatment of paediatric bipolar disorder. *The International Journal of Neuropsychopharmacology,* Vol.6, No.3, (September 2003), pp. 293-300, ISSN 1461-1457

Biederman, J. (2003). Editorial: Paediatric bipolar disorder coming of age. *Biological Psychiatry,* Vol.53, No.11, (1 June 2003) pp. 931-934, ISSN 0006-3223

Bostic, J.Q., Wilens, T., Spencer, T., & Biederman, J. (1997). Juvenile mood disorders and office psychopharmacology. *Pediatric Clinics of North America,* Vol.44, No.6, (December 1997), pp. 1487-1503, ISSN 0031-3955

Brotman, M. A., Schmajuk, M., Rich, B. A., Dickstein, D. P., Guyer, A. E., Costello, E. J., Egger, H.L., Angold, A., Pine, D.S., & Leibenluft, E. (2006). Prevalence, clinical correlates, and longitudinal course of severe mood dysregulation in children. *Biological Psychiatry,* Vol.60, No.9, (1 November 2006), pp. 991–997, ISSN 0006-3223

Carey, B. (2007, February 15). Debate over children and psychiatric drugs. *The New York Times,* 22.02.2007, Available from http://www.nytimes.com/2007/02/15/us/15bipolar.html

Carlson, G. (1984). Classification Issues of Bipolar Disorders in Childhood. *Psychiatric Developments,* Vol.2, No.4, (**=====, 1984), pp. 273–285, ISSN 0262-9283

Carlson, G. & Meyer, S. (2006). Phenomenology and Diagnosis of Bipolar Disorder in Children, Adolescents and Adults: Complexities and Developmental Issues. *Development and Psychopathology,* Vol.18, No.4, (October 2006), pp. 939–969, ISSN 0954-5794

Carlson, G.A., Findling, R.L., Post, R.M., Birmaher, B., Blumberg, H.P., Correll, C., DelBello, M.P., Fristad, M., Frazier, J., Hammen, C., Hinshaw, S.P., Kowatch, R., Leibenluft, E., Meyer, S.E., Pavuluri, M.N., Wagner, K.D., & Tohen, M. (2009). AACAP 2006 – advancing research in early-onset bipolar disorder: barriers and suggestions. *Journal of Child and Adolescent Psychopharmacology,* Vol.19, No.1, (February 2009), pp. 3-12, ISSN 1044-5463

Cook, A., Spinazzola, J., Ford, J., Lanktree, C., Blaustein, M., Cloitre, M., DeRosa, R., Hubbard, R., Kagan, R., Liautaud, J., Mallah, K., Olafson, E., & van der Kolk B. (2005). Complex trauma in children and adolescents. *Psychiatric Annals,* Vol.35, No.5, (May 2005), pp. 390-398, ISSN 0048-5713

Costello, E.J., Pine, D.S., Hammen, C., March, J.S., Plotsky, P.M., Weissman, M.M., Biederman, J., Goldsmith, H.H., Kaufman, J., Lewinsohn, P.M., Hellander, M., Hoagwood, K., Koretz, D.S., Nelson, C.A., & Leckman, J.F. (2002). Developmental and natural history of mood disorders. *Biological Psychiatry,* Vol.52, No.6, (15 September 2002), pp. 529-542, ISSN 0006-3223

Cramer, M. (2007, Feb 9). DSS seeking medical experts. *Boston Globe,* 22.02.2007, Available from http://www.boston.com/news/local/articles/2007/02/09/dss_seeking_medical_experts/

Craney, J.L. & Geller, B. (2003). A prepubertal and early adolescent bipolar disorder-I phenotype: review of phenomenology and longitudinal course. *Bipolar Disorders,* Vol.5, No.4, (August 2003), pp. 243-256, ISSN 1398-5647

Cytryn, L. (2003). Recognition of childhood depression: Personal reminiscences. *Journal of Affective Disorders,* Vol.77, No.1, (October 2003), pp. 1-9, ISSN 0165-0327

DelBello, M. (2009). The neurobiology of pediatric bipolar disorder. *Proceedings of American Academy of Child & Adolescent Psychiatry, 56th annual meeting,* 27 Oct – 1 Nov 2009, Honolulu, Hawaii.

Denton, W.H. (2007). Editorial: Issues for DSM-V: Relational diagnosis: an essential component of biopsychosocial assessment. *American Journal of Psychiatry,* Vol.164, No.8, (August 2007) pp. 1146-1147, ISSN 0002-953X

Dickstein, D.P., & Leibenluft, E. (2006). Emotion regulation in children and adolescents: Boundaries between normalcy and bipolar disorder. *Development and psychopathology,* Vol.18, No.4, (December 2006) pp. 1105-1131, ISSN 0954-5794

Dignam, P., Parry, P.I., & Berk, M. (2010). Detached from attachment: neurobiology and phenomenology have a human face. *Acta Neuropsychiatrica,* Vol.22, No.4, (August 2010) pp. 202-206, ISSN 0924-2708

Doyle, A.E., Biederman, J., Ferreira, M.A.R., Wong, P., Smoller, J.W., & Faraone, S.V. (2010). Suggestive linkage of the child behavior checklist juvenile bipolar disorder phenotype to 1p21, 6p21, and 8q21. *Journal of the American Academy of Child and Adolescent Psychiatry,* Vol.49, No.3, (March 2010), pp. 378-387, ISSN 0890-8567

Faedda, G.L., Baldessarini, J., Glovinsky, I.P., & Austin, N.B. (2004). Pediatric bipolar disorder: phenomenology and course of illness. *Bipolar Disorders,* Vol.6, No.4, (August 2004), pp. 305-313, ISSN 1398-5647

Faraone, S.V., Biederman, J., & Monuteaux, M.C. (2001). Attention deficit hyperactivity disorder with bipolar disorder in girls: further evidence for a familial subtype? *Journal of Affective Disorders,* Vol.64, No.1, (April 2001), pp. 19-26, ISSN 0165-0327

Frances, A. (2010, April 8). Psychiatric diagnosis gone wild: The "epidemic" of childhood bipolar disorder. *Psychiatric Times,* Available from http://www.psychiatrictimes.com/display/article/10168/1551005# ISSN 0893-2905

Frances, A. (2011, July 22). DSM approves new fad diagnosis for child psychiatry: antipsychotic use likely to rise. *Psychiatric Times,* Available from http://www.psychiatrictimes.com/display/article/10168/1912195 ISSN 0893-2905

Frazier, J.A., Breeze, J.L., Papadimitriou, G., Kennedy, D.N., Hodge, S.M., Moore, C.M., Howard, J.D., Rohan, M.P., Caviness, V.S., & Makris, N. (2007). White matter abnormalities in children with and at risk for bipolar disorder. *Bipolar Disorders,* Vol.9, No.8, (December 2007), pp. 799-809, ISSN 1398-5647

Freedman, R. & 25 fellow signatories. (2009). Editorial Board Commentary: Conflict of interest - an issue for every psychiatrist. *American Journal of Psychiatry,* Vol.166, No.3, (March 2009), pp. 274-277, ISSN 0002-953X

Ghaemi S. & Martin A. (2007). Defining the boundaries of childhood bipolar disorder. *American Journal of Psychiatry,* Vol.164, No.2, (February 2007), pp. 185–188, ISSN 0002-953X

Geller, B., Sun, K., Zimerman, B., Frazier, J., & Williams M. (1995). Complex and Rapid Cycling in Bipolar Children and Adolescents. *Journal of Affective Disorders,* Vol.34, No.4, (August 1995), pp. 259-268, ISSN 0165-0327

Geller, B. & Luby, J. (1997). Child and Adolescent Bipolar Disorder: A Review of the Past 10 Years. *Journal of the American Academy of Child and Adolescent Psychiatry,* Vol.36, No.9, (September 1997), pp. 1168 - 1176, ISSN 0890-8567

Geller, B., Bolhofner, K., Craney, J.L., Williams, M., DelBello, M.P., & Gundersen, K. (2000). Psychosocial functioning in a prepubertal and early adolescent bipolar disorder phenotype. *Journal of the American Academy of Child & Adolescent Psychiatry,* Vol.39, No.12, (December 2000), pp. 1543-1548, ISSN 0890-8567

Geller, B., Zimerman, B., Williams, M., DelBello, M.P., Bolhofner, K., Craney, J.L., Frazier, J., Beringer, L., & Nickelsburg, M.J. (2002). DSM-IV mania symptoms in a prepurbertal and early adolescent bipolar disorder phenotype compared to attention-deficit hyperactive and normal controls. *Journal of Child and Adolescent Psychopharmacology,* Vol.12, No.1, (March 2002), pp. 11-25, ISSN 1044-5463

Geller, B., Tillman, R., Craney, J.L., & Bolhofner, K. (2004). Four-year prospective outcome and natural history of mania in children with a prepubertal and early adolescent bipolar disorder phenotype. *Archives of General Psychiatry,* Vol.61, No.5, (May 2004), pp. 459-467, ISSN 0003-990X

Geller, B., Tillman, R., Badner, J.A., & Cook, Jr E.H. (2005). Are the arginine vasopressin V1a receptor microsatellites related to hypersexuality in children with a prepubertal and early adolescent onset bipolar disorder phenotype? *Bipolar Disorders,* Vol.7, No.6, (December 2005), pp. 610-616, ISSN 1398-5647

Geller, B., Harms, M.P., Wang, L., Tillman, R., DelBello, M.P., Bolhofner, K., & Csernansky, J.G. (2009). Effects of age, sex, and independent life events on amygdale and nucleus accumbens volumes in child bipolar I disorder. *Biological Psychiatry,* Vol.65, No.5, (1 March 2009), pp. 432-437, ISSN 0006-3223

Harpold, T.L., Wozniak, J., Kwon, A., Gilbert, J., Wood, J., Smith, L., & Biederman, J. (2005). Examining the association between pediatric bipolar disorder and anxiety disorders in psychiatrically referred children and adolescents. *Journal of Affective Disorders,* Vol.88, No.1, (September 2005), pp. 19-26, ISSN 0165-0327

Harris, J. (2005). Child & adolescent psychiatry: the increased diagnosis of "juvenile bipolar disorder": what are we treating? *Psychiatric Services,* Vol.56, No.5, (May 2005), pp.529-531, ISSN 1075-2730

Healy, D. (2008). *Mania: a short history of bipolar disorder,* John Hopkins Biographies of Disease, John Hopkins University Press, ISBN: 978-0-8018-8822-9, Maryland

Healy, D. (2010). From mania to bipolar disorder, In: *Bipolar Disorder: Clinical and Neurobiological Foundations,* Lakshmi N Yatham & Mario Maj, pp. 1-7, John Wiley & Sons, ISBN: 978-0-470-72198-8, Chichester

Henin, A., Biederman, J., Mick, E., Sachs, G.S., Hirshfeld-Becker, D.R., Siegel, R.S., McMurrich, S., Grandin, L., & Nierenberg, A.A. (2005). Psychopathology in the offspring of parents with bipolar disorder: a controlled study. *Biological Psychiatry,* Vol.58, No.7, (1 October 2005), pp. 554-561, ISSN 0006-3223

Hirshfeld-Becker, D.R., Biederman, J., Calltharp, S., Rosenbaum, E.D., Faraone, S.V., & Rosenbaum, J.F. (2003). Behavioral inhibition and disinhibition as hypothesized precursors to psychopathology implications for pediatric bipolar disorder. *Biological Psychiatry,* Vol.53, No.11, (1 June 2003), pp. 985-999, ISSN 0006-3223

Hirshfeld-Becker, D.R., Biederman, J., Henin, A., Faraone, S.V., Dowd, S.T., DePetrillo, L.A., Markowitz, S.M., & Rosenbaum, J.F. (2006). Psychopathology in the young offspring of parents with bipolar disorder: a controlled pilot study. *Psychiatry Research,* Vol.145, No.2-3, (7 December 2006), pp. 155-167, ISSN 0165-1781

Joshi, G. & Wilens, T. (2009). Comorbidity in Pediatric Bipolar Disorder. *Child and adolescent psychiatric clinics of North America,* Vol.18, No.2, (April 2009), pp. 291-319, ISSN 1056-4993

Kaplan, S.L. (2011). *Your Child Does NOT Have Bipolar Disorder: How Bad Science and Good Public Relations Created the Diagnosis,* Praeger (Childhood in America) ISBN 978-0-313-38134-8, Santa Barbara, California

Kowatch, R.A., Fristad, M., Birmaher, B., Wagner, K.D., Findling, R.L., & Hellander, M. (2005). Treatment guidelines for children and adolescents with bipolar disorder. *Journal of the American Academy of Child and Adolescent Psychiatry,* Vol.44, No.3, (March, 2005), pp. 215-235, ISSN 0890-8567

Kuhn, T. (1962). *The structure of scientific revolutions,* University of Chicago Press, ISBN 0-226-45808-3, Chicago

Leibenluft, E. (2011). Severe mood dysregulation, irritability and the diagnostic boundaries of bipolar disorder in youths. *American Journal of Psychiatry,* Vol.168, No.2, (February 2011), pp.129-142, ISSN 0002-953X

Levin, E.C. (2009). The challenges of treating developmental trauma disorder in a residential agency for youth. *Journal of the American Academy of Psychoanalysis and Dynamic Psychiatry,* Vol.37, No.3, (Fall 2009), pp. 519-538, ISSN 1546-0371

Levin, E.C. & Parry, P.I. (2011). Conflict of interest as a possible factor in the rise of pediatric bipolar disorder. *Adolescent Psychiatry,* Vol.1, No.1, (January 2011), pp.61-66, ISSN 2210-6766

Liu, H.Y.; Potter, M.P.; Woodworth, K.Y.; Yorks, D.M.; Petty, C.R.; Wozniak, J.R.; Faraone, S.V. & Biederman, J. (2011). Pharmacologic treatments for pediatric bipolar disorder: a review and meta-analysis. *Journal of the American Academy of Child and Adolescent Psychiatry,* Vol.50, No.8, (August 2011), pp. 749-762, ISSN 0890-8567

Luby, J.L. & Belden, A.C. (2008). Clinical characteristics of bipolar vs. unipolar depression in preschool children: An empirical investigation. *Journal of Clinical Psychiatry,* Vol.69, No.12, (December 2008), pp. 1960-1969, ISSN 0160-6689

Luby, J.L. & Navsaria, N. (2010). Pediatric bipolar disorder: evidence for prodromal states and early markers. *Journal of Child Psychology and Psychiatry,* Vol.51, No.4, (April 2010), pp. 459-471, ISSN 0021-9630

Marchand, W.R., Wirth, L., & Simon, C. (2005). Adverse life events and pediatric bipolar disorder in a community mental health setting. *Community Mental Health Journal,* Vol.41, No.1, (February 2005), pp. 67-75, ISSN 0010-3853

McClellan, J. (2005). Commentary: treatment guidelines for child and adolescent bipolar disorder. *Journal of the American Academy of Child and Adolescent Psychiatry,* Vol.44, No.3, (March 2005), pp. 236-239, ISSN 0890-8567

McClure, E.B., Kubiszyn, T., & Kaslow, N.J. (2002). Advances in the diagnosis and treatment of childhood mood disorders. *Professional Psychology Research and Practice,* Vol.33, No.2, (April 2002), pp. 125-134, ISSN 0735-7028

McCrory, E., DeBrito, S.A., & Viding, E. (2011). The impact of childhood maltreatment: a reveiew of neurobiological and genetic factors. *Frontiers in Child and Neurodevelopmental Psychiatry,* Vol.2:48 Published online 28 July 2011, pp.1-14, ISSN 1664-0640

McDonnell, M.A. & Wozniak, J. (2008). *Is Your Child Bipolar?*, Bantam Books, ISBN 978-0-553-38462-8, New York, NY

Meyer, S.E., Carlson, G.A., Wiggs, E.A., Ronsaville, D.S., Martinez, P.E., Klimes-Dougan, B., Gold, P.W., & Radke-Yarrow, M. (2006). A prospective high-risk study of the association among maternal negativity, apparent frontal lobe dysfunction, and the development of bipolar disorder. *Development and psychopathology*, Vol.18, No.2, (June 2006), pp. 573–589, ISSN 0954-5794

Miklowitz, D.J. & Cicchetti, D. (2006). Towards a developmental psychopathology of bipolar disorder. *Development and Psychopathology*, Vol.18, No.4, (December 2006), pp. 935-938, ISSN 0954-5794

Miklowitz, D.J., Biukians, A., & Richards, J.A. (2006). Early-onset bipolar disorder: A family treatment perspective. *Development and psychopathology*, Vol.18, No.4, (December 2006), pp. 1247-1265, ISSN 0954-5794

Moore, C.M., Biederman, J., Wozniak, J., Mick, E., Aleardi, M., Wardrop, M., Dougherty, M., Harpold, T., Hammerness, P., Randall, E., Lyoo, K., & Renshaw, P.F. (2007a). Mania, glutamate/glutamine and risperidone in pediatric bipolar disorder: A proton magnetic resonance spectroscopy study of the anterior cingulate cortex. *Journal of Affective Disorders*, Vol.99, No.1, (April 2007), pp. 19-25, ISSN 0165-0327

Moore, C.M., Frazier, J.A., Glod, C.A., Breeze, J.L., Dieterich, M., Finn, C.T., Frederick, B.D., & Renshaw, P.F. (2007b). Glutamine and glutamate levels in children and adolescents with bipolar disorder: A 4.0-T proton magnetic resonance spectroscopy study of the anterior cingulate cortex. *Journal of the American Academy of Child and Adolescent Psychiatry*, Vol.46, No.4, (April 2007), pp. 524-534, ISSN 0890-8567

Nottelman, E. (2001). National Institute of Mental Health research roundtable on prepubertal bipolar disorder. *Journal of the American Academy of Child and Adolescent Psychiatry*, Vol.40, No.8, (August 2001), pp. 871-878, ISSN 0890-8567

Papolos, D., & Papolos, J. (2000). *The Bipolar Child: the Definitive and Reassuring Guide to Childhood's Most Misunderstood Disorder*, Broadway Books, ISBN 978-0767928601, New York, NY

Parens, E. & Johnston, J. (2010). Controversies concerning the diagnosis and treatment of bipolar disorder in children. *Child and Adolescent Psychiatry and Mental Health*, Vol.4, No.3, (March 2010), pp. 9-23, ISSN 1753-2000

Paris, J. (2009). The Bipolar Spectrum: A Critical Perspective. *Harvard Review of Psychiatry*, Vol.17, No.3, (June 2009), pp. 206-213, ISSN 1067-3229

Parry, P., & Allison, S. (2008). Pre-pubertal paediatric bipolar disorder: a controversy from America. *Australasian Psychiatry*, Vol.16, No.2, (January 2008), pp.80 - 84, ISSN 1039-8562

Parry, P., Allison, S., & Furber, G. (2008). Results of the survey of faculty members' views of paediatric bipolar disorder. *Faculty of child and adolescent psychiatry RANZCP eBulletin* – Nov 2008. Available from http://www.ranzcp.org/members/collegestructure/boards/fcap.asp

Parry, P., Furber, G., & Allison, S. (2009). The paediatric bipolar hypothesis: the view from Australia and New Zealand. *Child and Adolescent Mental Health*, Vol.14, No.3, (September 2009), pp. 140-147, ISSN 1475-357X

Parry, P. & Levin, E. (2011). Pediatric Bipolar Disorder in an Era of "Mindless Psychiatry". *Journal of Trauma and Dissociation,* Published online ahead of print 22 June 2011, http://www.tandfonline.com/doi/full/10.1080/15299732.2011.597826 ISSN 1529-9732

Pavuluri, MN., Birmaher, B., & Naylor, MW. (2005). Paediatric bipolar disorder: A review of the past 10 years. *Journal of the American Academy of Child and Adolescent Psychiatry,* Vol.44, No.7, (July 2005), pp. 846–871, ISSN 0890-8567

Pavuluri, M.N., Passarotti, A.M., Harral, E.M., & Sweeney, J.A. (2009). An fMRI study of the neural correlates of incidental versus directed emotion processing in pediatric bipolar disorder. *Journal of the American Academy of Child and Adolescent Psychiatry,* Vol.48, No.3, (March 2009), pp. 308-319, ISSN 0890-8567

Pavuluri, M.N. (2009). How to tell ADHD and pediatric bipolar apart. *Proceedings of American Academy of Child & Adolescent Psychiatry, 56th annual meeting,* 27 Oct – 1 Nov 2009, Honolulu, Hawaii.

Petti, T., Reich, W., Todd, R.D., Joshi, P., Galvin, M., Reich, T., DePaulo Jr, J.R., & Nurnberger Jr, J. (2004). Psychosocial variables in children and teens of extended families identified through bipolar affective disorder probands. *Bipolar Disorders,* Vol.6, No.2, (April 2004), pp. 106-114, ISSN 1398-5647

Post, R.M. & Leverich, G.S. (2006). The role of psychosocial stress in the onset and progression of bipolar disorder and its comorbidities: The need for earlier and alternative modes of therapeutic intervention. *Development and psychopathology,* Vol.18, No.4, (December 2006), pp. 1181-1211, ISSN 0954-5794

Post, R.M., Luckenbaugh, D.A., Leverich, G.S., Altshuler, L.L., Frye, M.A., Suppes, T., Keck, P.E., McElroy, S.L., Nolan, W.A., Kupka, R., Gruze, H., & Walden, J. (2008). Incidence of childhood-onset bipolar illness in the USA and Europe. *British Journal of Psychiatry,* Vol.192, No.2, (February 2008), pp150 – 151 ISSN 0007-1250

Robbins, B.D., Higgins, M., Fisher, M., & Over, K. (2011). Conflicts of interest in research on antipsychotic treatment of pediatric bipolar disorder, temper dysregulation disorder, and attenuated psychotic symptoms syndrome: exploring the unholy alliance between big pharma and psychiatry. *Journal of Psychological Issues in Organizational Culture,* Vol.1, No.4, (January 2011), pp.32-49, ISSN 2041-8426

Rucklidge, J. (2006). Psychosocial functioning of adolescents with and without paediatric bipolar disorder. *Journal of Affective Disorders,* Vol.91, No.2-3, (April 2006), pp. 181-188, ISSN 0165-0327

Schenkel, L.S., West, A.E., Harral, E.M., Patel, N.B., & Pavuluri, M.N. (2008). Parent-child interactions in pediatric bipolar disorder. *Journal of Clinical Psychology,* Vol.64, No.4, (April 2008), pp. 422-437, ISSN 0021-9762

Schore, A. (2002). Dysregulation of the right brain: a fundamental mechanism of traumatic attachment and the psychopathogenesis of posttraumatic stress disorder. *Australian and New Zealand Journal of Psychiatry,* Vol.36, No.1, (February 2002), pp. 9-30, ISSN 0004-8674

Schore, A. (2003a). *Affect dysregulation and disorders of the self,* Norton, ISBN 0-393-70406-8, New York, NY

Schore, A. (2003b). *Affect regulation and repair of the self,* Norton, ISBN 0-393-70407-6, New York, NY

Silva, RR., Matzner F., Diaz J., Singh S., & Dummitt ES. (1999). Bipolar disorder in children and adolescents: a guide to diagnosis and treatment. *CNS Drugs,* Vol.12, No.6, (December 1999), pp. 437-450, ISSN 1172-7047

Soutullo, C.A., Escamilla-Canales, I., Wozniak, J., Gamazo-Garrán, P., Figueroa-Quintana, A., & Biederman, J. (2009). Pediatric bipolar disorder in a Spanish sample: Features before and at the time of diagnosis. *Journal of Affective Disorders,* Vol.118, No.1-3, (November 2009), pp. 39-47, ISSN 0165-0327

Spencer, T., Biederman, J., & Wilens, T. (1999). Attention-deficit/hyperactivity disorder and comorbidity. *Pediatric Clinics of North America,* Vol.46, No.5, (October 1999), pp. 915-927, ISSN 0031-3955

Steinbuchel, P.H., Wilens, T.E., Adamson, J.J., & Sgambati, S. (2009). Posttraumatic stress disorder and substance use disorder in adolescent bipolar disorder. *Bipolar disorders,* Vol.11, No.2, (March 2009), pp. 198-205, ISSN 1398-5647

Stringaris, A., Cohen, P., Pine, D. S., & Leibenluft, E. (2009). Adult outcomes of adolescent irritability: A 20-year community follow-up. *American Journal of Psychiatry,* Vol.166, No.9, (September 2009), pp. 1048–1054, ISSN 0002-953X

Stringaris, A. (2011). Irritability in children and adolescents: a challenge for DSM5. *European Child and Adolescent Psychiatr,y* Vol.20, No.2, (February 2011), pp.61-66, ISSN 1018-8827

Tillman, R., Geller, B., Nickelsburg, M.J., Bolhofner, K., Craney, J.L., DelBello, M.P., & Wigh, W. (2003). Life events in a prepubertal and early adolescent bipolar disorder phenotype compared to attention-deficit hyperactive and normal controls. *Journal of Child and Adolescent Psychopharmacology,* Vol.13, No.3, (September 2003), pp. 243-251, ISSN 1044-5463

Van der Kolk, B., & Courtois, C. (2005). Editorial comments: Complex developmental trauma. *Journal of Traumatic Stress,* Vol.18, No.5, (October 2005), pp. 385–388, ISSN 0894-9867

Van Meter, A.R., Moreira, A.L., & Youngstrom, E.A. Meta-analysis of epidemiologic studies of pediatric bipolar disorder. (2011). *Journal of Clinical Psychiatry,* May 31 ** Epub ahead of print. ISSN 0160-6689

Wilens, T.E., Biederman, J., Forkner, P., Ditterline, J., Morris, M., Moore, H., Galdo, M., Spencer, T.J., & Wozniak, J. (2003). Patterns of comorbidity and dysfunction in clinically referred preschool and school-age children with bipolar disorder. *Journal of Child and Adolescent Psychopharmacology,* Vol.13, No.4, (December 2003), pp. 495-505, ISSN 1044-5463

Wilens, T.E., Biederman, J., Adamson, J.J., Henin, A., Sgambati, S., Gignac, M., Sawtelle, R., Santry, A., & Monuteaux, M.C. (2008). Further evidence of an association between adolescent bipolar disorder with smoking and substance use disorders: A controlled study. *Drug and Alcohol Dependence,* Vol.95, No.3, (1 June 2008), pp. 188-198, ISSN 0376-8716

Wozniak, J., Biederman, J., Kiely, K., Ablon, J. S., Faraone, S. V., Mundy, E., & Mennin, D. (1995). Mania-like Symptoms Suggestive of Childhood-onset Bipolar Disorder in Clinically Referred Children. *Journal of the American Academy of Child and Adolescent Psychiatry,* Vol.34, No.7 (July 1995), pp. 867-876, ISSN 0890-8567

Wozniak, J., Crawford, M.H., Biederman, J., Faraone, S.V., Spencer, T.J., Taylor, A., & Blier, H.K. (1999). Antecedents and complications of trauma in boys with ADHD: Findings from a longitudinal study. *Journal of the American Academy of Child and Adolescent Psychiatry,* Vol.38, No.1, (January 1999), pp. 48-55, ISSN 0890-8567

Wozniak, J. (2003). Pediatric bipolar disorder: The new perspective on severe mood dysfunction in children. *Journal of Child and Adolescent Psychopharmacology,* Vol.13, No.4, (December 2003). pp. 449-451, ISSN 1044-5463

Part 4

Psychosocial Approaches

9

Psychosocial Functioning in Bipolar Disorder from a Social Justice Perspective

Emily Manove, Lauren M. Price and Boaz Levy
Department of Counseling and School Psychology, University of Massachusetts, Boston, USA

1. Introduction

Previous research on psychosocial functioning in bipolar disorder (BD) has uncovered various factors that exacerbate disability over the course of illness, including illness severity (Burdick, Goldberg and Harrow, 2010; Haro et al., 2011; Mur, Portella, Martinez-Aran, Pifarre, & Vieta, 2009), anxiety (Boylan et al., 2004), substance use (Jaworski, Dubertret, Adès, & Gorwood, 2011), and cognitive impairment (Martino, Igoa, Marengo, Scapola, & Strejilevich, 2011). Accordingly, current efforts toward improving functional outcome in BD focus primarily on pharmacological and behavioral interventions that decrease mood symptoms (Fountoulakis & Vieta, 2008; Miklowitz, Goodwin, Bauer, & Geddes, 2008), attenuate anxiety (Rakofsky & Dunlop, 2011), reduce substance use (Weiss et al., 2009), and increase cognitive functioning (Deckersbach et al., 2010; Goldberg & Roy Chengappa, 2009). Whereas this approach addresses the neuro-psychological aspect of functional disability in BD, it largely ignores its social components.

Disability is inherently contextual, and is always defined by the interaction between the person and the environment. In other words, psychosocial functioning varies with the person's abilities, but also with the particular characteristics of the person's social environment. In theory, changes in any given society can shift the functional variability among its members, and therefore carry the potential to decrease the number of people who experience severe forms of psychosocial disability.

Applied to BD, this notion relates to removing social barriers to functioning in mainstream settings such as stigma and discrimination. Currently available therapies can reduce the symptoms of BD to some extent; however, the consequent potential for psychosocial growth remains latent in a social environment that limits the person's opportunities to obtain psychiatric care (Boyd, Linsenmeyer, Woolhandler, Himmelstein, & Nardin, 2011), medical care (Corrigan, 2004) and reputable social standing (Corrigan, Markowitz, & Watson, 2004; Corrigan & Wassel, 2008).

This chapter focuses on the contribution of stigma and discrimination to functional disability in BD. We explore the mechanisms of stigma's effects on functioning, which may involve increasing painful emotions (notably shame) and stress, reducing treatment seeking and adherence, and limiting social and vocational opportunities through both consumer avoidance and public exclusion.

We advance a two-pronged approach to treatment. The first involves therapy directly targeting the negative effects of internalized stigma on the psychological health of people with BD. The second requires providers to publicly align with federal and state governments, advocacy groups and consumers in making the issues of stigma and discrimination visible. These efforts should seek to reduce the ill effects of stigma and discrimination through educational programs, media campaigns, litigation, and reform of service systems.

These two prongs will likely have synergistic effects. With one alleviating internal distress and the other removing external obstacles, they may converge to attenuate the debilitating effects of stigma and discrimination on people with BD. This approach may even shift the functional trajectory of BD from psychosocial decline to growth over the course of illness despite the recurrence of mood symptoms.

2. Functional impairment in BD

Longitudinal, cross-sectional and qualitative studies, now summarized in multiple reviews, have documented substantial levels of social and occupational impairment in BD, even during euthymia (Bonnin et al., 2010; Burdick, Goldberg, & Harrow, 2010; Coryell et al., 1993; Dickerson et al., 2010; Huxley & Baldessarini, 2007; Latalova, Prasko, Diveky, & Velartova, 2011; Martino et al., 2011; McMorris, Downs, Panish, & Dirani, 2010; Michalak, Yatham, Maxwell, Hale, & Lam, 2007; Michalak et al., 2011; Montoya et al., 2010; Pope, Dudley, & Scott, 2007; Rosa et al., 2010). Longitudinal studies show that between 90 to 98% of people receiving treatment for an acute episode of BD achieve syndromal recovery within six months to two years (Bonnin et al., 2010; Dickerson et al., 2010; Huxley & Baldessarini, 2007; Tohen, Waternaux, & Tsuang, 1990; Tohen et al., 2000). However, only 30 to 40% of consumers demonstrate functional recovery after a significant decrease in acute symptoms (Huxley & Baldessarini, 2007; Montoya et al., 2010; Tohen et al., 2000). Even when full syndromal remission is achieved, only about 50% of people with BD recover premorbid functionality (Montoya et al., 2010; Tohen et al., 2000). Disability rates in euthymic BD are similarly high (Gutierrez-Rojas, Jurado, & Gurpegui, 2011; Huxley & Baldessarini, 2007; Montoya et al.; Rosa et al., 2010; Tohen et al., 1990), indicating that acute symptoms are only partially responsible for functional impairment.

This failure to return to premorbid levels of functioning is especially concerning given that while normal psychosocial functioning prior to a first episode is possible (Reichenberg et al., 2002), premorbid functioning in BD tends to be below average (Rietschel et al., 2009; Cannon et al., 1997). A vulnerable premorbid baseline suggests that further decline in functioning after illness onset can result in the type of severe life-long disability often observed in BD.

The effects of disability in BD pervade all areas of functioning (Coryell et al., 1993; Elgie & Morselli, 2007; Pope et al., 2007; Sanchez-Moreno et al., 2009; Sanchez-Moreno et al., 2010). People with BD have more familial conflict and fewer enjoyable social interactions or leisure activities than people without psychiatric diagnoses (Bauwens, Tracy, Pardoen, Vander Elst, & Mendlewicz, 1991; Bauwens, Pardoen, Staner, Dramaix, & Mendlewicz, 1998; Elgie & Morselli, 2007; Sanchez-Moreno et al., 2009; Shapira et al., 1999). In a five-year study following the outcome of an acute mood episode, people with BD were only half as likely as the general population to be married (Coryell et al., 1993). People suffering from BD report significantly more illness-related obstruction of their social lives, even in comparison to people with chronic medical conditions including rheumatoid arthritis and renal failure (Robb, Cooke, Devins, Young, & Joffe, 1997). These studies further suggest that the disruption to social adjustment and development in BD may extend beyond the direct effects of acute symptoms on social engagement.

Beyond the challenges to social adjustment, limited occupational functioning is also widespread in BD, including during euthymia (Bonnin et al., 2010; Burdick et al., 2010; Dickerson et al., 2010; Zimmerman et al., 2010). Despite studies reporting people with BD to have more years of college and more B.A. degrees than is average in the general population of the U.S. (Kupfer et al., 2002), they demonstrate a much higher degree of unemployment, disability rates, workplace absenteeism, and other occupational impairments (Bowden, 2005; Latalova et al., 2011; Lloyd & Waghorn, 2007; McMorris et al., 2010; Reed, Goetz, Vieta, Bassi, & Haro, 2010; Rosa et al., 2010; Sanchez-Moreno et al., 2009). On average, about 60% of people with BD who are in syndromal remission and have either some college education or a college degree, are unemployed (compared to the average of 6% in the general population), and up to 88% suffer from reduced occupational functioning (Bearden et al., 2011; Elgie & Morselli, 2007; Hirschfeld, Lewis, & Vornick, 2003; Huxley & Baldessarini, 2007; Kessler et al., 2006). These studies indicate that in BD education and competency show much weaker correlations with professional or vocational development than expected in the general population.

Occupational impairment in BD is likely related to the difficulties in maintaining consistent engagement at work. For example, people with BD miss an average of up to 65.5 workdays a year due to absenteeism and presenteeism for illness-related reasons - up to more than one day per week (Gardner et al., 2006; Kessler et al., 2006). These problems can result in placement of people with BD in less skilled, entry-level, poorly paid positions despite superior education and intellectual abilities. Such placement may then lead to greater work dissatisfaction, which has a negative impact on achieving longer-term vocational goals (Baldwin & Marcus, 2011; Tse & Walsh, 2001).

Although work difficulties and underemployment in BD may be attributed to the direct effects of the illness, it is also quite possible that an inhospitable work environment contributes to absenteeism and the tendency to employ people with BD in capacities below their skill level. There is some evidence that despite the therapeutic advancements in BD, psychosocial functioning in this population may be declining. In a 2000 replication (Hirschfeld et al., 2003) of the 1992 National Depressive & Manic-Depressive Association survey (Lish, Dime-Meenan, Whybrow, Price, & Hirschfeld, 1994), people with BD reported that their illness had a more severe impact on their social, occupational and overall functioning than it had in 1992. These studies specifically indicated that rates of unemployment, underemployment and disability in BD had risen (Hirschfeld et al., 2003) during a period in which the stigmatization of people with mental illness (MI) also increased (Corrigan, Roe, & Tsang, 2011; Mehta, Kassam, Leese, Butler, & Thornicroft, 2009). Functional decline in the face of therapeutic progress suggests the presence of potentially important undetected factors that contribute to psychosocial disability in BD. Some of these factors may be external to the treated person, such as those related to social barriers rooted in discrimination.

3. Predictors of psychosocial functioning in BD

A volume of longitudinal studies has identified several predictors of poor psychosocial adjustment in BD. A more severe illness history - including a younger age at onset (Hays, Krishnan, George, & Blazer, 1998) and a greater number of mood episodes (MacQueen et al., 2000) and psychiatric hospitalizations (Altshuler et al., 2007) - yields reduced functioning even during euthymia (Martinez-Aran et al., 2004; Zubieta, Huguelet, O'Neil, & Giordani, 2001). In addition, common comorbidities of BD, such as anxiety (Boylan et al., 2004; Kauer-Sant Anna et al., 2007) and substance use disorders (SUD; Jaworski et al., 2011; Lagerberg et al., 2010), disrupt identity development (Michalak et al., 2011), increase symptoms and social avoidance (Jaworski et al., 2011; Otto et al., 2006), interfere with interpersonal engagement (Elgie & Morselli, 2007), exacerbate cognitive impairment (Latvala et al., 2009; Levy & Weiss, 2009; Silva & Leite, 2000), and generally reduce functioning (Goldstein & Levitt, 2008; Sanchez-Moreno et al., 2009). Finally, emotion dysregulation, sleep disturbances, and an elevated rate of medical comorbidities all predict reduced psychosocial functioning in BD (M.L. Phillips, Ladouceur, & Drevets, 2008 (Deckersbach, T., Hölzel, B. K., Eisner, L. R., Stange, J. P., Peckham, A. D., Dougherty, D. D., Nierenberg, A. A., 2011).

In euthymic BD, two of the strongest predictors of social and occupational functioning are subsyndromal depressive symptoms and cognitive impairment (Bearden et al., 2011; Bonnin et al., 2010; Dickerson et al., 2010; Martínez-Arán et al., 2011; Martino et al., 2009; Tabares-Seisdedos et al., 2008). Even very mild subsyndromal depression leads to reduced functioning including poorer vocational performance (Bearden et al., 2011; Gitlin, Mintz, Sokolski, Hammen, & Altshuler, 2010). Likewise, cognitive deficits (notably in verbal memory, processing speed and executive functioning) are significantly associated with worse social and occupational functioning (Bearden et al., 2011; Dickerson et al., 2010).

These findings focused current efforts for improving psychosocial functioning in BD on reducing subsyndromal depressive symptoms (M. Bauer et al., 2009) and cognitive dysfunction (Bearden et al., 2011; Deckersbach et al., 2010; Martínez-Arán et al., 2011), as well as enhancing stress management and interpersonal skills (Frank et al., 2000; Martínez-Arán et al., 2011). The bulk of this research emphasizes pharmacological interventions (Nivoli et al., 2011), followed by psychosocial treatments, such as psychotherapy (Ball et al., 2006; Hollon & Ponniah, 2010; Miklowitz, George, Richards, Simoneau, & Suddath, 2003; Miklowitz et al., 2008) and psychoeducation (Colom et al., 2009), as well as cognitive (Deckersbach et al., 2010) and functional remediation (Martínez-Arán et al., 2011). More broadly, the current overall approach to improving psychosocial adjustment in BD focuses on changing internal factors, with less consideration given to reducing environmental barriers to functioning.

The view that functioning can be fully restored in BD by treating internal factors such as depressive symptoms and cognitive dysfunction may be restrictive. Even if all internal factors that predict lower functioning were identified and successfully treated, people with BD may still not achieve functional recovery due to environmental influences that include level of expressed emotion (i.e., criticism or stigmatization of person with BD) in families (Elgie & Morselli, 2007; McMurrich & Johnson, 2009), stigma (Michalak et al., 2011; Perlick et al., 2001; Vazquez et al., 2011), discrimination (Corrigan et al., 2004; Elgie & Morselli, 2007; Michalak et al., 2007; Tse & Yeats, 2002), and lack of social support (Elgie & Morselli, 2007; Michalak, Yatham, Kolesar, & Lam, 2006; Michalak et al., 2011).

This notion is consistent with qualitative studies showing people with BD describing stigma, discrimination, reduction in social (especially family) support, shame and identity loss as the greatest challenges they face in dealing with their illness and functional recovery (Elgie & Morselli, 2007; Michalak et al., 2006; Michalak et al., 2007; Michalak et al., 2011; Proudfoot et al., 2009). People with BD describe social validation and support as the most important components that determine their quality of life (QOL; Michalak et al., 2006; Michalak et al., 2007; Michalak et al., 2011). In fact, many people with BD consider social inclusion to be more important than symptom resolution or clinical remission (Michalak et al., 2006).

Given the importance of social inclusion to recovery in BD, as well as the role of stigma in reducing treatment seeking and adherence (Berk et al., 2010; N. Rusch, Corrigan, Wassel, et al., 2009), a more effective investment in treatment may involve a distribution of resources that extends to both sides of the psychosocial equation. Improving functioning in BD may require the simultaneous development and implementation of interventions that target both internal consumer characteristics and external societal barriers to functioning.

4. The stigma of mental illness

"Stigma" comes from a sixteenth-century Greek word meaning a mark that was branded or tattooed onto people to denote a devalued social status - for example, that of a criminal or a slave (Hinshaw, 2007). Likewise, in most countries, MI is associated with harmful social devaluation (Brohan, Gauci, Sartorius, & Thornicroft, 2011). In the U.S., both the Surgeon General and the Department of Defense have identified current stigma against people with MI as one of the most serious barriers to the nation's mental health (Department of Defense Task Force on Mental Health, 2007; Surgeon General, 1999). There is also evidence that stigma against people with MI has grown since 1950, despite greater scientific insights into the phenomenon of MI and the relatively widespread acceptance of these insights by the public (Australian Journal of Pharmacy, 2011; Corrigan & Wassel, 2008; Corrigan et al., 2011; Mehta et al., 2009; Pescosolido, Martin, Lang, & Olafsdottir, 2008; Silton, Flannelly, Milstein, & Vaaler, 2011; Star, 1952, 1955). Recent surveys finding an upward trend in stigmatizing people with MI have inspired large-scale investigations of this stigma and its consequences.

The scientific inquiry into stigma and related constructs, however, is quite complex. While the word "stigma" has grown to be colloquially associated with prejudicial attitudes and discriminatory conduct, it remains a challenging construct to isolate, define and measure for the purpose of scientific research (Corrigan et al., 2011; B. G. Link, Yang, Phelan, & Collins, 2004). Despite these challenges, stigma instruments have been developed and validated for use with people who suffer from depression (Gabriel & Violato, 2010; Interian et al., 2010; Kanter, Rusch, & Brondino, 2008), psychotic disorders (B. G. Link et al., 2004), and SUD (Luoma, O'Hair, Kohlenberg, Hayes, & Fletcher, 2010). In BD, important progress in this area has been made with the development of qualitative literature describing widespread consumer experiences of stigma (Michalak et al., 2006; Michalak et al., 2011; Proudfoot et al., 2009).

Theoretical progress in stigma research has advanced several models that describe its impact on people with MI (Corrigan, 2004; Goffman, 1963; Jones et al., 1984; B. G. Link, Mirotznik, & Cullen, 1991). In the last fifteen years, research has generally converged to support a cognitive-behavioral model of stigma's operation (Corrigan et al., 2011; Crocker, Major, & Steele, 1998; Goffman, 1963; Jones et al., 1984; B. G. Link et al., 1991).

According to the model delineated by Corrigan and colleagues (Corrigan, 2004; Corrigan & Wassel, 2008), stigma is a contextual and relational psychosocial phenomenon whose impact on people suffering from MI can be described using three primary constructs: structural, or institutional, stigma (macro level), public, or enacted, stigma (meso level), and internalized, or self, stigma (micro level) (Corrigan, 2004). This model also describes a fourth construct known as perceived stigma that refers to an individual's perception of structural and public stigma (Corrigan & Watson, 2002).

4.1 Structural stigma

Institutional or structural stigma includes policies and practices of private and governmental institutions that intentionally or unintentionally restrict rights or limit opportunities for people with MI (Corrigan et al., 2004; Corrigan & Wassel, 2008; Michalak et al., 2011). Laws that directly restrict the rights of people with MI are examples of intentional structural discrimination that inhibits functioning (Corrigan et al., 2004; Corrigan, Watson, Heyrman, et al., 2005). In the private sector, lack of equal insurance coverage for mental health treatment (compared to non-psychiatric treatment) and the increasingly negative portrayals of MI in the media are both examples of intentional structural stigma (Corrigan et al., 2004). An example of unintentional structural stigma includes workplace rules that do not allow for scheduling flexibility (even in the absence of a clear need for rigid schedules) such that a person with BD's need for maintenance treatment appointments during business hours is not accommodated (Schultz et al., 2011; Michalak et al., 2007).

In the U.S., private and public institutions play critical roles in perpetuating stigma against people with MI across multiple domains (Krupa, Kirsh, Cockburn, & Gewurtz, 2009; Krupa, 2011a, 2011b). In the legal realm, the Americans with Disabilities Act (Americans with Disabilities Act, 1990) and Fair Housing Act (Fair Housing Amendments Act, 1988) have largely failed to protect people with MI, as indicated by judicial rulings over many years that greatly favored employers (Petrila, 2009) and landlords (Carter, 2010; Swanson, Burris, Moss, Ullman, & Ranney, 2006). In the area of health care, current government policies allow extensive discrimination against people with MI by health insurance companies (Boyd et al., 2011), fail to support deinstitutionalization of psychiatric patients with adequate community alternatives (Corrigan, Watson, Gracia, et al., 2005; Heginbotham, 1998; Phelan & Link, 1998), and provide insufficient funding for effective dissemination of evidence-based psychiatric practices (McHugh, K.R., & Barlow, D.H., 2010). With respect to civil rights, people with MI experience restrictions in obtaining firearms, voting, holding elective office, serving on juries, or gaining full parental rights over their children (Burton, 1990; Corrigan, Watson, Heyrman, et al., 2005; Hemmens, Miller, Burton, & Milner, 2002). People with MI also suffer a great deal of harm from the media (Corrigan, Watson, Gracia, et al., 2005). Currently, the government provides only limited support for anti-MI-discrimination campaigns, and allows the national media to portray people with MI as violent and immoral without penalty (Corrigan, Watson, Gracia, et al., 2005; Corrigan et al., 2011; Heginbotham,

1998; Krupa et al., 2009; Phelan & Link, 1998). Finally, even Social Security Administration (SSA) disability benefits may be structured in a way that incentivizes disability rather than rehabilitation for people who suffer from MI (Drake, Skinner, Bond, & Goldman, 2009; Elinson, Houck, & Pincus, 2007). All of these forms of structural stigma may compromise the psychosocial adjustment of people with MI, beyond the functional limitations imposed by psychiatric symptoms.

4.2 Public stigma

Public stigma operates at the level of interpersonal interactions through a three-part cognitive-behavioral process (Corrigan et al., 2011). First, contextual cues known as "marks" associated with a given person signal to a perceiver (who is in a position of power via the marked person) that a larger group stereotype may be a relevant cognitive response to the marked person. Second, a stereotype is invoked as a cognitive short-cut to allow the perceiver to quickly access a summary of characteristics of that larger group. Finally, the perceiver responds with behavior based on the assumption that these stereotypical group characteristics are true with respect to the marked person (Corrigan, 2004; Corrigan & Wassel, 2008; B. G. Link et al., 2004; Pescosolido et al., 2008).

In this context, the term "prejudice" refers to the adoption of a stereotype as truth with respect to individual members of a stigmatized group, as in step two of the model (Pescosolido et al., 2008). "Discrimination" is defined as a behavior resulting from prejudice that negatively impacts members of the stigmatized group, or step three in the model (Corrigan, 2004 (*American Psychologist*)). The three most common relevant stereotypes found in research literature are that people with MI are dangerous and incompetent, as well as blameworthy with respect to their illness (Corrigan & Wassel, 2008).

The four major categories of cues or "marks" that trigger the use of MI stereotypes are labels, psychiatric symptoms, social skill deficits, and physical appearance (Corrigan, 2004). Several studies demonstrated that a person labeled mentally ill is likely to face prejudiced and discriminatory responses even in the absence of visible psychiatric symptoms or atypical behavior (Link, Cullen, Frank, & Wozniak, 1987; Link, Phelan, Bresnahan, Stueve, & Pescosolido, 1999; Martinez at al., 2011). Additional studies have found that even if a potential stigmatizer or stigmatized person does not endorse negative societal stereotypes of people with MI, either party's knowledge of these stereotypes may increase anxiety and discomfort during an interaction (Hinshaw, 2007; B. G. Link, Struening, Neese-Todd, Asmussen, & Phelan, 2001). This discomfort, in turn, may raise anxiety levels and future interactional avoidance for people with MI, even when they are not aware that the other party has knowledge of their illness (Hinshaw, 2007). Thus, the pervasive and harmful nature of public stigma can negatively impact interpersonal relationships even in the absence of individual prejudice.

4.3 Perceived and internalized stigma

Perceived stigma refers to the degree to which a person with MI is aware of structural and public stigma, whether through personal experience, the media, or other means (Corrigan & Wassel, 2008; Livingston & Boyd, 2010). Perceived stigma describes people's awareness that stigma exists, but not their reaction to it, and is thus deemed a separate construct from internalized stigma (Livingston & Boyd, 2010; N. Rusch, Corrigan, Wassel, et al., 2009; N. Rusch, Todd, Bodenhausen, Olschewski, & Corrigan, 2010). Internalized stigma is broadly defined by feelings of shame, self-criticism, identity changes and maladaptive (most notably avoidant) behavior that may develop in response to perceived stigma (Brohan et al., 2011; Corrigan & Wassel, 2008; Livingston & Boyd, 2010; Manos, Rusch, Kanter, & Clifford, 2009; Michalak et al., 2011; N. Rusch, Lieb, Bohus, & Corrigan, 2006).

5. The impact of stigma on psychosocial functioning in BD

These four forms of stigma - structural, public, perceived and internalized - interact with each other in complex ways to decrease psychosocial functioning in BD. Understanding their individual and combined contributions to these negative outcomes among people with BD is a vital step towards promoting functional adjustment in this population.

On a macro level, structural stigma probably affects functioning in BD through the same mechanisms that limit psychosocial adjustment in MI more generally. In the absence of institutional protection, funding and support, structural stigma directly circumscribes the opportunities of people with BD to improve their clinical, social and vocational functioning (Baldwin & Marcus, 2011; Boyd et al., 2011; Corrigan, Watson, Heyrman, et al., 2005; Krupa, 2007; Michalak et al., 2007; Michalak et al., 2011). Thus, people with BD may struggle to function in part because they attempt to negotiate a social environment that is not designed to accommodate their limitations (e.g. cognitive impairment) and special needs (e.g. flexible work schedules and occupational supports), and thus deprives them of opportunities that are available to others.

On the meso level, people with BD encounter functional challenges that emanate from public discrimination. Studies suggest that public stigma against people with BD is prevalent and intense (Australian Journal of Pharmacy, 2011). A 2004 survey of 1200 American adults by the Depression and Bipolar Support Alliance (DBSA) found that a large number of respondents reported serious doubts about the ability of people with BD to keep responsible jobs, hold public office, and maintain positive relationships (Depression and Bipolar Support Alliance, 2008). Employers reported particular concerns as to the potential effects of emotional instability and atypical behavior on clients and co-workers, in addition to absenteeism, poor work performance, and an inability to tolerate stress (Schultz, Milner, Hanson, & Winter, 2011). People with BD therefore appear to suffer unemployment at least partly because employers refuse to hire or retain them (Michalak et al., 2007; Michalak et al., 2011). This notion is consistent with research findings in BD and MI generally indicating that stigma-related constructs are stronger predictors of employment outcomes than diagnosis, employment history or symptoms (Baldwin & Marcus, 2011; Krupa, 2011b; Larson et al., 2011; Michalak et al., 2007; Tse & Walsh, 2001; Tse, 2002; Tse & Yeats, 2002). Unemployment and underemployment in BD therefore may not emanate solely or even primarily from the limiting effects of psychiatric symptoms. Instead, occupational difficulties in BD may be compounded to a considerable degree by social barriers rooted in stigma and discrimination.

Public stigma may also hinder psychosocial adjustment in BD within the close family circle. There is evidence that "expressed emotion" - which consists of emotional over-involvement, as well as hostile, critical, and blaming attitudes of a family toward its members with MI -

is high in families of people with BD (McMurrich & Johnson, 2009). Studies further show that expressed emotion predicts depression and relapse in people with BD (Eisner & Johnson, 2008; McMurrich & Johnson, 2009), which independently correlate with each other and with poor overall functional outcomes (M. Bauer et al., 2009; Bonnin et al., 2010). Conversely, greater social support predicts better social (Perlick et al., 2001) and occupational adjustment in BD (Gutierrez-Rojas et al., 2011). Changes in public stigma within the family therefore may help to improve psychosocial adjustment in BD.

On a micro level, studies consistently show that people with BD suffer from high levels of perceived and internalized stigma (Brohan et al., 2011). Research indicates that many people with BD report experiencing shame about their diagnosis (Michalak et al., 2011) as well as illness-related social and occupational rejection or exclusion (Michalak et al., 2006; Michalak et al., 2007; Michalak et al., 2011; Proudfoot et al., 2009; Ward, 2011).

Both internalized and perceived stigma have been shown to be associated with functional outcome in BD (Brohan et al., 2011; Vasquez et al., 2011). The presence of significant internalized stigma is associated with reduced self-esteem, and lower self-efficacy (Brohan et al., 2011). Internalized stigma is also related to an overall reduction in QOL (Alonso et al., 2009; Brohan et al., 2011), delayed treatment seeking, poor treatment adherence, and greater depressive symptoms (Brohan et al., 2011; Sirey et al., 2001).

Micro-level stigma may play an important role in the mechanisms that decrease psychosocial functioning in BD. As qualitative findings imply, perceived and internalized stigma evoke marked distress in people with a BD diagnosis (Michalak et al., 2006; Michalak et al., 2011; Ward, 2011). More specifically, recent studies found that heightened stress from stigma was associated with increased social anxiety and shame (N. Rusch, Corrigan, Powell, et al., 2009). While anxiety has been shown to predict worse outcomes in BD (Boylan et al., 2004), shame predicts depression, avoidance and psychopathology generally (Andrews, Qian, & Valentine, 2002; Highfield, Markham, Skinner, & Neal, 2010; J.P. Tangney, 1995), which also decrease functional adjustment.

The automatic nature of internalized stigma makes it particularly destructive. There is evidence that the primary response to stigma is an involuntary aversive emotional experience rather than a conscious cognitive reaction (N. Rusch, Corrigan, Powell, et al., 2009). Thus, the impact of internalized stigma on functional adjustment in BD may operate primarily through the mediating factors of shame and social anxiety (Andrews et al., 2002; Brohan et al., 2011; Cheung, Gilbert, & Irons, 2004; B. Link et al., 1987; B. G. Link, 1987; Michalak et al., 2011; Vazquez et al., 2011), rather than through maladaptive cognitions such as endorsed stereotypes (N. Rusch et al., 2010; N. Rusch, Corrigan, Todd, & Bodenhausen, 2011). Shame, avoidance, social isolation and depressive symptoms all contribute to lower QOL in BD and heightened risk of relapse (Cohen et al., 2004; Bonnin et al., 2010; Michalak et al., 2011; Perlick et al., 2001). Collectively, these stigma-related factors probably decrease psychosocial development in BD.

In broader conceptualization, it may be useful to understand the debilitating roles that shame and social anxiety play in stigma's operation across its various levels. More specifically, the different levels of stigma are probably looped together in detrimental ways that exacerbate their individual effects. In this respect, insight into the connection between public and internalized stigma might be particularly informative, with the leading hypothesis being that public stigma and discrimination often create internalized stigma, which manifests as shame and social anxiety.

In support of this line of reasoning, shame and social anxiety are known as key outcomes of social devaluation by the public (L. C. Rusch, Kanter, Angelone, & Ridley, 2008; N. Rusch, Corrigan, Powell, et al., 2009; N. Rusch et al., 2010). These emotions may be triggered when people perceive themselves to be at risk of social exclusion due to not meeting societal norms or losing social rank (Cheung et al., 2004; Gilbert & Procter, 2006). Further, ongoing criticism or stigmatization of a person may result in the internalization of these devaluations as trait or internal shame, key elements of which are chronic self-invalidation, self-criticism and self-blame (Cheung et al., 2004; Gilbert, 1992, 2002; Gilbert & Procter, 2006; Gilbert et al., 2010; Miklowitz & Johnson, 2009). Trait shame is strongly associated with avoidance (J. P. Tangney & Dearing, 2002) and social anxiety (Gilbert & Irons, 2004), and all three are known as core emotional and behavioral components of internalized stigma (Birchwood et al., 2007; Perlick et al., 2001; N. Rusch et al., 2006; N. Rusch et al., 2010). Internalized stigma therefore may emerge as a function of public stigma (Livingston & Boyd, 2010; Schneider, Beeley, & Repper, 2011) in a manner that compromises psychosocial adjustment (Brohan et al., 2011; Vasquez et al., 2011). Interventions aimed at improving psychosocial functioning in BD thus need to address all levels of stigma to be effective.

6. Implications for care

The conventional approach to improving functional adjustment in BD focuses primarily on pharmacological interventions to decrease symptoms (Balanza-Martinez et al., 2010). These efforts have been traditionally supported by psychosocial treatments that are designed to increase practical knowledge about the illness and its management (e.g., group psychoeducation; Colom et al., 2009), as well as to improve interpersonal and coping skills (Frank et al., 2005). Although studies support the efficacy of these interventions, they are clearly insufficient in fully addressing the problem of psychosocial impairment in BD. Further improvement to functional outcome in BD may therefore come from addressing factors such as stigma and discrimination that originate in the larger social milieu rather than within the individual consumer.

6.1 Bringing about structural changes

On a structural level, treatment providers and mental health care administrators need to advocate for affirmative litigation, proper funding of services, effective dissemination of evidence-based practices, and incentives for work that do not, in the near term, jeopardize the financial support and stability derived from maintaining disability status. These efforts may reduce discrimination in housing and employment. They may also equalize health insurance coverage and access to care for people with BD compared to those with non-psychiatric diagnoses. People with BD need major structural changes in the larger social context to create environments that accommodate their special needs. The absence of these changes currently limits their ability to participate in mainstream social functions.

Accommodations in the workplace should rank high in priorities for reform, due to the pernicious effects of stigma on employment in BD, and the drastic effects of unemployment on mental and physical health generally (Bambra, 2010; Bowden, 2005; Bush, Drake, Xie,

McHugo, & Haslett, 2009 Michalak et al., 2007). Qualitative studies (Michalak et al., 2007; Tse & Yeats, 2002) indicate that augmenting employment in BD requires interventions in four key areas: illness management, social support, and employer-employee relations, as well as in combating broader societal barriers such as stigma and disincentives to working embedded in disability regulations (Elinson et al., 2007; Tse & Yeats, 2002). Evidence-based supported employment (SE) programs that include early intervention after diagnosis, integrated clinical care and vocational services (Cook et al., 2005; Cook et al., 2008), training in goal-setting, cognitive and social skills (Arbesman & Logsdon, 2011; Cook, 2006; Cook et al., 2008; Krupa, 2007; Krupa et al., 2009), a collaborative and individualized approach to rapid employment placement without lengthy pre-placement training, and especially, high-intensity on-the-job support of unlimited duration (G. R. Bond & Kukla, 2011; Cook et al., 2005; Evans & Bond, 2008; McHugo, Drake, & Becker, 1998), address the first three of these areas.

SE programs have been found to be highly effective in improving employment outcomes for people with severe MI, including BD (Arbesman & Logsdon, 2011; Tse & Yeats, 2002). Importantly, as Tse & Yeats (2002) note, the absence of time limits in SE programs is likely critical in helping people with BD achieve their long-term vocational goals. People with BD may need to be guided from working in low-level positions as they initially adjust to employment, towards taking on increasingly challenging work or education over time (Drake et al., 2009; Tse & Yeats, 2002). From a broader structural viewpoint, a widespread implementation of supportive programs and workplace accommodations for people with BD will require affirmative litigation, adequate funding and a change in the social climate.

Beyond the wide-scale implementation of SE, further structural reforms that may increase employment in BD should address the process of shifting consumers from the protective service of disability benefits towards enrollment in SE programs. Even if SE programs were sufficiently funded and disseminated, a remaining obstacle to this shift involves SSA disability benefits recipients' concerns about the loss of income and health insurance benefits in the short and long-term were they to begin working (Drake et al., 2009; Elinson et al., 2007; Tse & Yeats, 2002). In 2011, SSA disability beneficiaries face powerful disincentives to employment as, after a brief trial work period, they risk losing their cash benefits immediately and public health insurance several months thereafter, if they gross over 1000 dollars per month (Drake et al., 2009; Elinson et al., 2007). Not surprisingly, in BD, very few employed people receive disability benefits, and most experience reduced health insurance coverage and access to mental health care compared to those receiving benefits (Elinson et al., 2007).

The vocational reform under discussion is guided by the notion that BD is a form of major mental illness, with recurrent mood and/or psychotic episodes that periodically preclude continuous attendance at work. The instability that inheres in this condition necessitates external supports to assure even minimal long-term financial security. In the absence of such security, the decision to forego the benefits of disability for employment opportunities seems unreasonably risky. For this reason, allowing people with BD to reap the health and economic benefits of employment while reducing the tremendous cost to the nation of lost productivity due to BD, requires a structural mechanism that ensures consumers larger rewards from working without the risk of losing financial security due to relapse.

6.2 Reducing public stigma

Attenuating the detrimental effects of public stigma may also be important for improving functional adjustment in BD. In this regard, there is evidence that family-focused therapy (FFT; Miklowitz et al., 2003), which aims to mitigate expressed emotion and augment social support from family members, can be highly beneficial. Studies suggest that FFT improves clinical outcomes in BD even over long periods of time (Rea et al., 2003), including by delaying time to relapse and lessening residual symptoms (Miklowitz et al., 2003), which are highly correlated with functional adjustment (M. Bauer et al., 2009).

While the focus on family members is essential, the efforts to reduce the ill effects of public stigma should also include larger social circles. In this realm, advocates have traditionally supported the use of protests (Corrigan et al., 2011), anti-stigma mass media campaigns (Schneider et al., 2011), and educational programs that contradict negative stereotypes (Corrigan et al., 2011). However, there is no evidence that these interventions lead to lasting reductions in stigma, and they sometimes may have opposite effects (Corrigan & Wassel, 2008; Masuda, Hayes, et al., 2009). As with internalized stigma, the stigmatization of people with MI by others appears to be an automatic emotional process occurring outside awareness, which then triggers stereotypical cognitions (N. Rusch et al., 2011). Further efforts to alter or suppress these unwanted cognitions often paradoxically result in reinforcing them (Masuda, Hayes, et al., 2009). Thus, educational interventions that aim to decrease the frequency of stereotypical cognitions without addressing the underlying emotional process that produces stigma, may in fact trigger an intensified emotional reaction of heightened stigma. This reaction may then unintentionally increase the occurrence of stereotypical thoughts (Masuda, Hayes, et al., 2009; Masuda, Price, Anderson, Schmertz, & Calamaras, 2009). The same type of resistance to change was observed in educational interventions aimed at altering racial prejudice (McKown, 2005; Paluck & Green, 2009). At the same time, extensive research in social psychology has developed models and programs involving advocacy and education to combat public stigmatization of historically oppressed groups that have had some success (Buhin & Vera, 2009). These models may inform the development of interventions to reduce public stigma in BD.

Models that come from the field of clinical psychology may also be helpful in reducing public stigma. There is some evidence that mindfulness-based interventions can decrease the stigmatizing of people with MI by college students (Masuda, Hayes, et al., 2009). In this study, after an acceptance and commitment therapy (ACT) intervention of only 2.5 hours, stigmatization of MI was reduced even at a one-month follow-up, compared to an educational control intervention. Mindfulness-based interventions may be practical to implement in educational settings, as part of a curriculum that aims to promote social justice.

6.3 Individual counseling

Addressing the broader social context of stigma, with its structural and public components, should be accompanied by counseling methods that reduce the effects of stigma at an individual level. One potentially useful approach to alleviating the shame, social anxiety and avoidance that result from stigma in BD may be based on mindfulness- (Luoma, Kohlenberg, Hayes, Bunting, & Rye, 2008) and compassion-focused (Gilbert & Procter, 2006) therapies. Mindfulness-based interventions increase nonjudgmental awareness and acceptance of thoughts and feelings, which allow people to better regulate these internal experiences (Luoma et al., 2008). Thus, these therapies may be suitable for decreasing the chronic, automatic self-criticism that emanates from internalized stigma and impedes functioning (Gilbert et al., 2010). Some studies show that mindfulness-based cognitive therapy (MBCT; Weber et al., 2010; Williams et al., 2008) raised levels of treatment adherence while diminishing stress, anxiety, depression and relapse in BD, all of which improved functioning (Bonvalot et al., 2010; Miklowitz & Johnson, 2009; Richardson, 2010). ACT has been found to diminish internalized stigma in SUD, a common comorbidity of BD (Luoma et al., 2008), warranting research to examine the effects of this intervention on stigma-related thoughts and feelings in BD.

Recent research also points to the potential usefulness of compassion-focused therapy (CFT; Gilbert & Procter, 2006; Highfield et al., 2010) in improving functioning in BD. CFT trains people in countering chronic self-criticism by practicing self-compassion (Gilbert & Procter, 2006). Studies show that CFT can promote self-soothing and decrease shame, social anxiety, depression and self-criticism in personality as well as mood disorders (Gilbert & Procter, 2006; Pauley & McPherson, 2010; Van Dam, Sheppard, Forsyth, & Earleywine, 2010). In BD, it may also promote affect regulation (Galvez et al., 2011; Gruber et al., 2009; Lowens, 2010). Given CFT's potential to diminish shame and self-criticism, future research may therefore wish to examine the effects of CFT on internalized stigma in BD.

Use of empowerment-based interventions to counter internalized stigma also deserves consideration (Brohan et al., 2011; Corrigan et al., 2011). Broadly defined, empowerment refers to perceived and actual control over the central domains of life, such as personal security, financial stability, social networks, employment and health (Corrigan et al., 2011). To empower people with BD in the face of heavy internal and public stigma, treatment providers and researchers need to give greater weight and attention to consumer experiences including self-management strategies (Ilic et al., 2011), as well as to the potential use of consumer-led support services (Repper & Carter, 2011) and a strengths-based collaborative care treatment model (M. S. Bauer et al., 2006). While research on empowerment in BD remains scarce, Brohan et al. (2011) found that empowerment was negatively correlated with internalized stigma in BD. Empowerment interventions may therefore carry the potential to decrease the ill effects of stigma and improve functioning in BD.

Finally, psychoeducation for consumers with BD that explicitly acknowledges the realities of stigma and discrimination and helps consumers share experiences and improve strategies for navigating and coping with stigma, may serve a three-part function. First, such psychoeducation may reduce individual shame with respect to stigma by naming it as part of unjust systemic inequalities. Second, helping people with BD learn strategies for negotiating stigma may be of practical value to them. Third, such psychoeducational programs could provide a mechanism for people with BD who are managing their illness and related stigma successfully to impart their experiences to others with BD, in a way that empowers the community of people with BD as a whole. Such consumer-led psychoeducation and empowerment is taking place in some psychiatric rehabilitation programs, as well as in consumer support centers known as "recovery learning centers," that are being created in growing numbers. However, significantly increased resources must be devoted to the peer-led recovery movement if it going to play a dominant role in combating stigma and improving functioning in BD.

In sum, diminishing the effects of stigma on functioning in BD requires efforts on multiple levels. Treatment providers, researchers and mental health care administrators need to join with consumers and others in advocating for equality, support and accommodation in employment, housing and social opportunities, as well as in health insurance coverage for people with BD. To reduce public and internalized stigma, mindfulness- and compassion-based therapies might be considered along with more traditional empirically validated psychosocial interventions.

7. Conclusions

Tremendous advances in pharmacological and psychosocial treatments now allow many people with BD to achieve syndromal remission within two years of an acute episode, and sometimes much sooner (Huxley & Baldessarini, 2007; Tohen et al., 2000). However, the reduction in symptoms is often not accompanied by functional recovery (Montoya et al., 2010; Tohen et al., 2000). Given the limited ability of identified predictors of functioning in euthymic BD to account for the variance in functional outcomes, researchers have begun to look more closely at contextual factors that may restrict functioning. Stigma and discrimination have been identified as important contextual factors that contribute to psychosocial disability in BD (Vazquez et al., 2011).

Different forms of stigma interact with each other and the illness in a series of vicious cycles that push people with BD towards social and vocational marginalization (Brohan et al., 2011; Michalak et al., 2006, 2011; Vasquez et al., 2011). Reversing the impact of stigma on functioning in BD will likely require simultaneous interventions on multiple fronts. First, advocacy is needed to reform laws and regulations that restrict the rights and opportunities of people with BD. Second, funding is required for supported employment and social programs that empower them. Third, a major effort at reducing public stigma and discrimination against BD, both in the family and society at large, is required for developing the necessary social support and employment opportunities for psychosocial adjustment. Finally, counseling in BD needs to include methods for addressing stigma-related shame, self-criticism, social anxiety and avoidance, which carry detrimental clinical and psychosocial effects.

From a broad-based perspective, the presence of powerful societal barriers to functioning limits current efforts to improve functional outcomes in BD, which mostly focus on interventions that address internal factors. An inhospitable social climate hampers psychosocial development in ways that are unaffected by psychopharmacology or psychotherapy. For this reason, significant improvement to functioning in BD requires balancing investment into interventions aimed at removing societal barriers to functional adjustment and alleviating the ill effects of these barriers on the individual, with more traditional forms of treatment.

8. References

Alonso, J., Buron, A., Rojas-Farreras, S., de Graaf, R., Haro, J. M., de Girolamo, G., Vilagut, G. (2009). Perceived stigma among individuals with common mental disorders. *Journal of Affective Disorders, 118*(1-3), 180-186.

Altshuler, L., Tekell, J., Biswas, K., Kilbourne, A. M., Evans, D., Tang, D., & Bauer, M. S. (2007). Executive function and employment status among veterans with bipolar disorder. *Psychiatric Services, 58*(11), 1441.

Americans with Disabilities Act of 1990, 42 U.S.C.A. §§ 12101 (1990) *et seq*. (West 1993).

Andrews, B., Qian, M., & Valentine, J. D. (2002). Predicting depressive symptoms with a new measure of shame: The Experience of Shame Scale. *British Journal of Clinical Psychology, 41*(Pt 1), 29-42.

Arbesman, M., & Logsdon, D. W. (2011). Occupational therapy interventions for employment and education for adults with serious mental illness: A systematic review. *American Journal of Occupational Therapy, 65*, 238-246.

Australian Journal of Pharmacy. (2011). Practice updates: Rising misconceptions about mental illness. *AJP: The Australian Journal of Pharmacy, 92*(1091), 29.

Balanza-Martinez, V., Selva, G., Martinez-Aran, A., Prickaerts, J., Salazar, J., Gonzalez-Pinto, A., Tabares-Seisdedos, R. (2010). Neurocognition in bipolar disorders -- A closer look at comorbidities and medications. *European Journal of Pharmacology, 626*(1), 87-96.

Baldwin, M. L., & Marcus, S. C. (2011). Stigma, discrimination, and employment outcomes among persons with mental health disabilities. In I. Z. Schultz & E. S. Rogers (Eds.), *Work accommodation and retention in mental health* (pp. 53-69). New York: Springer Science+Business Media.

Ball, J. R., Mitchell, P. B., Corry, J. C., Skillecorn, A., Smith, M., & Malhi, G. S. (2006). A randomized controlled trial of cognitive therapy for bipolar disorder: Focus on long-term change. *Journal of Clinical Psychiatry, 67*(2), 277-286.

Bambra, C. (2010) 'Yesterday once more? unemployment and health in the 21st century.', *Journal of epidemiology and community health, 64*, 213-215.

Bauer, M., Glenn, T., Grof, P., Rasgon, N. L., Marsh, W., Sagduyu, K., Whybrow, P. C. (2009). Frequency of subsyndromal symptoms and employment status in patients with bipolar disorder. *Social Psychiatry and Psychiatric Epidemiology, 44*(7), 515-522.

Bauer, M. S., McBride, L., Williford, W. O., Glick, H., Kinosian, B., Altshuler, L., Sajatovic, M. (2006). Collaborative care for bipolar disorder: Part I. Intervention and implementation in a randomized effectiveness trial. *Psychiatric Services, 57*(7), 927-936.

Bauwens, F., Tracy, A., Pardoen, D., Vander Elst, M., & Mendlewicz, J. (1991). Social adjustment of remitted bipolar and unipolar out-patients. A comparison with age- and sex-matched controls. *The British Journal of Psychiatry, 159*(2), 239.

Bauwens, F., Pardoen, D., Staner, L., Dramaix, M., & Mendlewicz, J. (1998). Social adjustment and the course of affective illness: A one year controlled longitudinal study involving bipolar and unipolar outpatients. *Depression and Anxiety, 8*(2), 50-57.

Bearden, C. E., Shih, V. H., Green, M. F., Gitlin, M., Sokolski, K. N., Levander, E., Altshuler, L. L. (2011). The impact of neurocognitive impairment on occupational recovery of clinically stable patients with bipolar disorder: A prospective study. *Bipolar Disorders, 13*(4), 323-333.

Berk, L., Hallam, K. T., Colom, F., Vieta, E., Hasty, M., Macneil, C., & Berk, M. (2010). Enhancing medication adherence in patients with bipolar disorder. *Human Psychopharmacology, 25*(1), 1-16.

Birchwood, M., Trower, P., Brunet, K., Gilbert, P., Iqbal, Z., & Jackson, C. (2007). Social anxiety and the shame of psychosis: A study in first episode psychosis. *Behaviour Research and Therapy, 45*(5), 1025-1037.

Bond, G. R., Xie, H., & Drake, R. E. (2007). Can SSDI and SSI beneficiaries with mental illness benefit from evidence-based supported employment? *Psychiatric Services, 58*(11), 1412.

Bond, G. R., & Kukla, M. (2011). Impact of follow-along support on job tenure in the individual placement and support model. *Journal of Nervous and Mental Disorders, 199*(3), 150-155.

Bonnin, C. M., Martinez-Aran, A., Torrent, C., Pacchiarotti, I., Rosa, A. R., Franco, C., Vieta, E. (2010). Clinical and neurocognitive predictors of functional outcome in bipolar euthymic patients: A long-term, follow-up study. *Journal of Affective Disorders, 121*(1-2), 156-160.

Bonvalot, T., Placines, B., Bouché, C., Le Moigne, L., Boldi, I., Millet, B., & Thym, B. (2010). P01-13 - The bipolar patients observance and participation in cognitive behavioral therapy groups tend to be improved through mindfulness meditation. *European Psychiatry, 25*(Suppl 1), 232-233.

Bowden, C. L. (2005). Bipolar disorder and work loss. *The American Journal of Managed Care, 11*(3 Suppl), S91-94.

Boyd, J. W., Linsenmeyer, A., Woolhandler, S., Himmelstein, D. U., & Nardin, R. (2011). The crisis in mental health care: A preliminary study of access to psychiatric care in Boston. *Annals of Emergency Medicine, 58*(2), 218.

Boylan, K. R., Bieling, P. J., Marriott, M., Begin, H., Young, L. T., & MacQueen, G. M. (2004). Impact of comorbid anxiety disorders on outcome in a cohort of patients with bipolar disorder. *Journal of Clinical Psychiatry, 65*(8), 1106-1113.

Brohan, E., Gauci, D., Sartorius, N., & Thornicroft, G. (2011). Self-stigma, empowerment and perceived discrimination among people with bipolar disorder or depression in 13 European countries: The GAMIAN-Europe study. *Journal of Affective Disorders, 129*(1-3), 56-63.

Buhin, L., & Vera, E. M. (2009). Preventing racism and promoting social justice: Person-centered and environment-centered interventions. *Journal of Primary Prevention, 30*, 43-59.

Burdick, K. E., Goldberg, J. F., & Harrow, M. (2010). Neurocognitive dysfunction and psychosocial outcome in patients with bipolar I disorder at 15-year follow-up. *Acta Psychiatrica Scandinavica, 122*(6), 499-506.

Burton, V. S., Jr. (1990). The consequences of official labels: A research note on rights lost by the mentally ill, mentally incompetent, and convicted felons. *Community Mental Health Journal, 26*(3), 267-276.

Bush, P. W., Drake, R. E., Xie, H., McHugo, G. J., & Haslett, W. R. (2009). The long-term impact of employment on mental health service use and costs for persons with severe mental illness. *Psychiatric Services, 60*(8), 1024-1031.

Cannon, M., Jones, P., Gilvarry, C., Rifkin, L., McKenzie, K., Foerster, A., & Murray, R. (1997). Premorbid social functioning in schizophrenia and bipolar disorder: Similarities and differences. *American Journal of Psychiatry, 154*(11), 1544-1550.

Carter, M. P. (2010). How evictions from subsidized housing routinely violate the rights of persons with mental illness. *Northwestern Journal of Law and Social Policy, 5*, 118-148.

Caspi, A., Sugden, K., Moffitt, T. E., Taylor, A., Craig, I. W., Harrington, H., Poulton, R. (2003). Influence of life stress on depression: Moderation by a polymorphism in the 5-HTT gene. *Science, 301*(5631), 386-389.

Cheung, M., Gilbert, P., & Irons, C. (2004). An exploration of shame, social rank and rumination in relation to depression. *Personality and Individual Differences, 36*(5), 1143-1153.

Cohen, A. N., Hammen, C., Henry, R. M., & Daley, S. E. (2004). Effects of stress and social support on recurrence in bipolar disorder. *Journal of Affective Disorders, 82*(1), 143-147.

Colom, F., Vieta, E., Sanchez-Moreno, J., Palomino-Otiniano, R., Reinares, M., Goikolea, J. M., . . . Martinez-Aran, A. (2009). Group psychoeducation for stabilised bipolar disorders: 5-year outcome of a randomised clinical trial. *British Journal of Psychiatry, 194*(3), 260-265.

Cook, J. A., Lehman, A. F., Drake, R., McFarlane, W. R., Gold, P. B., Leff, H. S., Grey, D. D. (2005). Integration of psychiatric and vocational services: A multisite randomized, controlled trial of supported employment. *American Journal of Psychiatry, 162*, 1948-1956.

Cook, J. A. (2006). Employment barriers for persons with psychiatric disabilities: A report for the President's New Freedom Commission. *Psychiatric Services, 57*(10), 1391-1405.

Cook, J. A., Blyler, C. R., Leff, H. S., McFarlane, W. R., Goldberg, R. W., Gold, P. B., Razzano, L. A. (2008). The Employment Intervention Demonstration Program: Major findings and policy implications. *Psychiatric Rehabilitation Journal, 31*(4), 291-295.

Corrigan, P. W., & Watson, A. C. (2002). The paradox of self-stigma and mental illness. *Clinical Psychology: Science and Practice, 9*(1), 35-53.

Corrigan, P. W. (2004). How stigma interferes with mental health care. *American Psychologist, 59*(7), 614-625.

Corrigan, P. W., Markowitz, F. E., & Watson, A. C. (2004). Structural levels of mental illness stigma and discrimination. *Schizophrenia Bulletin, 30*(3), 481-491.

Corrigan, P. W., Watson, A. C., Gracia, G., Slopen, N., Rasinski, K., & Hall, L. L. (2005). Newspaper stories as measures of structural stigma. *Psychiatric Services, 56*(5), 551-556.

Corrigan, P. W., Watson, A. C., Heyrman, M. L., Warpinski, A., Gracia, G., Slopen, N., & Hall, L. L. (2005). Structural stigma in state legislation. *Psychiatric Services, 56*(5), 557-563.

Corrigan, P. W., & Wassel, A. (2008). Understanding and influencing the stigma of mental illness. *Journal of Psychosocial Nursing and Mental Health Services, 46*(1), 42-48.

Corrigan, P. W., Roe, D., & Tsang, H. W. (2011). *Challenging the stigma of mental illness: Lessons for therapists and advocates*. West Sussex, UK: John Wiley & Sons Ltd.

Coryell, W., Scheftner, W., Keller, M., Endicott, J., Maser, J., & Klerman, G. L. (1993). The enduring psychosocial consequences of mania and depression. *American Journal of Psychiatry, 150*(5), 720.

Crocker, J., Major, B., & Steele, C. (1998). Social stigma. In D. T. Gilbert, S. T. Fiske & G. Lindzey (Eds.), *Handbook of social psychology* (4th ed., pp. 504-553). Boston & New York: McGraw-Hill; Distributed exclusively by Oxford University Press.

Deckersbach, T., Hölzel, B. K., Eisner, L. R., Stange, J. P., Peckham, A. D., Dougherty, D. D., Nierenberg, A. A. (2011). Mindfulness-based cognitive therapy for non-remitted patients with bipolar isorder. *CNS Neuroscience & Therapeutics*, doi: 10.1111/j.1755-5949.2011.00236.x

Deckersbach, T., Nierenberg, A. A., Kessler, R., Lund, H. G., Ametrano, R. M., Sachs, G., Dougherty, D. (2010). RESEARCH: Cognitive rehabilitation for bipolar disorder: An open trial for employed patients with residual depressive symptoms. *CNS Neuroscience and Therapeutics, 16*(5), 298-307.

Denicoff, K. D., Ali, S. O., Mirsky, A. F., Smith-Jackson, E. E., Leverich, G. S., Duncan, C. C., Post, R. M. (1999). Relationship between prior course of illness and neuropsychological functioning in patients with bipolar disorder. *Journal of Affective Disorders, 56*(1), 67-73.

Department of Defense Task Force on Mental Health. (2007). *An achievable vision: Report of the Department of Defense Task Force on Mental Health.* Falls Church, VA: Defense Health Board.

Depression and Bipolar Support Alliance. (2008). Workplace Stigma & Bipolar Disorder: DBSA Speaks to ABC News Retrieved 5 September 2011, from http://www.dbsalliance.org/site/PageServer?pagename=media_eupdate0308_complete

Dickerson, F. B., Origoni, A., Stallings, C., Khushalani, S., Dickinson, D., & Medoff, D. (2010). Occupational status and social adjustment six months after hospitalization early in the course of bipolar disorder: A prospective study. *Bipolar Disorders, 12*(1), 10-20.

Drake, R. E., Skinner, J. S., Bond, G. R., & Goldman, H. H. (2009). Social security and mental illness: Reducing disability with supported employment. *Health Affairs, 28*(3), 761.

Eisner, L. R., & Johnson, S. L. (2008). An acceptance-based psychoeducation intervention to reduce expressed emotion in relatives of bipolar patients. *Behavior Therapy, 39*(4), 375-385.

Elgie, R., & Morselli, P. L. (2007). Social functioning in bipolar patients: The perception and perspective of patients, relatives and advocacy organizations - A review. *Bipolar Disorders, 9*(1-2), 144-157.

Elinson, L., Houck, P., & Pincus, H. A. (2007). Working, receiving disability benefits, and access to mental health care in individuals with bipolar disorder. *Bipolar Disorders, 9*(1-2), 158-165.

Evans, L. J., & Bond, G. R. (2008). Expert ratings on the critical ingredients of supported employment for people with severe mental illness. *Psychiatric Rehabilitation Journal, 31*(4), 318-331.

Fair Housing Amendments Act of 1988, 42 U.S.C.A. §§ 3601-3619 *et seq.* (West 1988).

Fountoulakis, K. N., & Vieta, E. (2008). Treatment of bipolar disorder: A systematic review of available data and clinical perspectives. *International Journal of Neuropsychopharmacology, 11*(7), 999-1029.

Frank, E., Hlastala, S., Ritenour, A., Houck, P., Tu, X. M., Monk, T. H., Kupfer, D. J. (1997). Inducing lifestyle regularity in recovering bipolar disorder patients: Results from the Maintenance Therapies in Bipolar Disorder protocol. *Biological Psychiatry, 1997*(41).

Frank, E., Swartz, H. A., & Kupfer, D. J. (2000). Interpersonal and social rhythm therapy: Managing the chaos of bipolar disorder. *Biological Psychiatry, 48*(6), 593-604.

Frank, E., Kupfer, D. J., Thase, M. E., Mallinger, A. G., Swartz, H. A., Fagiolini, A. M., Monk, T. (2005). Two-year outcomes for interpersonal and social rhythm therapy in individuals with bipolar I disorder. *Archives of General Psychiatry, 62*, 996-1004.

Gabriel, A., & Violato, C. (2010). The development and psychometric assessment of an instrument to measure attitudes towards depression and its treatments in patients suffering from non-psychotic depression. *Journal of Affective Disorders, 124*(3), 241-249.

Galvez, J. F., Thommi, S., & Ghaemi, S. N. (2011). Positive aspects of mental illness: A review in bipolar disorder. *Journal of Affective Disorders, 128*(3), 185-190.

Gardner, H. H., Kleinman, N. L., Brook, R. A., Rajagopalan, K., Brizee, T. J., & Smeeding, J. E. (2006). The economic impact of bipolar disorder in an employed population from an employer perspective. *Journal of Clinical Psychiatry, 67*(8), 1209-1218.

Gilbert, P. (1992). *Depression: The evolution of powerlessness.* Hove, U.K.: Erlbaum.

Gilbert, P. (2002). Body shame: A biopsychosocial conceptualisation and overview, with treatment implications. In P. Gilbert & J. Miles (Eds.), *Body shame: Conceptualisation, research and treatment* (pp. 3-54). London: Brunner-Routledge.

Gilbert, P., & Irons, C. (2004). A pilot exploration of the use of compassionate images in a group of self-critical people. *Memory, 2004*(12), 4.

Gilbert, P., & Procter, S. (2006). Compassionate mind training for people with high shame and self-criticism: Overview and pilot study of a group therapy approach. *Clinical Psychology & Psychotherapy, 13*(6), 353-379.

Gilbert, P., McEwan, K., Irons, C., Bhundia, R., Christie, R., Broomhead, C., & Rockliff, H. (2010). Self-harm in a mixed clinical population: The roles of self-criticism, shame, and social rank. *British Journal of Clinical Psychology, 49*(Pt 4), 563-576.

Gitlin, M., Mintz, J., Sokolski, K., Hammen, C., & Altshuler, L. (2010). Subsyndromal depressive symptoms after symptomatic recovery from mania are associated with delayed functional recovery. *Journal of Clinical Psychiatry.*

Goffman, E. (1963). *Stigma: Notes on the management of spoiled identity.* Englewood Cliffs, N.J.,: Prentice-Hall.

Goldberg, J. F., & Chengappa, K.N.R. (2009). Identifying and treating cognitive impairment in bipolar disorder. *Bipolar Disorders, 11*, 123-137.

Goldstein, B. I., & Levitt, A. J. (2008). The specific burden of comorbid anxiety disorders and of substance use disorders in bipolar I disorder. *Bipolar Disorders, 10*(1), 67-78.

Gruber, J., Culver, J. L., Johnson, S. L., Nam, J. Y., Keller, K. L., & Ketter, T. A. (2009). Do positive emotions predict symptomatic change in bipolar disorder? *Bipolar Disorders, 11*(3), 330-336.

Gutierrez-Rojas, L., Jurado, D., & Gurpegui, M. (2011). Factors associated with work, social life and family life disability in bipolar disorder patients. *Psychiatry Research, 186*(2-3), 254-260.

Haro, J. M., Reed, C., Gonzalez-Pinto, A., Novick, D., Bertsch, J., & Vieta, E. (2011). 2-Year course of bipolar disorder type I patients in outpatient care: Factors associated with remission and functional recovery. *European Neuropsychopharmacology, 21*(4), 287-293.

Harvey, A. G., Schmidt, D. A., Scarna, A., Semler, C. N., & Goodwin, G. M. (2005). Sleep-related functioning in euthymic patients with bipolar disorder, patients with insomnia, and subjects without sleep problems. *American Journal of Psychiatry, 162*(1), 50.

Hays, J. C., Krishnan, K. R., George, L. K., & Blazer, D. G. (1998). Age of first onset of bipolar disorder: Demographic, family history, and psychosocial correlates. *Depression and Anxiety, 7*(2), 76-82.

Heginbotham, C. (1998). U.K. mental health policy can alter the stigma of mental illness. *Lancet, 352*(9133), 1052-1053.

Hemmens, C., Miller, M., Burton, V. S., Jr., & Milner, S. (2002). The consequences of official labels: An examination of the rights lost by the mentally ill and mentally incompetent ten years later. *Community Mental Health Journal, 38*(2), 129-140.

Highfield, J., Markham, D., Skinner, M., & Neal, A. (2010). An investigation into the experience of self-conscious emotions in individuals with bipolar disorder, unipolar depression and non-psychiatric controls. *Clinical Psychology & Psychotherapy, 17*(5), 395-405.

Hinshaw, S. P. (2007). *The mark of shame: Stigma of mental Illness and an agenda for change.* New York: Oxford University Press.

Hirschfeld, R. M., Lewis, L., & Vornick, L. A. (2003). Perception and impact of bipolar disorder: How far have we really come? Results of the National Depressive and Manic-Depressive Association 2000 Survey of individuals with BD. *Journal of Clinical Psychiatry, 64*(161-174).

Hollon, S. D., & Ponniah, K. (2010). A review of empirically supported psychological therapies for mood disorders in adults. *Depression and Anxiety, 27*(10), 891-932.

Huxley, N., & Baldessarini, R. (2007). Disability and its treatment in bipolar disorder patients. *Bipolar Disorders, 9*(1-2), 183-196.

Ilic, M., Reinecke, J., Bohner, G., Rottgers, H. O., Beblo, T., Driessen, M., Corrigan, P. W. (2011). Protecting self-esteem from stigma: A test of different strategies for coping with the stigma of mental illness. *International Journal of Social Psychiatry*, 1-12.

Interian, A., Ang, A., Gara, M. A., Link, B. G., Rodriguez, M. A., & Vega, W. A. (2010). Stigma and depression treatment utilization among Latinos: Utility of four stigma measures. *Psychiatric Services, 61*(4), 373-379.

Jaworski, F., Dubertret, C., Adès, J., & Gorwood, P. (2011). Presence of co-morbid substance use disorder in bipolar patients worsens their social functioning to the level observed in patients with schizophrenia. *Psychiatry Research, 185*(1-2), 129-134.

Johnson, S. L., Meyer, B., Winett, C., & Small, J. (2000). Social support and self-esteem predict changes in bipolar depression but not mania. *Journal of Affective Disorders, 58*(1), 79-86.

Jones, E., Farina, A., Hastorf, A., Markus, H., Miller, D., & Scott, R. (1984). *Social stigma: The psychology of marked relationships*. New York: W. H. Freeman.

Kanter, J. W., Rusch, L. C., & Brondino, M. J. (2008). Depression self-stigma: A new measure and preliminary findings. *Journal of Nervous and Mental Disorders, 196*(9), 663-670.

Kauer-Sant Anna, M., Frey, B. N., Andreazza, A. C., Cereser, K. M., Gazalle, F. K., Tramontina, J., . . . Kapczinski, F. (2007). Anxiety comorbidity and quality of life in bipolar disorder patients. *Canadian Journal of Psychiatry, 52*(3), 175.

Kessler, R. C., Akiskal, H. S., Ames, M., Birnbaum, H., Greenberg, P., Hirschfeld, R. M., Wang, P. S. (2006). Prevalence and effects of mood disorders on work performance in a nationally representative sample of U.S. workers. *American Journal of Psychiatry, 163*(9), 1561-1568.

Krishnan, K. R. R. (2005). Psychiatric and medical comorbidities of bipolar disorder. *Psychosomatic Medicine, 67*, 1-8.

Krupa, T. (2007). Interventions to improve employment outcomes for workers who experience mental illness. *Canadian Journal of Psychiatry, 52*(6), 339.

Krupa, T., Kirsh, B., Cockburn, L., & Gewurtz, R. (2009). Understanding the stigma of mental illness in employment. *Work, 33*(4), 413-425.

Krupa, T. (2011a). Approaches to improving employment outcomes for people with serious mental illness. In I. Z. Schultz & E. S. Rogers (Eds.), *Work accommodation and retention in mental health* (pp. 219-231). New York: Springer Science+Business Media.

Krupa, T. (2011b). Employment and serious mental health disabilities. In I. Z. Schultz & E. S. Rogers (Eds.), *Work accommodation and retention in mental health* (pp. 91-101). New York: Springer Science+Business Media.

Kupfer, D. J., Frank, E., Grochocinski, V. J., Cluss, P. A., Houck, P. R., & Stapf, D. A. (2002). Demographic and clinical characteristics of individuals in a bipolar disorder case registry. *Journal of Clinical Psychiatry, 63*(2), 120-125.

Lagerberg, T. V., Andreassen, O. A., Ringen, P. A., Berg, A. O., Larsson, S., Agartz, I., Melle, I. (2010). Excessive substance use in bipolar disorder is associated with impaired functioning rather than clinical characteristics, a descriptive study. *BMC Psychiatry, 10*, 9.

Larson, J. E., Ryan, C. B., Wassel, A. K., Kaszynski, K. L., Ibara, L., Glenn, T. L., & Boyle, M. G. (2011). Analyses of employment incentives and barriers for individuals with psychiatric disabilities. *Rehabilitation Psychology, 56*(2), 145-149.

Latalova, K., Prasko, J., Diveky, T., & Velartova, H. (2011). Cognitive impairment in bipolar disorder. *Biomedical Papers, 155*(1).

Latvala, A., Castaneda, A. E., Perälä, J., Saarni, S. I., Aalto Setälä, T., Lönnqvist, J., . . . Tuulio Henriksson, A. (2009). Cognitive functioning in substance abuse and dependence: A population based study of young adults. *Addiction, 104*(9), 1558-1568.

Levy, B., & Weiss, R. D. (2009). Cognitive functioning in bipolar and co-occurring substance use disorders: A missing piece of the puzzle. *Harvard Review of Psychiatry, 17*(3), 226-230.

Lim, L., Nathan, P., O'Brien-Malone, A., & Williams, S. (2004). A qualitative approach to identifying psychosocial issues faced by bipolar patients. *Journal of Nervous and Mental Disorders, 192*(12), 810-817.

Link, B., Cullen, F. T., Frank, J., & Wozniak, J. F. (1987). The social rejection of former mental patients: Understanding why labels matter. *American Journal of Sociology, 92*, 1461-1500.

Link, B. G. (1987). Understanding labeling effects in the area of mental disorders: An assessment of the effects of expectations of rejection. *American Sociological Review*, 96-112.

Link, B. G., Mirotznik, J., & Cullen, F. T. (1991). The effectiveness of stigma coping orientations: Can negative consequences of mental illness labeling be avoided? *Journal of Health and Social Behavior, 32*(3), 302-320.

Link, B. G., Phelan, J. C., Bresnahan, M., Stueve, A., & Pescosolido, B. A. (1999). Public conceptions of mental illness: Labels, causes, dangerousness, and social distance. *American Journal of Public Health, 89*(9), 1328-1333.

Link, B. G., Struening, E. L., Neese-Todd, S., Asmussen, S., & Phelan, J. C. (2001). Stigma as a barrier to recovery: The consequences of stigma for the self-esteem of people with mental illnesses. *Psychiatric Services, 52*(12), 1621.

Link, B. G., Yang, L. H., Phelan, J. C., & Collins, P. Y. (2004). Measuring mental illness stigma. *Schizophrenia Bulletin, 30*(3), 511-541.

Lish, J. D., Dime-Meenan, S., Whybrow, P. C., Price, R. A., & Hirschfeld, R. M. (1994). The National Depressive and Manic-depressive Association (DMDA) survey of bipolar members. *Journal of Affective Disorders, 31*(4), 281-294.

Livingston, J. D., & Boyd, J. E. (2010). Correlates and consequences of internalized stigma for people living with mental illness: A systematic review and meta-analysis. *Social Science & Medicine, 71*(12), 2150-2161.

Lloyd, C., & Waghorn, G. (2007). The importance of vocation in recovery for young people with psychiatric disabilities. *British Journal of Occupational Therapy, 70*(2), 50-59.

Lowens, I. (2010). Compassion-focused therapy for people with bipolar disorder. *International Journal of Cognitive Therapy, 3*(2), 172-185.

Luoma, J. B., Kohlenberg, B. S., Hayes, S. C., Bunting, K., & Rye, A. K. (2008). Reducing self-stigma in substance abuse through acceptance and commitment therapy: Model manual development, and pilot outcomes. *Addiction Research and Theory, 16*, 149-165.

Luoma, J. B., O'Hair, A. K., Kohlenberg, B. S., Hayes, S. C., & Fletcher, L. (2010). The development and psychometric properties of a new measure of perceived stigma toward substance users. *Substance Use & Misuse, 45*(1-2), 47-57.

MacQueen, G. M., Young, L. T., Robb, J. C., Marriott, M., Cooke, R. G., & Joffe, R. T. (2000). Effect of number of episodes on wellbeing and functioning of patients with bipolar disorder. *Acta Psychiatrica Scandinavica, 101*(5), 374-381.

Manos, R. C., Rusch, L. C., Kanter, J. W., & Clifford, L. M. (2009). Depression self-stigma as a mediator of the relationship between depression severity and avoidance. *Journal of Social and Clinical Psychology, 28*(9), 1128-1143.

Martinez, A. G., Piff, P. K., Mendoza-Denton, R., & Hinshaw, S. P. (2011). The power of a label: Mental illness diagnoses, ascribed humanity, and social rejection. *Journal of Social and Clinical Psychology, 30*(1), 1-23.

Martinez-Aran, A., Vieta, E., Reinares, M., Colom, F., Torrent, C., Sanchez-Moreno, J., Salamero, M. (2004). Cognitive function across manic or hypomanic, depressed, and euthymic states in bipolar disorder. *American Journal of Psychiatry, 161*(2), 262.

Martínez-Arán, A., Torrent, C., Solé, B., Bonnín, C. M., Rosa, A. R., Sánchez-Moreno, J., & Vieta, E. (2011). Functional remediation for bipolar disorder. *Clinical Practice and Epidemiology in Mental Health: CP & EMH, 7*, 112.

Martino, D. J., Marengo, E., Igoa, A., Scápola, M., Ais, E. D., Perinot, L., & Strejilevich, S. A. (2009). Neurocognitive and symptomatic predictors of functional outcome in bipolar disorders: A prospective 1 year follow-up study. *Journal of Affective Disorders, 116*(1-2), 37-42.

Martino, D. J., Igoa, A., Marengo, E., Scapola, M., & Strejilevich, S. A. (2011). Neurocognitive impairments and their relationship with psychosocial functioning in euthymic bipolar II disorder. *Journal of Nervous and Mental Disorders, 199*(7), 459-464.

Masuda, A., Hayes, S. C., Lillis, J., Bunting, K., Herbst, S. A., & Fletcher, L. B. (2009). The relation between psychological flexibility and mental health stigma in acceptance and commitment therapy: A preliminary process investigation. *Behavior and Social Issues, 18*, 25-40.

Masuda, A., Price, M., Anderson, P. L., Schmertz, S. K., & Calamaras, M. R. (2009). The role of psychological flexibility in mental health stigma and psychological disress for the stigmatizer. *Journal of Social and Clinical Psychology, 28*(10), 1244-1262.

McHugh, K.R., & Barlow, D.H. (2010). The dissemination and implementation of evidence-based psychological treatments. *American Psychologist, 65(2),* 73-84. DOI: 10.1037/a0018121

McHugo, G. J., Drake, R. E., & Becker, D. R. (1998). The durability of supported employment effects. *Psychiatric Rehabilitation Journal, 22*, 55-61.

McKown, C. (2005). Applying ecological theory to advance the science and practice of school-based prejudice reduction interventions. *Educational Psychology, 40*, 177-189.

McMorris, B. J., Downs, K. E., Panish, J. M., & Dirani, R. (2010). Workplace productivity, employment issues, and resource utilization in patients with bipolar I disorder. *Journal of Medical Economics, 13*(1), 23-32.

McMurrich, S. L., & Johnson, S. L. (2009). The role of depression, shame-proneness, and guilt-proneness in predicting criticism of relatives towards people with bipolar disorder. *Behavior Therapy, 40*(4), 315-324.

Mehta, N., Kassam, A., Leese, M., Butler, G., & Thornicroft, G. (2009). Public attitudes towards people with mental illness in England and Scotland, 1994-2003. *British Journal of Psychiatry, 194*(3), 278-284.

Michalak, E. E., Yatham, L. N., Kolesar, S., & Lam, R. W. (2006). Bipolar disorder and quality of life: A patient-centered perspective. *Quality of Life Research, 15*(1), 25-37.

Michalak, E. E., Yatham, L. N., Maxwell, V., Hale, S., & Lam, R. W. (2007). The impact of bipolar disorder upon work functioning: A qualitative analysis. *Bipolar Disorders, 9*(1-2), 126-143.

Michalak, E. E., Livingston, J. D., Hole, R., Suto, M., Hale, S., & Haddock, C. (2011). 'It's something that I manage but it is not who I am': Reflections on internalized stigma in individuals with bipolar disorder. *Chronic Illness, 0*(0), 1-16.

Miklowitz, D. J., George, E. L., Richards, J. A., Simoneau, T. L., & Suddath, R. L. (2003). A randomized study of family-focused psychoeducation and pharmacotherapy in the outpatient management of bipolar disorder. *Archives of General Psychiatry, 60*(9), 904-912.

Miklowitz, D. J., Goodwin, G. M., Bauer, M. S., & Geddes, J. R. (2008). Common and specific elements of psychosocial treatments for bipolar disorder: A survey of clinicians participating in randomized trials. *Journal of Psychiatric Practice, 14*(2), 77-85.

Miklowitz, D. J., & Johnson, S. L. (2009). Social and familial factors in the course of bipolar disorder: Basic processes and relevant interventions. *Clinical Psychology, 16*(2), 281-296.

Montoya, A., Tohen, M., Vieta, E., Casillas, M., Chacon, F., Polavieja, P., & Gilaberte, I. (2010). Functioning and symptomatic outcomes in patients with bipolar I disorder in syndromal remission: A 1-year, prospective, observational cohort study. *Journal of Affective Disorders, 127*(1-3), 50-57.

Mur, M., Portella, M. J., Martinez-Aran, A., Pifarre, J., & Vieta, E. (2009). Influence of clinical and neuropsychological variables on the psychosocial and occupational outcome of remitted bipolar patients. *Psychopathology, 42*(3), 148-156.

Nivoli, A. M., Colom, F., Murru, A., Pacchiarotti, I., Castro-Loli, P., Gonzalez-Pinto, A., Vieta, E. (2011). New treatment guidelines for acute bipolar depression: A systematic review. *Journal of Affective Disorders, 129*(1-3), 14-26.

Otto, M. W., Simon, N. M., Wisniewski, S. R., MIklowitz, D. J., Kogan, J. N., Reilly-Harrington, N. A., Pollack, M. H. (2006). Prospective 12-month course of bipolar disorder in out-patients with and without comorbid anxiety disorders. *British Journal of Psychiatry, 189*, 20-25.

Paluck, E. L., & Green, D. P. (2009). Prejudice reduction: What works? A critical look at evidence from the field and the laboratory. *Annual Review of Psychology, 60*, 339-367.

Pauley, G., & McPherson, S. (2010). The experience and meaning of compassion and self compassion for individuals with depression or anxiety. *Psychology and Psychotherapy: Theory, Research and Practice, 83*(2), 129-143.

Perlick, D. A., Rosenheck, R. A., Clarkin, J. F., Sirey, J. A., Salahi, J., Struening, E. L., & Link, B. G. (2001). Stigma as a barrier to recovery: Adverse effects of perceived stigma on social adaptation of persons diagnosed with bipolar affective disorder. *Psychiatric Services, 52*(12), 1627.

Perlick, D. A., Miklowitz, D. J., Link, B. G., Struening, E., Kaczynski, R., Gonzalez, J., Rosenheck, R. A. (2007). Perceived stigma and depression among caregivers of patients with bipolar disorder. *British Journal of Psychiatry, 190*, 535-536.

Pescosolido, B. A., Martin, J. K., Lang, A., & Olafsdottir, S. (2008). Rethinking theoretical approaches to stigma: A Framework Integrating Normative Influences on Stigma (FINIS). *Social Science & Medicine, 67*(3), 431-440.

Petrila, J. (2009). Congress restores the Americans with Disabilities Act to its original intent. *Psychiatric Services, 60*, 878-879.

Phelan, J. C., & Link, B. G. (1998). The growing belief that people with mental illnesses are violent: The role of the dangerousness criterion for civil commitment. *Social Psychiatry and Psychiatric Epidemiology, 33*(Suppl 1), S7-S12.

Pope, M., Dudley, R., & Scott, J. (2007). Determinants of social functioning in bipolar disorder. *Bipolar Disorders, 9*(1-2), 38-44.

Proudfoot, J. G., Parker, G. B., Benoit, M., Manicavasagar, V., Smith, M., & Gayed, A. (2009). What happens after diagnosis? Understanding the experiences of patients with newly-diagnosed bipolar disorder. *Health Expectations, 12*(2), 120-129.

Rakofsky, J. J., & Dunlop, B. W. (2011). Treating nonspecific anxiety and anxiety disorders in patients with bipolar disorder: A review. *Journal of Clinical Psychiatry, 72*(1), 81-90.

Raune, D., Kuipers, E., & Bebbington, P. E. (2004). Expressed emotion at first-episode psychosis: Investigating a carer-appraisal model. *British Journal of Psychiatry, 184*, 321-326.

Rea, M. M., Tompson, M. C., Miklowitz, D. J., Goldstein, M. J., Hwang, S., & Mintz, J. (2003). Family-focused treatment versus individual treatment for bipolar disorder: Results of a randomized clinical trial. *Journal of Consulting and Clinical Psychology, 71*(3), 482-492.

Reed, C., Goetz, I., Vieta, E., Bassi, M., & Haro, J. M. (2010). Work impairment in bipolar disorder patients -- Results from a two-year observational study (EMBLEM). *European Psychiatry, 25*(6), 338-344.

Reichenberg, A., Weiser, M., Rabinowitz, J., Caspi, A., Schmeidler, J., Mark, M., & Kaplan, Z. (2002). A population-based cohort study of premorbid intellectual, language, and behavioral functioning in patients with schizophrenia, schizoaffective disorder, and nonpsychotic bipolar disorder. *American Journal of Psychiatry, 159*(12), 2027-2035.

Repper, J., & Carter, T. (2011). A review of the literature on peer support in mental health services. *Journal of Mental Health, 20*, 392-411.

Richardson, T. H. (2010). Psychosocial interventions for bipolar disorder: A review of recent research. *Journal of Medical Science, 10*(6), 143-152.

Rietschel, M., Georgi, A., Schmael, C., Schirmbeck, F., Strohmaier, J., Boesshenz, K. V.,. Schulze, T. G. (2009). Premorbid adjustment: A phenotype highlighting a distinction rather than an overlap between schizophrenia and bipolar disorder. *Schizophrenia Research, 110*(1-3), 33-39.

Robb, J. C., Cooke, R. G., Devins, G. M., Young, L. T., & Joffe, R. T. (1997). Quality of life and lifestyle disruption in euthymic bipolar disorder. *Journal of Psychiatric Research, 31*(5), 509-517.

Rosa, A. R., Bonnin, C. M., Vazquez, G. H., Reinares, M., Sole, B., Tabares-Seisdedos, R., . Vieta, E. (2010). Functional impairment in bipolar II disorder: Is it as disabling as bipolar I? *Journal of Affective Disorders, 127*(1-3), 71-76.

Rusch, L. C., Kanter, J. W., Angelone, A. F., & Ridley, R. C. (2008). The impact of In Our Own Voice on stigma. *American Journal of Psychiatric Rehabilitation, 11*(4), 373-389.

Rusch, N., Lieb, K., Bohus, M., & Corrigan, P. W. (2006). Self-stigma, empowerment, and perceived legitimacy of discrimination among women with mental illness. *Psychiatric Services, 57*(3), 399-402.

Rusch, N., Corrigan, P. W., Powell, K., Rajah, A., Olschewski, M., Wilkniss, S., & Batia, K. (2009). A stress-coping model of mental illness stigma: II. Emotional stress responses, coping behavior and outcome. *Schizophrenia Research, 110*(1-3), 65-71.

Rusch, N., Corrigan, P. W., Wassel, A., Michaels, P., Larson, J. E., Olschewski, M., Batia, K. (2009). Self-stigma, group identification, perceived legitimacy of discrimination and mental health service use. *British Journal of Psychiatry, 195*(6), 551-552.

Rusch, N., Todd, A. R., Bodenhausen, G. V., Olschewski, M., & Corrigan, P. W. (2010). Automatically activated shame reactions and perceived legitimacy of discrimination: A longitudinal study among people with mental illness. *Journal of Behavior Therapy and Experimental Psychiatry, 41*(1), 60-63.

Rusch, N., Corrigan, P. W., Todd, A. R., & Bodenhausen, G. V. (2011). Automatic stereotyping against people with schizophrenia, schizoaffective and affective disorders. *Psychiatry Research, 186*(1), 34-39.

Sajatovic, M., Smith, D., Singer, E., Meyer, W. J., Cassidy, K. A., & Jenkins, J. H. (2009). Subjective experience of bipolar disorder, treatment adherence and stigma. In D. Roth & W. J. Lutz (Eds.), New research in mental health, 2006-2007 (Vol. 18, pp. 47-51). Columbus, OH: Ohio Department of Mental Health.

Sanchez-Moreno, J., Martinez-Aran, A., Tabarés-Seisdedos, R., Torrent, C., Vieta, E., & Ayuso-Mateos, J. (2009). Functioning and disability in bipolar disorder: An extensive review. *Psychotherapy and Psychosomatics, 78*(5), 285-297.

Sanchez-Moreno, J., Martinez-Aran, A., Gadelrab, H. F., Cabello, M., Torrent, C., Bonnin Cdel, M., Vieta, E. (2010). The role and impact of contextual factors on functioning in patients with bipolar disorder. *Disability and Rehabilitation, 32*(Suppl 1), S94-S104.

Schneider, J., Beeley, C., & Repper, J. (2011). Campaign appears to influence subjective experience of stigma. *Journal of Mental Health, 20*(1), 89-97.

Schultz, I. Z., Milner, R. A., Hanson, D. B., & Winter, A. (2011). Employer attitudes towards accommodations in mental health disability. In I. Z. Schultz & E. S. Rogers (Eds.), *Work accommondation and retention in mental health* (pp. 325-340). New York: Springer Science+Business Media.

Shapira, B., Zislin, J., Gelfin, Y., Osher, Y., Gorfine, M., Souery, D., Lerer, B. (1999). Social adjustment and self-esteem in remitted patients with unipolar and bipolar affective disorder: A case-control study. *Comprehensive Psychiatry, 40*(1), 24-30.

Silton, N. R., Flannelly, K. J., Milstein, G., & Vaaler, M. L. (2011). Stigma in America: Has anything changed? Impact of perceptions of mental illness and dangerousness on the desire for social distance: 1996 and 2006. *Journal of Nervous and Mental Disorders, 199*(6), 361-366.

Silva, F. T., & Leite, J. R. (2000). Physiological modifications and increase in state anxiety in volunteers submitted to the Stroop Color-Word Interference Test: A preliminary study. *Physiology & Behavior, 70*(1-2), 113-118.

Sirey, J. A., Bruce, M. L., Alexopoulos, G. S., Perlick, D. A., Friedman, S. J., & Meyers, B. S. (2001). Stigma as a barrier to recovery: Perceived stigma and patient-rated severity of illness as predictors of antidepressant drug adherence. *Psychiatric Services, 52*(12), 1615.

Star, S. A. (1952). *What the public thinks about mental health and mental illness.* Paper presented at the Annual Meeting of the National Association for Mental Health.

Star, S. A. (1955). *The public's ideas about mental illness.* Paper presented at the Annual Meeting of the National Association for Mental Health.

Surgeon General, U. S. (1999). Mental health: A report of the Surgeon General. Retrieved August 26, 2011, from http://www.surgeongeneral.gov/library/mentalhealth/home.html

Swanson, J., Burris, S., Moss, K., Ullman, M., & Ranney, L. M. (2006). Justice disparities: Does the ADA enforcement system treat people with psychiatric disabilities fairly? *Maryland Law Review, 66*(94), 94-139.

Tabares-Seisdedos, R., Balanza-Martinez, V., Sanchez-Moreno, J., Martinez-Aran, A., Salazar-Fraile, J., Selva-Vera, G., Vieta, E. (2008). Neurocognitive and clinical predictors of functional outcome in patients with schizophrenia and bipolar I disorder at one-year follow-up. *Journal of Affective Disorders, 109*(3), 286-299.

Tangney, J. P. (1995). Recent advances in the empirical study of shame and guilt. *American Behavioral Scientist, 38*(8), 1132.

Tangney, J. P., & Dearing, R. L. (2002). *Shame and guilt.* New York: Guilford.

Tohen, M., Waternaux, C. M., & Tsuang, M. T. (1990). Outcome in mania. A 4-year prospective follow-up of 75 patients utilizing survival analysis. *Archives of General Psychiatry, 47*(12), 1106-1111.

Tohen, M., Hennen, J., Zarate Jr, C. M., Baldessarini, R. J., Strakowski, S. M., Stoll, A. L., Cohen, B. M. (2000). Two-year syndromal and functional recovery in 219 cases of first-episode major affective disorder with psychotic features. *American Journal of Psychiatry, 157*(2), 220.

Tse, S. S., & Walsh, A. E. S. (2001). How does work work for people with bipolar affective disorder? *Occupational Therapy International, 8*(3), 210-225.

Tse, S. S. (2002). Practice guidelines: Therapeutic interventions aimed at assisting people with bipolar affective disorder achieve their vocational goals. *Work, 19*(2), 167-179.

Tse, S. S., & Yeats, M. (2002). What helps people with bipolar affective disorder to succeed in employment: A grounded theory approach. *Work, 19*(1), 47–62.

Uzelac, S., Jaeger, J., Berns, S., & Gonzales, C. (2006). Premorbid adjustment in bipolar disorder: Comparison with schizophrenia. *Journal of Nervous and Mental Disorders, 194*(9), 654-658.

Van Dam, N. T., Sheppard, S. C., Forsyth, J. P., & Earleywine, M. (2010). Self-compassion is a better predictor than mindfulness of symptom severity and quality of life in mixed anxiety and depression. *Journal of Anxiety Disorders, 25*(1), 123-130.

Vazquez, G. H., Kapczinski, F., Magalhaes, P. V., Cordoba, R., Lopez Jaramillo, C., Rosa, A. R., Tohen, M. (2011). Stigma and functioning in patients with bipolar disorder. *Journal of Affective Disorders, 130*(1-2), 323-327.

Ward, T. D. (2011). The lived experience of adults with bipolar disorder and comorbid substance use disorder. *Issues in Mental Health Nursing, 32*(1), 20-27.

Weber, B., Jermann, F., Gex-Fabry, M., Nallet, A., Bondolfi, G., & Aubry, J. M. (2010). Mindfulness-based cognitive therapy for bipolar disorder: A feasibility trial. *European Psychiatry, 25*(6), 334-337.

Weiss, R. D., Griffin, M. L., Jaffee, W. B., Bender, R. E., Graff, F. S., Gallop, R. J., & Fitzmaurice, G. M. (2009). A "community-friendly" version of integrated group therapy for patients with bipolar disorder and substance dependence: A randomized controlled trial. *Drug and Alcohol Dependence, 104*(3), 212-219.

Williams, J. M., Alatiq, Y., Crane, C., Barnhofer, T., Fennell, M. J., Duggan, D. S., Goodwin, G. M. (2008). Mindfulness-based cognitive therapy (MBCT) in bipolar disorder: Preliminary evaluation of immediate effects on between-episode functioning. *Journal of Affective Disorders, 107*(1-3), 275-279.

Zimmerman, M., Galione, J., Chelminski, I., Young, D., Dalrymple, K., & Ruggero, C. (2010). Sustained unemployment in psychiatric outpatients with bipolar disorder: Frequency and association with demographic variables and comorbid disorders. *Bipolar Disorders, 12*(7), 720-726.

Zubieta, J. K., Huguelet, P., O'Neil, R. L., & Giordani, B. J. (2001). Cognitive function in euthymic bipolar I disorder. *Psychiatry Research, 102*(1), 9-20.

10

Star Shots: Stigma, Self-Disclosure and Celebrity in Bipolar Disorder

Wendy Cross[1] and Ken Walsh[2]
[1]*Monash University*
[2]*University of Wollongong*
Australia

1. Introduction

Recent public disclosure by Catherine Zeta Jones of her treatment for Bipolar 2 disorder and the list of other celebrities past and present who have also experienced mental health problems prompted us to explore the phenomenon of the self-disclosure of mental illness by such well known public figures given the risk of such disclosure. We asked these questions:

What does this mean for deepening understanding?

What does this mean for reducing stigma?

What is the role of self-disclosure?

2. Bipolar disorder

What a creature of strange moods [Winston Churchill] is – always at the top of the wheel of confidence or at the bottom of an intense depression. (Lord Beaverbrook (1879-1964), quoted in Manchester, 1984:24)

The above reference to Sir Winston Churchill's periods of fluctuating mood from elevation to depression sums up the most common presentation of the mood disorder generally known in Western society as Bipolar Disorder. The American Psychiatric Association Diagnostic and Statistical Manual of Mental Disorders Fourth Edition (Text Revision), DSM-IV-TR, divides mood disorders into two basic categories: Depressive Disorders (also known as Unipolar Depression) and Bipolar Disorders (previously referred to as Manic Depression). These disorders are characterized by experiences of elevated or depressed mood. Whilst all people experience fluctuations in mood, people with mood disorders experience a sense of a loss of control of their moods and high degrees of subjective distress (Sadock et al., 2003).

The depressive episodes that accompany all mood disorders are not merely episodes of sadness. They are severe, distressing and prolonged episodes of depressed mood or loss of interest or pleasure that represent a change in previous functioning and significantly interfere with the person's ability to function socially, interpersonally or occupationally (American Psychiatric Association [APA], 2000). In addition to depressed mood and/or loss of interest or pleasure, people with depressive episodes may complain of weight loss or gain, fatigue, restlessness, problems thinking or concentrating and may have recurring thoughts of death or suicide (indeed it has been estimated that up to 24% of Bipolar Disorder suffers will attempt suicide (Rihmer & Kiss, 2002)).

Manic episodes are not just periods of subjective wellbeing but are intense and prolonged episodes of elevated, expansive or irritable mood which, as with depressive episodes, are severe enough to cause impairments to occupational, relational and social functioning. In addition to the elevated or irritable mood, people may also experience grandiosity or inflated self-esteem, decreased need for sleep, racing thoughts and excessive involvement in pleasurable pursuits such as buying sprees and sexual activity (APA, 2000).

People who experience depressive episodes alone are said to have a Depressive Disorder and those who experience both elevated mood (manic episodes) and depressive episodes are said to have Bipolar Disorder (APA, 2000).

Whilst the above descriptions represent the broad parameters generally accepted as defining mood disorders (both depressive and bipolar), it needs to be acknowledged that each individual experiences mood disorders in different ways. In Bipolar Disorder, some people will experience more depressive than manic episodes whilst for others it will be the opposite. Some may have severe manic episodes whilst for others it may be the depressive symptoms that predominate. Episodes may occur following a psychosocial stressor whilst for other people they may appear out to the blue and can last from weeks to months, often with long periods in between where the person can be symptom free.

Falret first characterized Bipolar Disorder as a distinct illness in 1851. He termed the cycle of manic and depressive episodes with intervening symptom free periods "folie circulaire" or circular madness. Later Kraepelin (1899) coined the term "Manic Depressive Insanity" (Angst & Sellaro, 2000). However, descriptions of Bipolar Disorder date back to ancient times. Mood disorders describing melancholia and mania are mentioned by Homer and by Hippocrates (Marneros, 2008).

Bipolar Disorder today (whilst not common), is still a significant burden on the individual and society (Das Gupta & Guest, 2002). Some studies have shown a lifetime prevalence of between 1% and 5% for Bipolar 1 Disorder (Emilien et al., 2007; Marneros, 2008). A large national survey in Australia reported a 12 month prevalence of 0.5% for Bipolar Disorder with a more equal gender ratio when compared to major depression (Mitchell et al., 2004). An earlier study by Bebbington and Ramana (1995) based on first admission rates, concluded that the incidence of Bipolar Disorder was around 8–10 per 100 000 (a morbid risk of the order of 0.5%).

Given that the mean age at onset for a first Manic Episode is the early 20s, with some cases starting in childhood and adolescence, and given the high suicide risk, cost to the community is consequently high. Indeed Das Gupta and Guest (2002) estimated the annual direct and indirect costs of Bipolar Disorder in the UK to be around £2 billion. Of this, only 10% was attributed to the direct cost of treatment. However, this may only be the tip of the iceberg. There is evidence to suggest a high degree of misdiagnosis of BD with some studies showing an average gap of 8–10 years between the onset of symptoms and the correct diagnosis (Emilien et al., 2007; Firipis, 2010).

2.1 Treatment and care

Typically the treatment of BD falls into two categories: medication and psychological treatments. As has been noted, people with BD always have some degree of depressive symptoms. However, treatment with antidepressants alone can precipitate the onset of hypomania or mania. Consequently, treatment with a mood stabilizer (with or without an antidepressant) is usually the first line of treatment (Emilien et al., 2007). Lithium and Valproate are the most common mood stabilizers in use today. Antidepressants are also commonly used in some cases of BD although they may, as mentioned earlier, precipitate switching from depression to mania in some people. Atypical antipsychotics have also proved to be effective but these can have unpleasant or dangerous side effects as can the mood stabilizers such as Lithium.

In cases where people with severe Bipolar Disorder do not respond to pharmacotherapy, there is evidence that Electro Convulsive Therapy (ECT) can be an effective treatment for acute mania and bipolar depression. However, whilst it is still considered safe and effective, it is usually used as a last treatment option (Loo et al., 2011; Taylor, 2007).

Psychological treatments such as cognitive therapy, family focused therapy and psycho-education can reduce the risk of relapse and are an important adjunct to pharmacotherapy (Emilien et al., 2007; Sorensen et al., 2006).

There would appear to be a need to find more efficacious drug treatments that have fewer side effects and work across the range of symptoms and phases of the disorder. As well, more research is needed into non-pharmacological treatments that could reduce the use of medications and assist in avoiding hospitalization.

2.2 Experience

The lived experience of Bipolar Disorder is in stark contrast to the measured words of the diagnostic criteria that cannot convey the pain the sufferer experiences, nor the disruption the illness causes to their life and those around them.

I was terrified that I could feel so wrecked inside, terrified that something was horribly wrong with me and terrified that I didn't know how to tell anyone that I was slipping into a dark place. I felt tortured by my mind and I thought I could not contain myself or be contained by those I loved. (Anonymous, quoted in Licinio, 2005:828)

As has been discussed, BD can run a fluctuating course. It is not surprising that the illness can be misdiagnosed or the diagnosis delayed, sometimes for years. All people experience fluctuations of mood, often in response to external stressors. However, with people who suffer from BD, the stressors can coincide with the cyclic nature of the illness and be exacerbated by the illness. Also, the behaviours associated with the illness may in turn precipitate the external stressor making a feedback loop which further exacerbates the episode of illness.

I have this belief that stress sometimes brings out an illness. Maybe you have a predisposition just like you have a predisposition to diabetes or varicose veins, and you if don't get those variable things that cause, that would impact on them, you won't get the disease. (Anonymous, quoted in Sajatovic et al., 2008:721)

The wit, humour, creativity and energy that often accompany the elevated mood states of BD can spiral into grandiosity, insomnia, irritability and poor decision-making. The depressed mood can spiral down into deep depression and thoughts of suicide and suicide attempts. All of this has a negative effect on personal relationships, work, finances and physical and psychological well-being.

Diagnosis, especially after years of suffering, often comes as a relief. Whilst the accompanying self-disclosure to family, friends and work colleagues can be beneficial, it may also be accompanied by stigmatisation that adds further stress to an already stressful situation.

2.3 Diagnosis

As mentioned, many people experience a delay in diagnosis or misdiagnosis. After several episodes of elevated or depressed mood of varying severity, the person or their family come to a realisation that something is wrong. The individual may feel guilty for their behavioural excesses whilst in a manic state or be tormented by depression. When they finally do seek help they may be misdiagnosed with Unipolar Depression when presenting in a depressed state, or misdiagnosed with a psychotic disorder or anxiety when presenting in a manic or hypomanic state (Delmas et al., 2011).

When the diagnosis is finally made it may be accompanied by a sense of relief, as expressed here by Neil Cole, a former politician:

I must say I was then, as I am today, grateful for his diagnosis. He had done for me what no-one had done for me over the previous 15 years of trying; namely to establish what on earth was wrong with me! (Cole, 2004:671)

2.4 Treatment

Once the illness has been diagnosed, treatment can commence. As mentioned earlier, the first line of treatment will generally be pharmacological, usually a mood stabiliser and/or antidepressants. In some cases atypical antipsychotics will also be used. In most cases people with BD will experience initial relief from their symptoms.

The anger that used to overcome me, and the irritability, had done much damage to a lot of people. Worse still was a perception that my temper was very bad, which meant it was best not to get near me. It upset me when I lost my temper. Since my treatment with lithium it is hardly like what it was; in fact it's nothing like it was at all. (Cole, 2004:671)

Unfortunately treatment, like the illness, can also run a fluctuating course. After the initial relief drug treatment usually brings, there are often relapses and sometimes these are precipitated by the medications themselves. As has been discussed, antidepressants may precipitate a manic or hypomania episode.

My new antidepressants were good in stopping depression, but I couldn't sleep. Then as often happens when I don't sleep or I am on an antidepressant that over-stimulates me, I went high. I bought a house I couldn't afford, gave away money, became extremely anxious... (Cole, 2004:673)

It may be some time (perhaps years) before the right combination of medications can be tailored to the needs of the individual. Many people with Bipolar Disorder will need to grapple with an acceptance of the diagnosis and the possibility that they will be on medication for the rest of their lives.

Every night my husband and I embark on the ritual of laying out the pills just so, arranged by color. This is a constant reminder that I am bipolar and that I will be taking these meds for the rest of my life. I lay out the next morning's dose in a silver heart-shaped pillbox that has my initial on one side and 'Love Always, J' on the other. This pillbox is all about my husband's wholehearted acceptance of me as a bipolar woman. (Anonymous, quoted in Licinio, 2005:827)

Whilst medication and medication adherence is still the mainstay of treatment, factors effecting social supports and coping strategies also have an impact on rates of relapse. This realisation has seen a greater acceptance in the medical community of psychosocial approaches to treatment, including Cognitive Behaviour Therapy and Psycho-education amongst others (Sorensen et al., 2006). A strong therapeutic alliance with a trusted professional and strong family supports can assist treatment adherence and build resilience and other constructive changes in people with Bipolar Disorder (Lazarus & Lazarus, 1991; McEnany, 2005).

2.5 Coming out

However, the path to diagnosis, acceptance of the illness and treatment is not an easy one. Not least because it entails some form of self-disclosure to a health professional, family member or friend.

Self disclosure (SD) is an important part of building identity and relationships. It is a means by which we express thoughts and feelings, build intimacy within personal relationships and develop a sense of self (Chaudoir & Fisher, 2010). However, for people with a stigmatising mental illness such as Bipolar Disorder, it is of potential benefit and also a great risk as it may invoke the prejudice of others as following quotes illustrate.

He was admitted to the Psychiatric Department. It was hurtful to experience the gossip. Behind shop shelves, at Sunday parties: 'He's crazy'! Our friends just vanished. When important people abandon you, it really gets to you and undermines your identity, and I feel as if I'm the one being abandoned by them. And then; just silence. I was left sitting alone with the children, who were terrified. (Anonymous, quoted in Tranvåg & Kristoffersen, 2008:9)

The social impact of the illness as a general rule, it can either make or break friendships. It can make people encourage you and be supportive or they can totally leave you alone. (Anonymous, quoted in Sajatovic et al., 2008:721)

With this very real risk in mind, it is not surprising that people may try to hide their illness.

You close up and isolate yourself, you hide the disease so you try to look presentable when you're out, talk presentably, seem interested, you don't want to go to social things because you don't know if you can hide it anymore. (Anonymous, quoted in Sajatovic et al., 2008:721)

The fear and experience of stigma for people with a mental illness can be profound and have far reaching consequences.

3. Stigma and mental illness

The term "stigma" has its origins in Greece (Ng, 1997). It meant to mark the body with a burn or cut to signify that the person is shameful. With time, this meaning evolved to refer to a mark of disgrace with a negative connotation invoking rejection, stereotyping and discrimination (Ng, 1997). Goffman (1974) comprehensively discusses the interpersonal and intrapersonal forces of stigma resulting in a damaged sense of self and identity that is profoundly compromised. The stigmatised person is undesirably different and prone to be segregated and discriminated against. Indeed they are more likely to experience social exclusion across a number of domains (Sayce, 1998). In a qualitative study of self-disclosure and mental illness (Cross, 2009) one person with mental illness stated:

When I first found out I actually got more depressed because I didn't want to be mentally ill. I didn't want to think that there was something mentally wrong with me. It was just abhorrent to me the feeling; and I either was going to be depressed about it forever or had to come to terms with it. (Cross, 2009:133)

And another:

How I felt about my mental illness? I hated having it, I felt like I wasn't a full citizen, that my rights were taken away from me, that I had to take medication for the rest of my life. I didn't want to take drugs because it might have side-effects. (Cross, 2009:134)

Stigma is common for people with a mental illness with three out of four people reporting it. The labelling of people according to their illness stereotypes them and creates prejudice and discrimination (Corrigan et al., 2009). Stigmatised people with mental illness experience humiliation, guilt, despondency, despair, media misrepresentation and sensationalism and impaired help-seeking. Their families are also touched by stigma, giving rise to embarrassment, which in turn leads to avoidance and isolation (ARAFMI NSW, 2005; Corrigan et al., 2006; Hayes et al., 2008).

[N]ot many people understood mental illness and I think my family were embarrassed in a sense. Like my Mum had manic depression and … nobody understood her… None of her children understood, my grandfather didn't understand and people just kind of kept her away from society because she was different. That wasn't my experience but I still was afraid that people were embarrassed about my illness. (Cross, 2002:203)

Highet et al. (2006) showed that while 90% of respondents in their survey believe mental health is a significant issue in Australia, Australians do not have a clear understanding of mental illness. Another Australian study found that:

- almost 25% of people thought that depression was a sign of personal weakness and would not employ a person with depression
- 30% would not vote for a politician with depression
- 42% thought people with depression were unpredictable
- 20% said that if they had depression they would not tell anyone
- almost 66% thought people with schizophrenia were unpredictable and a 25% considered them dangerous (SANE Australia, 2009)

Francis et al. (2001) stated that surveys of the general population as well as consumers of mental health services indicate that a significant percentage believe that media representations (in both news and entertainment) of mental illness are mainly negative, biased and unbalanced. Further, this negativity contributes to less tolerant attitudes and for consumers had detrimental impact on their mental health (Ferriman, 2000; Granello et al., 1999; National Mental Health Association, 2000). Positive portrayals had little influence on attitudes (Domino, 1983; Thornton & Wahl, 1996; Wahl & Lefkowits, 1989).

3.1 Self-stigma

Self-stigma occurs when people internalise the negative messages and this leads to low self esteem, poor self-efficacy, despair, vulnerability and self loathing (Corrigan et al., 2009). Self-stigma results in avoidance of contact with others. In particular, the avoidance of mental health and social services, the consequences of which are further social isolation and loss of support networks. In the end, the person experiences a deterioration in mental health status and possibly an increased risk of suicide (Commonwealth of Australia, 2009).

Well … I'm sort of knocking myself all the time. I'm beginning to look beyond that, bit by bit. I'm the one that's stopping myself stigmatising myself, no-one else. I'm a very beautiful person people love. So I have nothing to be afraid of in that regard. (Cross, 2002:204).

So, given all this negativity around mental illness why do people like Catherine Zeta Jones disclose their Bipolar Disorder? Surely this is a great risk?

According to SANE Australia (2009), we all have a role in creating a mentally healthy community that supports recovery and social inclusion and reduces discrimination. The more sharing the facts about mental health and illness and talking openly about personal experiences of mental illness the better because when mental illness remains out of sight, the community continues to believe that it is shameful and needs to be hidden.

Stephen Fry, another celebrity with Bipolar Disorder, has gone to great lengths to promote understanding of the illness. He uses his fame, in his words, "to fight the stigma and to give a clearer picture of a mental illness most people know little about" (The Independent, 2006). He has taken this further with the BBC2 documentary, *The Secret Life of the Manic Depressive.* In this documentary he seeks advice from experts and connects with other celebrities with Bipolar Disorder, including comedian Tony Slattery, singer Robbie Williams and actors Richard Dreyfuss and Carrie Fisher as well as ordinary people to shed light on this life-long debilitating disorder.

We feel as if we "know" these people because of their celebrity and that we invite them into our lives and homes through a variety of media. Our intimacy with them enables a relationship that embraces self-disclosures that would normally remain private. That they experience mental illness and can still look stunning, land starring roles, have robust relationships and beautiful children can have a very positive effect on the stigma that is associated with mental illness. The media portrays them far more positively than many other people with mental illness.

Indeed, research has shown that people with mental illness tend to be depicted in the media (both news and entertainment media) as objects of mockery, violent, unpredictable, bungling and inept (Thornicroft, 2006). This is confirmed by Australian studies (Hazelton, 1997) though recent Australian research suggests reporting may be improving (Francis et al., 2001).

Francis et al. (2001) reviewed the literature regarding media portrayal of mental health and illness, how it influences community attitudes towards mental illness, and the impact of mental health promotion in the media. They found that community attitudes are influenced by the negative coverage of mental illness (both fictional and non-fictional) and where the media is a major source of information, people are more likely to have disapproving attitudes towards people with a mental illness. Importantly, there appears to be an imbalance between the reporting of positive and negative mental health issues. Mental health promotional campaigns are thought to influence community attitudes more optimistically.

High profile celebrities break the stereotype and as a consequence are treated with respect and dignity. They speak on behalf of others to create an enabling environment that promotes support, decreases labelling and discrimination and challenges our assumptions. Of course there are exceptions to this where some celebrities have had to not only face the problem of mental illness but the unrelenting public scrutiny and ridicule from some sections of the media (Britney Spears [singer], Robert Downey Jr. [actor]); all of which exacerbates the stigma and shame surrounding mental illness. They would probably not welcome the exposure.

I actually think it depends on the personality. It really depends on whether you care about what other people think and whether you trust the [person] you are talking to. I know I've made some terrible mistakes. I haven't been the best person in the world. I wouldn't like to be a politician or a movie star, because they are under scrutiny of journalists all the time. Like I was reading about [a singer] who committed suicide. I thought well every part of his life is scrutinised and [judged]. I don't think I'd pass either if someone were to judge me, I wouldn't pass the moral test. (Cross, 2002:197)

It is important therefore to understand the role of self-disclosure.

4. Self-disclosure

Self-disclosure is defined as the process of sharing personal information, thoughts and feelings with others. Although Heidegger and Laing (Rosenfeld, 2000) both considered self-disclosure within their philosophical views it was Sydney Jourard (1964) who was the first to examine the conceptual basis of self-disclosure. Jourard claimed that in order to be healthy (or to have a healthy personality) a person must be courageous enough to be themselves with others. That is, through word and behaviour, to self-disclose.

Self-disclosure follows in the phenomenological traditions of Husserl (1859–1938) and Heidegger (1889–1976) where the concept of self-disclosure is closely related to the existentialist tenet of sharing "here and now" feelings and the humanistic tenet of authenticity. The authentic being acknowledges the truth of their feelings, needs and drives and communicates this with others. It involves being "genuine" and "real" which indicates openness with the world and an acceptance of that world *as it is*. This is the truth of self-disclosure (Chelune, 1979).

Spontaneity in self-disclosure relates to the here and now element and reflects a self-confluence to which an authentic being is attuned. There is an element of risk-taking in the act of self-disclosure. In a strong healthy interpersonal relationship the message should be accepted and this should encourage greater self-disclosure in both amount and level of intimacy (Jourard, 1964).

Mental health is distinguished by two common attributes. They are the individual's psychological end states and his/her ability to cope with and respond to stressful situations (Toukmanian & Brouwers, 1998). As previously stated, Jourard (1971:32) claimed that self-disclosure is "a means of ultimately achieving a healthy personality". However, early research only weakly supported Jourard's assertion. Cozby (1973) reviewed 11 of these early studies and could find positive associations in only three, whereas five actually demonstrated a negative association.

Other writers have agreed with Jourard's belief. Moreover, they have made connections between Jourard's position and physical health. In an infrequently reported essay, Jourard (1959:503-504) suggested that an individual's lack of self-disclosure could lead to physical ill health:

In the effort to avoid becoming known a person provides for himself a cancerous kind of stress which is subtle and unrecognised but nonetheless effective in producing ... [a] wide array of physical ills.

Tardy (2000) revisited Jourard's hypothesis and concluded that several studies have demonstrated the link between self-disclosure (as either trait or behaviour), with health. Kowalski (1999) also identified many positive consequences of self-disclosure. These positive outcomes relate to physical and psychological well being, acquiring new insights into old problems, securing confirmation for thoughts and feelings and the satisfaction of engaging in meaningful relationships (Kowalski, 1999).

Self-disclosure is purposeful. The aim of disclosing may be to establish or strengthen a relationship, to confess, to inform or to reciprocate. An individual might disclose for catharsis, for self-clarification, self-validation, reciprocity, impression formation, relationship maintenance and enhancement, social influence, manipulation and possibly to remedy injustices or address stereotypes as occurs in celebrity disclosure (Rosenfeld, 2000).

The goal of the interaction influences self-disclosure and thus it is important to examine the goals that lead to self-disclosure. Goals differ when the target is a friend than when the target is a stranger. Are disclosure patterns different when the goal is different? Given specific goals, are there differences in beliefs, resources and strategies in self-disclosure?

According to Archer (1987) the goals of self-disclosure are both intrapersonal and interpersonal. As an intrapersonal goal it is a form of self-expression whereby the individual may reveal his/her innermost feelings. It may also be used to clarify the individuals thoughts, beliefs and for social validation. As an interpersonal goal, self-disclosure functions as a mechanism for social control where the individual may self-disclose in order to please others or to avoid reprisals from others. It is also used interpersonally to elicit self-disclosure from others and for the development of the relationship.

There are a number of other factors that moderate self-disclosure and a particularly important one in this context is that of the valence of the disclosure.

The valence of the disclosure relates to the value placed on the content by the speaker or the listener. Positive or negative values may be ascribed depending on whether the information perceived as "good" or "bad" and is judged on the perceived reaction of the listener.

Kowalski (1999) states that negative thoughts and feelings are more likely to remain concealed because they lead to embarrassment or humiliation or have serious implications. For example, some individuals will avoid unpleasant disclosures in an effort to prevent being a burden to the listener or upsetting them by negative disclosures. This conclusion was drawn in interviews with Holocaust survivors (Pennebaker et al., 1989). The researchers concluded that survivors typically did not disclose their experiences because they were attempting to protect the listener from physical and psychological distress. An alternate conclusion to be drawn from these studies, however, is that the survivors have experienced such trauma themselves and avoid disclosure to prevent reliving the trauma and exposing themselves to psychological distress. This is particularly evident in sufferers of Post Traumatic Stress Disorder (PTSD).

According to Dindia (1997), disclosure of stigmatising (negative valence) content such as information related to mental illness follows a dialectical process rather than a single act. That is, the information is revealed in pieces and built upon to create the whole story rather than being divulged in its entirety in the first instance. There appears to be an "unlayering" of information as the person reveals then conceals then reveals a bit more (Gard, 1990; Limandri, 1989; Marks et al., 1992). Limandri (1989:69) describes the nature of disclosure of stigmatising topics:

To perceive her/his condition as stigmatising, an individual experiences an underlying feeling of shame. The notion of hiding or concealment is intrinsic to and inseparable from the concept of shame. To avoid shame, the individual must avoid disclosure of the condition ... This, however, cannot always be avoided. In fact, the individual may need to disclose her/his stigmatising condition in order to receive necessary health care.

Interpersonal control will also moderate self-disclosure. Anderson and Randlet (1994) found that high self-monitoring and self-control was significantly related to satisfaction with disclosure. They concluded that high self-monitors experienced increased satisfaction because they had a greater sense of interpersonal control and self-disclosed in a situation-specific strategic way. That is, the high self-monitor judged the interpersonal situation and disclosed as they perceived the listener's responsiveness. Bryan, Dodson and Cullari (1997) found an inverse relationship between self-monitoring and self-disclosure. They found that males had significantly higher self-monitoring scores than females. This sense of interpersonal control is important in that the disclosure of areas of privacy potentially gives the listener power over the discloser and renders him/her susceptible to rejection and other negative consequences.

Limandri (1989:70) argues:

The circumstances for voluntary disclosure must yield sufficient anticipated reward to counterbalance the disesteem and rejection that may result. The stigmatised person struggles with the conflict of the need to reveal due to a concomitant stressor versus the need to conceal due to further stigmatisation.

Petronio, Ellemers, Giles and Gallois (1998) examine self-disclosure in light of privacy and argue that communication boundary management explains why people want to control who knows private information about them. Boundaries relate to groups, families, organizations and communities. Rules govern the opening up of the boundaries that allow others to gain access to personal/private information. Some rules are tightly controlled and others loosely. Moreover, disclosure routines apply for some sorts of information. For example, some rules for disclosure may have been utilised for so long that they are invoked as a matter of routine: *"I never tell people my salary"*. Petronio et al. (1998) claim that the ramifications of the disclosure will determine access to the information contained within it. Once the information is shared, both parties to the disclosure have a responsibility for the management of the information. Petronio (2000) states that boundary rules regulate the flow of information to others, are influenced by motivations and gender, can be altered by events or ritualised. Boundaries must also be coordinated because the individual co-owns different information with different people or groups. Regardless of the nature of the relationship, the information is private and originally belongs to the individual.

In the study by Cross (2002:203) some of the participants stated that they would have difficulty in disclosing:

Psychiatric symptoms,

I would try to and there would still probably be certain things that I would keep to myself. Maybe what I was hearing or seeing [hallucinations] – I might say yes that's happening but not describe it.

Confusing thoughts,

Because sometimes I've just been in a situation where it's just happened and I sometimes my thoughts override what I say so but in general, I would find it difficult to describe.

Taboos,

I found that it would be easier to say things that I'm happy about or that I'm fairly kind of neutral things to me in my life. What I care most about or the money and the religion and stuff and that I found that it would be more difficult to talk about negative aspects of my life – like what I felt bad about or anxious about or things like that.

Families,

I wouldn't talk about my sexual abuse. I wouldn't talk about my father at all or my family. I think it's little things that you just don't want to talk about, the person in yourself. But a lot of the people I talk to … it's confidential so I don't mind expressing how I feel about my father and stuff.

According to Chaudoir and Fisher (2010:236), people with a mental illness carry a "concealable stigmatized identity". This means that they hold personal information that is devalued by society though hidden from others. Other types of information include HIV status, sexual orientation, experiences of abuse or assault. They will be seen as non-stigmatised and as such be acutely aware of the stereotypes and discrimination afforded the hidden identity. Moreover, their sensitivity to negative social cues could make them more burdened by their identity and experience such distress that it impairs performance (Quinn et al., 2004). Choosing to disclose these types of information requires an assessment of the risk depending on the recipient of the information and reasons for the disclosure. The effect on the recipient of the disclosure and their response predicts whether the disclosure will produce a beneficial outcome or not. This is a complex decision to make.

Chaudoir and Fisher (2010) examine the interaction between internal and external factors involved in self-disclosure through the Disclosure Process Model (DPM). They assert that a feedback loop encompassing the antecedent conditions for disclosure (approach/avoidance goals) coupled with the disclosure event (content of the disclosure and response of the recipient) and mediating processes (alleviation of inhibition, social support and changes in social information) effects the long term outcomes for the individual, their relationships with others and broad social contexts. The interaction of these conditions through the DPM attempts to explain why self-disclosure might be beneficial or not.

Let's apply the example of celebrity (Catherine Zeta Jones; Stephen Fry and others) to the DPM. They have a concealable stigmatised identity. They made the decision to disclose this information at significant personal, professional and social risk. The reaction of the recipient(s) (general public, media) to this information was benign and they are admired for "coming out". Their disclosure led to others with mental illness to also reveal their problems with mental health and other related issues.

In addressing the mediating processes, we must consider three things: alleviation of inhibition, social support and changes in social information. Disclosure in alleviating inhibition permits celebrities to express hidden information leading to reduced suffering, and better overall wellbeing. Disclosure permits them to gain social support leading to greater liking, intimacy and trust. It also has the individual benefits with regard to improving their individual self-esteem. Disclosure contributes fresh information about their concealed Bipolar Disorder to the wider social context and impacts ensuing social exchanges based on the expectations of the community for regulated disclosure. This example is based on the assumption that both had the antecedent condition of a goal approach motivation. If they were motivated by avoidance goals, the outcomes might not have been so positive because they would have had more difficulty attending to positive cues, indeed, focussing on negative cues and the possibility of rejection thus potentially making the decision not to disclose.

More importantly, when celebrities disclose, they are in control of the disclosure event. Using Petronio's et al. (1998) explanation, celebrity self-disclosure in the context of privacy it can be seen why they would seek to want to control who knows private information about them. They can reveal as much or as little as suits them and when and in what situations thereby assuring themselves that sensationalism, "scoops", "tell-alls" are minimised. Boundary management, in this instance, is clear. They have set rules governing the opening of these boundaries to permit the general public to gain limited access to ordinarily personal/private information. In any case the information is private and originally belongs to that individual. The sharing of information induces all those party to the disclosure to be accountable for it (Petronio et al., 1998).

5. Summary

In this chapter we have attempted to shed light on the reasons why celebrities choose to disclose their diagnosis of Bipolar Disorder utilising self-disclosure literature and how this could contribute to reducing the stigma and social exclusion for people with mental illness.

Self disclosure is a complex and risk laden activity. However, where the factors around the antecedent conditions, the disclosing event and the mediating processes are right, the long-term outcomes for the individual and their relationships can be very positive. When celebrities can control these processes, disclosure by them may have a positive effect on society's attitudes to the hidden stigmatised identity. Whilst there is some risk of the disease becoming trendy (the naive "I want to be bipolar" phenomenon (Chan & Sireling, 2010)) or trivialised as the "disease du jour", it would appear that purposeful and deliberate celebrity self disclosure of the type modelled by Stephen Fry can be beneficial in reducing stigma. In this way celebrity self-disclosure can achieve benefits for others as well as the celebrities themselves.

Given that the World Health Organisation estimates that 25% of people will experience a mental illness across their lifetime and that it is the fastest growing occupational disability (World Health Organisation, 2001), there is a dire need for greater understanding of the stigmatisation of people with mental illness and how that can be mitigated. Mental Health organisations and governments are committed to increasing public awareness in the hope of reducing stigma and are undertaking numerous strategies in their quest for non-discrimination and social inclusion. If it takes a celebrity to influence others then we must acknowledge that this is a step in the right direction.

6. References

American Psychiatric Association [APA] (2000) *Diagnostic and Statistical Manual of Mental Disorders,* IV-TR ed., American Psychiatric Association, Washington, DC.

Angst, J. & Sellaro, R. (2000) Historical perspectives and natural history of bipolar disorder. *Biological Psychiatry,* 48, 445-457.

ARAFMI NSW (2005) Career Services Mapping Project, Report for NSW Government Department of Health

Archer, R. L. (1987) Commentary: self-disclosure, a very useful behavior. In *Self-Disclosure: Theory Research and Therapy,* Derlega, V. J. & Berg, J. H. (Eds.), pp. 329-342, Plenum, New York.

Bebbington, P. E. & Ramana, R. (1995) Epidemiology of bipolar affective disorder: a review. *Social Psychiatry and Psychiatric Epidemiology,* 30, 279-292.

Bryan, J., Dodson, D. & Cullari, S. (1997) The association of self-monitoring with self-disclosure. *Psychological Reports,* 80, 940-942.

Chan, D. & Sireling, L. (2010) 'I want to be bipolar' ... a new phenomenon. *The Psychiatrist,* 34, 103-105.

Chaudoir, S. R. & Fisher, J. D. (2010) The disclosure processes model: understanding disclosure decision making and postdisclosure outcomes among people living with a concealable stigmatized identity. *Psychological Bulletin,* 136, 236-256.

Chelune, G. K. (Ed.) (1979) *Self-Disclosure: Origins, patterns and implications of openness in interpersonal relationships,* Jossey-Bass, San Francisco.

Cole, N. (2004) A bipolar journey. *Australian and New Zealand Journal of Psychiatry,* 38, 671-673.

Commonwealth of Australia (2009) Suicide Risk Factors and Warning Signs: Response-ability fact sheet B6. Available from: http://www.responseability.org/client_images/78125.pdf

Corrigan, P. W., Larson, J. E., Hautamaki, J., Matthews, A., Kuwabara, S., Rafacz, J., Walton, J., Wassel, A. & O'Shaughnessy, J. (2009) What lessons do coming out as gay men or lesbians have for people stigmatized by mental illness? *Community Mental Health Journal,* 45, 366-374.

Corrigan, P. W., Watson, A. C. & Miller, F. E. (2006) Blame, shame, and contamination: the impact of mental illness and drug dependence stigma on family members. *Journal of Family Psychology,* 20, 239-246.

Cozby, P. C. (1973) Self-disclosure: a literature review. *Psychological Bulletin,* 79, 73-91.

Cross, W. M. (2002) *Culture, communication and self-disclosure: implications for psychiatric assessment and diagnosis.* University of New South Wales, Ph. D. thesis.

Cross, W. M. (2009) A phenomenological exploration of self-disclosure and mental illness. In *From Birth to Death: Clinical nursing and midwifery research across the lifespan,* O'Connor, M., Griffiths, D., Ives, G., Newton, J. & Tan, H. (Eds.), pp. 128-144, Monash University, Melbourne.

Das Gupta, R. & Guest, J. F. (2002) Annual cost of bipolar disorder to UK society. *British Journal of Psychiatry,* 180, 227-233.

Delmas, K., Proudfoot, J., Parker, G. & Manicavasagar, V. (2011) Recoding past experiences: a qualitiative study of how patients and family members adjust to the diagnosis of bipolar disorder. *The Journal of Nervous and Mental Disease,* 199, 136-139.

Dindia, K. (1997) Self-disclosure, self-identity, and relationship development: a transactional/dialectical perspective. In *Handbook of personal relationships: Theory, research and interventions,* 2nd ed., Duck, S. (Ed.), pp. 411-426, John Wiley, Chichester, UK.

Domino, G. (1983) Impact of the film 'One Flew Over the Cuckoo's Nest', on attitudes towards mental illness. *Psychological Reports,* 53.

Emilien, G., Septien, L., Brisard, C., Corruble, E. & Bourin, M. (2007) Bipolar disorder: how far are we from a rigorous definition and effective management? *Progress in Neuro-Psychopharmacology & Biological Psychiatry,* 31, 975-996.

Ferriman, A. (2000) The stigma of schizophrenia. *British Medical Journal,* 320, 522.

Firipis, M. (2010) Management of bipolar disorder. *Australian Pharmacist,* 29, 604-609.

Francis, C., Pirkis, J., Dunt, D. & Blood, R. W. (2001) Mental Health and Illness in the Media: A Review of the Literature, Commonwealth Deapartment of Health and Ageing, Canberra, ACT

Gard, L. (1990) Patient disclosure of human immunodeficiency virus (HIV) status to parents: clinical considerations. *Professional Psychology: Research and Practice,* 21, 252-256.

Goffman, E. (1974) *Stigma: Notes on the management of spoiled identity,* Jason Aronson, New York.

Granello, D., Pauley, P. & Carmichael, A. (1999) Relationship of the media to attitudes towards people with mental illness. *Journal of Humanistic Counseling Education and Development,* 38, 98-103.

Hayes, A., Gray, M. & Edwards, B. (2008) Social inclusion: origins, concepts and key themes. In *Australian Insititute of Family Studies,* prepared for the Social Inclusion Unit, Department of the Prime Minister and Cabinet, Australia.

Hazelton, M. (1997) Reporting mental health: a discourse analysis of mental health-related news in two Australian newspapers. *Australian and New Zealand Journal of Mental Health Nursing,* 6, 73-89.

Highet, N. J., Luscombe, G. M., Davenport, T. A., Burns, J. M. & Hickie, I. B. (2006) Positive relationships between public awareness activity and recognition of the impacts of depression in Australia. *Australian and New Zealand Journal of Psychiatry*, 40, 55-58.

Jourard, S. M. (1959) Health, personality and self-dsiclosure. *Journal of Mental Hygiene*, 43, 499-507.

Jourard, S. M. (1964) *The Transparent Self*, D. Van Nostrand, New York.

Jourard, S. M. (1971) *The Transparent Self*, 2nd ed., D. Van Nostrand, New York.

Kowalski, R. M. (1999) Speaking the unspeakable: self-disclosure and mental health. In *The Social Psychology of Emotional and Behavioural Problems*, Kowalski, R. M. & Leary, M. R. (Eds.), American Psychological Association, Washington, DC.

Lazarus, A. A. & Lazarus, C. N. (1991) Let us not forsake the individual nor ignore the data: a response to Bozarth. *Journal of Counseling & Development*, 69, 463-465.

Licinio, J. (2005) The experience of bipolar disorder: a personal perspective on the impact of mood disorder symptoms. *Molecular Psychiatry*, 10, 827-830.

Limandri, B. (1989) Disclosure of stigmatizing conditions: the discloser's perspective. *Archives of Psychiatric Nursing*, 3, 69-78.

Loo, C., Katalinic, N., Mitchell, P. B. & Greenberg, B. (2011) Physical treatments for bipolar disorder: a review of electroconvulsive therapy, stereotactic surgery and other brain stimulation techniques. *Journal of Affective Disorders*, 132, 1-13.

Manchester, W. (1984) *The Last Lion: William Spencer Churchill: Visions of Glory 1874-1932*, Sphere, London.

Marks, G., Bundek, N., Richardson, J., Ruiz, M., Maldonado, N. & Mason, J. (1992) Self-disclosure of HIV infection: preliminary results from a sample of Hispanic men. *Health Psychology*, 11, 300-306.

Marneros, A. (2008) Mood disorders: epidemiology and natural history. *Psychiatry*, 8, 52-55.

McEnany, G. W. (2005) Beyond pharmacotherapy: the art of patient care. *Advanced Studies in Nursing*, 3, 54-58.

Mitchell, P. B., Slade, T. & Andrews, G. (2004) Twelve-month prevalance and disability of DSM-IV bipolar disorder in an Australian general population survey. *Psychological Medicine*, 34, 777-785.

National Mental Health Association (2000) *Stigma Matters: Assessing the media's impact on public perception of mental illness*, National Mental Health Association, Chicago, IL.

Ng, C. H. (1997) The stigma of mental illness in Asian cultures. *Australian and New Zealand Journal of Psychiatry*, 31, 382-390.

Pennebaker, J. W., Barger, S. D. & Tiebout, J. (1989) Disclosure of traumas and health among Holocaust survivors. *Psychosomatic Medicine*, 51, 577-589.

Petronio, S. (2000) The boundaries of privacy: praxis of everyday life. In *Balancing the Secrets of Private Disclosures*, Petronio, S. (Ed.), Lawrence Erlbaum Associates, Mahwah, NJ.

Petronio, S., Ellemers, N., Giles, H. & Gallois, C. (1998) (Mis)communicating across boundaries. *Communication Research*, 25, 571-595.

Quinn, D. M., Kahng, S. K. & Crocker, J. (2004) Discreditable: Stignma effects of revealing a mental illness history on test performance. *Personality and Social Psychology Bulletin*, 30, 803-815.

Rihmer, Z. & Kiss, K. (2002) Bipolar disorders and suicidal behaviour. *Bipolar Disorders*, 4, 21-25.

Rosenfeld, L. B. (2000) Overview of the ways of privacy, secrecy, and disclosure are balanced in today's society. In *Balancing the Secrets of Private Disclosures,* Petronio, S. (Ed.), pp. 3-17, Lawrence Erlbaum Associates, Mahwah, NJ.

Sadock, B. J., Sadock, V. A. & Kaplan, H. I. (2003) *Kaplan and Sadock's Synopsis of Psychiatry: Behavioural Sciences, Clinical Psychiatry,* 9th ed., Lippincott Williams & Wilkins, Philadelphia, PA.

Sajatovic, M., Jenkins, J. H., Safavi, R., West, J. A., Cassidy, K. A., Meyer, W. J. & Calabrese, J. R. (2008) Personal and societal construction of illness among individuals with rapid-cycling bipolar disorder: a life-trajectory perspective. *American Journal of Geriatric Psychiatry,* 16, 718-726.

SANE Australia (2009) SANE Research Bulletin 10: Stigma, the media and mental illness, SANE Australia, Melbourne

Sayce, L. (1998) Stigma, discrimination and social exclusion: What's in a word? *Journal of Mental Health,* 7, 331-343.

Sorensen, J., Done, D. J. & Rhodes, J. (2006) A case series evaluation of a brief, psycho-education approach intended for the prevention of relapse in bipolar disorder. *Behavioural and Congnitive Psychotherapy,* 35, 93-107.

Tardy, C. H. (2000) Self-disclosure and health: revisiting Jourard's hypothesis. In *Balancing the Secrets of Private Disclosures,* Petronio, S. (Ed.), pp. 111-122, Lawrence Erlbaum Associates, Mahwah, NJ.

Taylor, S. (2007) Electroconvulsive therapy: a review of history, patient selection, technique, and medication management. *Southern Medical Journal,* 100, 494-498.

The Independent (2006) Mr Fry improves the mood of the nation. Accessed August 2011. Available from: http://www.independent.co.uk/opinion/leading-articles/leading-article-mr-fry-improves-the-mood-of-the-nation-416318.html

Thornicroft, G. (2006) Actions Speak Louder: Tackling stigma and discrimination against people with mental illness, Mental Health Foundation, London. Available from: http://www.mentalhealth.org.uk/publications/

Thornton, J. A. & Wahl, O. F. (1996) Impact of a newspaper article on attitudes towards mental illness. *Journal of Community Psychology,* 24, 17-25.

Toukmanian, S. G. & Brouwers, M. C. (1998) Cultural aspects of self-disclosure and psychotherapy. In *Cultural clinical psychology: theory research and practice,* Kazarian, S. S. & Evans, D. R. (Eds.), pp. 106-124, Oxford University Press, New York.

Tranvåg, O. & Kristoffersen, K. (2008) Experience of being the spouse/cohabitant of a person with bipolar affective disorder: a cumulative process over time. *Scandanavian Journal of Caring Science,* 22, 5-18.

Wahl, O. F. & Lefkowits, J. Y. (1989) Impact of a television film on attitudes towards mental illness. *American Journal of Community Psychology,* 17, 521-528.

World Health Organisation (2001) *World Health Report 2001 - Mental Health: New Understanding, New Hope,* World Health Organisation, Geneva.